STP

NATIONAL CURRICULUM
MATHEMATICS

8A

STP

NATIONAL CURRICULUM
MATHEMATICS

8A

L. BOSTOCK, B.Sc.

S. CHANDLER, B.Sc.

A. SHEPHERD, B.Sc.

E. SMITH, M.Sc.

 Nelson Thornes

First published in 1996 by
Stanley Thornes (Publishers) Ltd

Reprinted in 2002 by:
Nelson Thornes Ltd
Delta Place
27 Bath Road
CHELTENHAM
GL53 7TH
United Kingdom

11 12 / 20

A catalogue record of this book is available from the British Library.

ISBN 978 0 7487 2440 6

Artwork by Peters and Zabransky, Mark Dunn, Linda Jeffrey.

Front cover image produced using material kindly supplied by I LOVE LOVE CO, makers of The Happy Cube © Laureyssens/Creative City Ltd 1986/91.
Distributed in UK by: RIGHTRAC, 119 Sandycombe Road, Richmond Surrey TW9 2ER
Tel. 0181 940 3322.

The publishers are grateful to the following for granting permission to reproduce photographs or other copyright material:
Martyn Chillmaid: p. 24
Cordon Art, Baam Holland – © 1996 M.C. Escher/Cordon Art.
All rights reserved: p. 133

Typeset by Tech-Set, Gateshead, Tyne & Wear.
Printed in China

CONTENTS

Introduction

Summary 1   **1**

1 Working with Numbers   **17**
Positive and negative indices. Zero index. Rules of indices for multiplying and dividing. Standard form. Significant figures. Estimating and using a calculator.

2 Probability   **38**
Probability that an event does not happen. Possibility space for two events. Expected number of times an event will occur.

3 Multiplication and Division of Fractions   **48**
Multiplication and division of fractions and mixed numbers. Mixed operations.

4 Fractions and Percentages   **63**
One quantity as a fraction or percentage of another. Interchanging fractions, decimals and percentages. Percentage increase and decrease.

5 Ratio   **88**
Comparing sizes of quantities. Simplifying ratios. Ratios as fractions. Using ratios to find unknown quantities. Division in a given ratio. Map ratios. Direct proportion.

Summary 2   **105**

6 Polygons   **113**
Regular and irregular polygons. Sum of the interior angles and exterior angles. Tessellations.

7 Areas of Triangles and Parallelograms   **134**
Formulas for the area of a triangle and the area of a parallelogram. Finding square roots using a calculator. Compound shapes.

8 Scatter Graphs   **156**
Hypothesis testing. Drawing scatter graphs. Line of best fit. Positive and negative correlation. Questionnaires.

9 Circumference and Area of a Circle   **168**
Parts of a circle. Introducing π. Formulas for the circumference and area of a circle.

10 Formulas   **184**
Constructing formulas. Using brackets. Multiplication and division of directed numbers. Substituting numbers into formulas. Finding the nth term of a sequence. Using a spreadsheet to generate sequences.

11 Reflections, Translations and Rotations   **202**
Reflecting in a mirror line. Translations. Order of rotational symmetry. Centre of rotation. Rotation of an object about a point.

Summary 3   **220**

12 Linear Equations 232

Forming and solving equations involving brackets. Multiplication and division of algebraic fractions. Solving equations involving fractions. Forming and solving linear inequalities.

13 Straight Line Graphs 256

Equation of a line through the origin. Plotting a line from its equation. Gradient of a straight line. Equations of lines that do not pass through the origin. The equation $y = mx + c$. Intersecting lines. Using a graphics calculator. Lines parallel to the axes. Drawing conversion graphs.

14 Curved Graphs 278

Drawing a curve through given points. Constructing a table from a formula. Equation of a curve. Shape of curves with equations of the form $y = ax^2 + bx + c$.

15 Continuous Data 290

Discrete and continuous data. Rounding continuous values. Grouping continuous data and illustrating with bar charts. Range and modal class. Frequency polygons.

16 Simultaneous Equations 310

Forming and solving simultaneous equations algebraically and graphically.

17 Solving Equations 323

Forming polynomial equations. Simplifying expressions with brackets. Solving equations of the form $x^2 = 20$ by using a calculator. Finding solutions by trial and improvement, including using a spreadsheet and using graphs.

Summary 4 336

18 Volumes 350

Finding volumes of solids with constant cross-section (prisms), including cylinders. Density.

19 Enlargement 365

Finding the centre of enlargement. Positive and fractional scale factors.

20 Scale Drawing 376

Making and using scale drawings. Angles of elevation and depression. Three-figure bearings.

21 Pythagoras' Theorem 396

Right-angled triangles. Pythagoras' theorem. The 3,4,5 right-angled triangle. Perigal's dissection. Using Pythagoras' theorem for finding lengths in isosceles triangles and circles. Converse of Pythagoras' theorem.

22 Travel graphs 416

Finding distances from distance-time graphs. Relationship between distance, speed and time. Average speed. Using travel graphs.

Summary 5 441

Index 454

INTRODUCTION

To the pupil

This book continues to help you to learn, enjoy and progress through Mathematics in the National Curriculum. As well as a clear and concise text the book offers a wide range of practical and investigational work that is relevant to the mathematics you are learning.

Everyone needs success and satisfaction in getting things right. With this in mind we have divided many of the exercises into three types of questions.

The first type, identified by plain numbers, e.g. **15**, helps you to see if you understand the work. These questions are considered necessary for every chapter.

The second type, identified by an underline, e.g. **15**, are extra, but not harder, questions for quicker workers, for extra practice or for later revision.

The third type, identified by a coloured square, e.g. **15**, are for those of you who like a greater challenge.

Most chapters have a 'mixed exercise' after the main work of the chapter has been completed. This will help you to revise what you have done, either when you have finished the chapter or at a later date. All chapters end with some mathematical puzzles, practical and/or investigational work. For this work you are encouraged to share your ideas with others, to use any mathematics you are familiar with, and to try to solve each problem in different ways, appreciating the advantages and disadvantages of each method.

The book starts with a summary of the main results from Book 7A. After every five or six chapters you will find further summaries. These list the most important points that have been studied in the previous chapters and conclude with revision exercises that test the work you have studied up to that point.

At this stage you will find that you use a calculator more frequently but it is still unwise to rely on a calculator for work that you should do in your head. Remember, whether you use a calculator or do the working yourself, always estimate your answer and always ask yourself the question, 'Is my answer a sensible one?'

Mathematics is an exciting and enjoyable subject when you understand what is going on. Remember, if you don't understand something, ask someone who can explain it to you. If you still don't understand, ask again. Good luck with your studies.

To the teacher

This is the second book of the STP National Curriculum Mathematics series. It is based on the ST(P) Mathematics series but has been extensively rewritten and is now firmly based on the Programme of Study for Key Stages 3 and 4.

The A series of books aims to prepare pupils for about Level 8 at Key Stage 3 and for the higher tier at GCSE.

SUMMARY 1

**MAIN RESULTS
FROM BOOK 7A**

**OPERATIONS OF
$\times$, $\div$, $+$, $-$**

The sign in front of a number refers to that number only.
Three or more numbers can be added in any order,

e.g. $2 + 7 + 8 = 2 + 8 + 7 = 10 + 7 = 17$

Three or more numbers can be multiplied in any order,

e.g. $2 \times 7 \times 5 = 2 \times 5 \times 7 = 10 \times 7 = 70$

The order in which you add or subtract does not matter, so

$1 - 5 + 8$ can be calculated in the order $1 + 8 - 5$, i.e. $9 - 5 = 4$

When a calculation involves a *mixture of operations*, start by calculating anything inside brackets, then follow the rule 'do the multiplication and division first',

e.g. $2 + 3 \times 2 = 2 + 6$ and $(2 + 3) \times 2 = 5 \times 2$
$= 8$ $= 10$

**TYPES OF
NUMBER**

A *factor* of a number will divide into the number exactly.
A *multiple* of a number has that number as a factor,

e.g. 3 is a factor of 12 and 12 is a multiple of 3.

A *prime number* has only 1 and itself as factors, e.g. 7

Square numbers can be drawn
as a square grid of dots, e.g. 9:

Rectangular numbers can be
drawn as a rectangular grid of dots, e.g. 6:

Triangular numbers can be drawn
as a triangular grid of dots, e.g. 6:

INDICES

The small 2 in 3^2 is called an index and it shows how many 3s are multiplied together,

e.g. 3^5 means $3 \times 3 \times 3 \times 3 \times 3$

1

FRACTIONS

Equivalent fractions are formed by multiplying or by dividing the top and the bottom of a fraction by the same number,

e.g. $\frac{1}{2} = \frac{3}{6}$ (multiplying top and bottom of $\frac{1}{2}$ by 3)

and $\frac{5}{10} = \frac{1}{2}$ (dividing top and bottom of $\frac{5}{10}$ by 5)

To add or subtract fractions they must have the *same* denominator,

e.g.   to add $\frac{1}{2}$ to $\frac{1}{3}$ we must first change them into equivalent fractions with the same denominators,

i.e. $\frac{1}{2} + \frac{1}{3} = \frac{1 \times 3}{2 \times 3} + \frac{1 \times 2}{3 \times 2} = \frac{3}{6} + \frac{2}{6} = \frac{5}{6}$

DECIMALS

In decimal notation, numbers to the right of the decimal point represent tenths, hundredths, thousandths, ...,

e.g. $0.534 = \frac{5}{10} + \frac{3}{100} + \frac{4}{1000} = \frac{534}{1000}$

Decimals can be added or subtracted using the same methods as for whole numbers, provided that the decimal points are placed in line,

e.g.   to add 12.56 and 7.9, we can write them as   12.56
                                                       $+\underline{\;\;7.9}$
                                                           20.46

To multiply a decimal by 10, 100, 1000, ..., we move the point 1, 2, 3, ... places to the right,

e.g. $2.56 \times 10 = 25.6$,   and $2.56 \times 1000 = 2560$

To divide a decimal by 10, 100, 1000, ..., we move the point 1, 2, 3, ... places to the left,

e.g. $2.56 \div 10 = 0.256$,   and $2.56 \div 1000 = 0.002\,56$

To multiply decimals without using a calculator, first ignore the decimal point and multiply the numbers. Then add together the number of decimal places in the decimals being multiplied; this gives the number of decimal places in the answer,

e.g. $2.5 \times 0.4 = 1.00 = 1$ ($25 \times 4 = 100$)
          $[\,(\,1\,) + (\,1\,) = (\,2\,)\,]$

To divide by a decimal, move the point in *both* numbers to the right until the number we are dividing by is a whole number,

e.g. $2.56 \div 0.4 = 25.6 \div 4$

Now we can use ordinary division, keeping the decimal point in the same place,

e.g. $25.6 \div 4 = 6.4$ $4\overline{)25.6}$ with 6.4 above

To change a fraction to decimal notation, divide the bottom number into the top number,

e.g. $\frac{3}{8} = 3 \div 8 = 0.375$

ROUNDING NUMBERS

To round (i.e. to correct) a number to a specified place value, look at the figure in the next place: if it is 5 or more, add 1 to the specified figure, otherwise leave the specified figure as it is,

e.g. $1\,3\,7 = 140$ to the nearest 10,

$1\,3\,7 = 100$ to the nearest 100,

$2.\,5\,6\,4 = 3$ to the nearest whole number,

$2.\,5\,6\,4 = 2.6$ correct to 1 decimal place,

$2.\,5\,6\,4 = 2.56$ correct to 2 decimal places.

PERCENTAGES

A percentage of a quantity describes how many of the 100 equal-sized parts of the quantity are being considered,

e.g.   20% of an apple means 20 out of the 100 equal-sized parts of the apple,

i.e. $20\% = \frac{20}{100} = 0.2$

UNITS

Metric units of length in common use are kilometres, metres, centimetres and millimetres, where

$$1\,km = 1000\,m, \qquad 1\,m = 100\,cm, \qquad 1\,cm = 10\,mm$$

Metric units of mass are tonnes, kilograms, grams and milligrams, where

$$1\,tonne = 1000\,kg, \qquad 1\,kg = 1000\,g, \qquad 1\,g = 1000\,mg$$

Imperial units of length in common use are miles, yards (yd), feet (ft) and inches (in), where

$$1\,mile = 1760\,yd, \qquad 1\,yd = 3\,ft, \qquad 1\,ft = 12\,in$$

Imperial units of mass still in common use are
tons, hundredweights (cwt), stones, pounds (lb) and ounces (oz), where

$$1\,ton = 2240\,lb, \quad 1\,cwt = 112\,lb, \quad 1\,stone = 14\,lb, \quad 1\,lb = 16\,oz$$

For a rough conversion between metric and Imperial units use

$$1\,km \approx \tfrac{1}{2}\,mile, \quad 1\,yard \approx 1\,m, \quad 1\,kg \approx 2\,lb, \quad 1\,tonne \approx 1\,ton$$

For a *better approximation* use

$$5\,miles \approx 8\,km, \qquad 1\,in \approx 2.5\,cm, \qquad 1\,kg \approx 2.2\,lb$$

AREA

Area is measured by standard-sized squares.

$$1\,cm^2 = 10 \times 10\,mm^2 = 100\,mm^2 \quad \text{and}$$
$$1\,m^2 = 100 \times 100\,cm^2 = 10\,000\,cm^2$$
$$1\,km^2 = 1000 \times 1000\,m^2 = 1\,000\,000\,m^2$$

The *area of a square* = (length of a side)^2

The *area of a rectangle* = length × breadth

VOLUME AND CAPACITY

Volume is measured by standard sized cubes.

$$1\,cm^3 = 10 \times 10 \times 10\,mm^3 = 1000\,mm^3$$
$$1\,m^3 = 100 \times 100 \times 100\,cm^3 = 1\,000\,000\,cm^3$$

The capacity of a container is the volume of liquid it could hold.

The main metric units of capacity are the litre (l) and the millilitre (ml), where

$$1\,\text{litre} = 1000\,ml \quad \text{and} \quad 1\,\text{litre} = 1000\,cm^3$$

The main Imperial units of capacity are the gallon and the pint, where

$$1\,\text{gallon} = 8\,\text{pints}$$

Rough conversions between metric and Imperial units of capacity are given by

$$1\,\text{litre} \approx 1.75\,\text{pints} \quad \text{and} \quad 1\,\text{gallon} \approx 4.5\,\text{litres}$$

The *volume of a cuboid* = length × breadth × height

One complete revolution = 4 right angles = 360°.

1 *right angle* = 90°.

An *acute angle* is less than 90°.

An *obtuse angle* is larger than 90° but less than 180°.

A *reflex angle* is larger than 180°.

Vertically opposite angles are equal.

Angles on a straight line add up to 180°.

Two angles that add up to 180° are called *supplementary angles*.

Angles at a point add up to 360°.

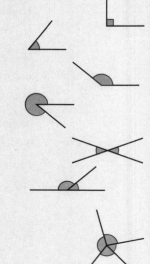

CONGRUENCE

Two figures are congruent when they are exactly the same shape and size. They are still congruent even if one is turned round or over with respect to the other.

TRIANGLES

The three angles in a triangle add up to 180°.

An *equilateral triangle* has all three sides equal and each angle is 60°.

An *isosceles triangle* has two sides equal and the angles at the base of these sides are equal.

QUADRILATERALS

A quadrilateral has four sides.
The four angles in a quadrilateral add up to 360°.

SPECIAL QUADRILATERALS

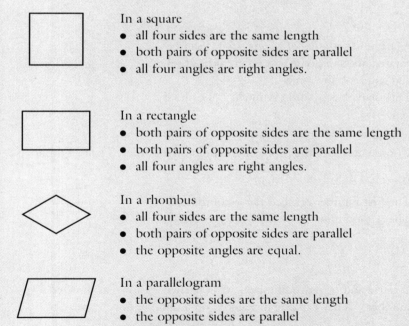

In a square
- all four sides are the same length
- both pairs of opposite sides are parallel
- all four angles are right angles.

In a rectangle
- both pairs of opposite sides are the same length
- both pairs of opposite sides are parallel
- all four angles are right angles.

In a rhombus
- all four sides are the same length
- both pairs of opposite sides are parallel
- the opposite angles are equal.

In a parallelogram
- the opposite sides are the same length
- the opposite sides are parallel
- the opposite angles are equal.

In a trapezium
- just one pair of opposite sides are parallel.

PARALLEL LINES When two parallel lines are cut by a transversal

the *corresponding angles* are equal

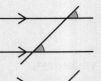

the *alternate angles* are equal

the *interior angles* add up to $180°$.

SYMMETRY A shape has *line symmetry* if, when folded along the line, one half fits exactly over the other half. This shape has one line of symmetry.

A shape has *rotational symmetry* if it can be rotated about a point to a different position and still look the same.
This shape has rotation symmetry of order 8.

COORDINATES Coordinates give the position of a point as an ordered pair of numbers,

e.g. $(2, 4)$.

The first number is called the x-coordinate.
The second number is called the y-coordinate.

A shape has rotation symmetry of order 8.

DIRECTED NUMBERS Positive and negative numbers are collectively known as directed numbers. They can be represented on a number line.

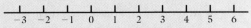

The rules for adding and subtracting directed numbers are

$$+(+a) = a \qquad +(-a) = -a$$
$$-(+a) = -a \qquad -(-a) = a$$

FORMULAS

A formula is a general rule for finding one quantity in terms of other quantities, e.g. the formula for finding the area of a rectangle is given by

$$\text{Area} = \text{length} \times \text{breadth}$$

When letters are used for unknown numbers, the formula can be written more concisely, i.e. the area, $A \text{ cm}^2$, of a rectangle measuring $l \text{ cm}$ by $b \text{ cm}$, is given by the formula

$$A = l \times b$$

SUMMARISING DATA

For a list of values,

- the *range* is the difference between the largest value and the smallest value
- the *mean* is the sum of all the values divided by the number of values
- the *median* is the middle value when they have been arranged in order of size, (when the middle of the list is halfway between two values, the median is the average of these two values)
- the *mode* is the value that occurs most frequently.

PROBABILITY

The probability that an event A happens is given by

$$\frac{\text{the number of ways in which } A \text{ can occur}}{\text{the total number of equally likely outcomes}}$$

When we perform experiments to find out how often an event occurs, the *relative frequency* of the event is given by

$$\frac{\text{the number of times the event occurs}}{\text{the number of times the experiment is performed}}$$

Relative frequency is used to give an approximate value for probability.

ALGEBRAIC EXPRESSIONS

Terms such as $5n$ mean $5 \times n = n + n + n + n + n$

Similarly ab means $a \times b$.

$2x + 5x$ are like terms and can be simplified to $7x$.

SOLVING EQUATIONS

An equation is a relationship between an unknown number, represented by a letter, and other numbers, e.g. $2x - 3 = 5$

Solving the equation means finding the unknown number.

Provided that we do the same thing to both sides of an equation, we keep the equality; this can be used to solve the equation,

e.g.   to solve $2x - 3 = 5$

First add 3 to both sides $2x - 3 + 3 = 5 + 3$

This gives $2x = 8$

Now divide each side by 2 $x = 4$

The exercises that follow are *not* intended to be worked through before starting the main part of this book. They are here for you to use when you need practice on the basic techniques.

**REVISION
EXERCISE 1.1
(Whole numbers)**

1 Write down

   a the value of the 4 in the number 1407

   b the number 3592 in words

   c the number two thousand seven hundred and eight in figures

2 Put the numbers 709, 970, 917, 794 and 799 in order with the smallest number first.

3 Write 1584   **a** correct to the nearest number of tens

              b correct to the nearest number of hundreds.

4 Without using a calculator, find

a $293 + 827$	**d** $834 - 392$	**g** $263 + 713 + 429$
b $421 + 943$	**e** $747 - 559$	**h** $612 - 385$
c $513 + 748$	**f** $638 - 145$	**i** $5210 - 3763 + 2140$

5 Without using a calculator, find

a 72×8	**g** $125 \div 5$	**m** 78×4
b 49×6	**h** $276 \div 6$	**n** $435 \div 5$
c 55×7	**i** $416 \div 8$	**p** $432 \div 6$
d 64×500	**j** $520 \div 20$	**q** $392 \div 8$
e 38×700	**k** $420 \div 70$	**r** 382×7
f 58×900	**l** $2550 \div 30$	**s** 561×9

6 Without using a calculator, find

a $8 \times 243 - 1492$	**d** $(18 + 15) \times 3 - 38$
b $4298 - 367 \times 8$	**e** $246 \div 23$, giving the remainder
c $36 \div (13 - 4) + 8$	**f** $1429 \div 54$, giving the remainder

7 a Write in index form   **i** $5 \times 5 \times 5 \times 5$ **ii** $7 \times 7 \times 7$

   b Find the value of

     i 3^4 **iii** $2^2 \times 3^2$ **v** $3^2 \times 5^3$

     ii $2^3 \times 3^2$ **iv** $2^3 \times 7^2$ **vi** $2^3 \times 5^2$

   c Express in index form   **i** 32   **ii** 125

   d Express 360 as the product of prime numbers in index form.

8 a Find the largest whole number that will divide exactly into

    i 42, 70 and 210   **ii** 72, 48, 120

b **i** Find the smallest whole number that 3, 5 and 6 will divide into exactly.

  ii Repeat part **i** for the numbers 6, 8 and 9.

9 a Write down the next square number after   **i** 36   **ii** 100

b Write down the prime numbers between 30 and 50.

c Which of the numbers 8, 11, 15, 22, 28, 30, 35, 36, 42, 45 are **i** prime numbers

      ii rectangular numbers

      iii triangular numbers?

10 a Estimate the value of each of the following multiplications, then use a calculator to find the exact value.

    i 53×472 **ii** 137×63 **iii** 86×952 **iv** 393×32

b Find, without using a calculator, the value of

    i $3 \times 4 \div 2 + 2 \times 5$

    ii $3 \times 5 \times 4 - 6 \times 5 \div 2$

    iii $12 \div 4 - 2 - 3 \times 2 + 7$

REVISION EXERCISE 1.2 (Fractions, decimals and percentages)

1 Fill in the missing numbers to make equivalent fractions.

a $\dfrac{3}{4} = \dfrac{}{28}$ **c** $\dfrac{5}{13} = \dfrac{25}{}$

b $\dfrac{11}{20} = \dfrac{44}{}$ **d** $\dfrac{4}{5} = \dfrac{}{60}$

2 Express   **a** $\dfrac{7}{20}$ as a decimal

        b 0.66 as a fraction in its lowest terms

        c 84% as a fraction in its lowest terms

        d 36% as a decimal

        e 0.54 as a percentage

3 Put either $>$ or $<$ between each of the following fractions.

a $\dfrac{9}{10}$ $\dfrac{7}{8}$ **b** $\dfrac{7}{10}$ $\dfrac{2}{3}$ **c** $\dfrac{1}{4}$ $\dfrac{3}{13}$ **d** $\dfrac{3}{7}$ $\dfrac{4}{9}$

4 a Change into an improper fraction

    i $3\frac{5}{7}$ **ii** $1\frac{4}{9}$ **iii** $5\frac{2}{3}$

b Give as a mixed number   **i** $\frac{38}{5}$ **ii** $\frac{14}{3}$ **iii** $\frac{54}{7}$

5 Find

a $\frac{2}{7} + \frac{3}{7}$ **g** $\frac{2}{3} + \frac{1}{4} + \frac{5}{6}$ **m** $\frac{5}{12} - \frac{1}{3}$

b $\frac{2}{5} + \frac{3}{10}$ **h** $\frac{3}{4} + \frac{1}{2} + \frac{1}{5}$ **n** $\frac{5}{8} - \frac{5}{12}$

c $\frac{3}{8} + \frac{5}{8}$ **i** $\frac{2}{3} + \frac{4}{5} + \frac{1}{6}$ **p** $\frac{1}{4} + \frac{3}{7} - \frac{3}{8}$

d $\frac{4}{5} + \frac{1}{7}$ **j** $\frac{11}{13} - \frac{2}{13}$ **q** $\frac{7}{8} - \frac{1}{2} + \frac{1}{4}$

e $\frac{5}{9} + \frac{2}{7}$ **k** $\frac{7}{12} - \frac{5}{12}$ **r** $\frac{11}{12} + \frac{1}{3} - \frac{3}{4}$

f $\frac{1}{4} + \frac{1}{3} + \frac{1}{2}$ **l** $\frac{3}{4} - \frac{1}{2}$ **s** $\frac{5}{9} - \frac{5}{18} + \frac{1}{2}$

6 Find

a $2\frac{3}{5} + 1\frac{3}{4}$ **c** $3\frac{2}{3} + 5\frac{5}{6}$ **e** $6\frac{7}{8} + 5\frac{2}{7}$

b $5\frac{3}{4} - 2\frac{1}{3}$ **d** $7\frac{6}{7} - 3\frac{1}{4}$ **f** $9\frac{4}{5} - 5\frac{7}{8}$

7 Find, without using a calculator

a $2.4 + 6.1$ **c** $5.1 - 1.874$ **e** $3 - 0.44$

b $13.72 + 4.96$ **d** $2.63 + 1.059 + 3.2$ **f** $9.29 - 5.7$

8 Find, without using a calculator

a 32.44×10 **e** 0.0849×100 **i** $0.392 \div 7$

b $62.93 \div 100$ **f** 1.72×300 **j** 0.0421×500

c 5.49×400 **g** $59 \div 200$ **k** $13.68 \div 30$

d $0.24 \div 6$ **h** $4.704 \div 8$ **l** $32.48 \div 20$

9 a Give 268 correct to **i** the nearest 10 **ii** the nearest 100.

b Give 6.498 correct to **i** the nearest whole number

ii 1 decimal place

iii 2 decimal places.

c Give 82.064 correct to **i** 1 decimal place **ii** 2 decimal places

10 Find, without using a calculator

a 0.005×7 **c** 160×1.5 **e** 0.05×4

b 0.6×0.34 **d** 0.4×0.8 **f** 300×1.8

11 Find, correct to 2 decimal places if necessary

a $6.09 \div 9$ **c** $0.0045 \div 0.5$ **e** $0.0143 \div 0.11$

b $13.2 \div 7$ **d** $0.5 \div 0.2$ **f** $0.0036 \div 0.07$

**REVISION
EXERCISE 1.3
(Metric and
   Imperial units)**

In questions **1** to **20** express each given quantity in terms of the units in brackets.

1 7240 m (km)

2 500 g (kg)

3 340 mm (cm)

4 87 cm (m)

5 0.04 kg (g)

6 1.09 m (cm)

7 2 m 45 cm (cm)

8 6450 mm (m)

9 0.0426 m (mm)

10 2.4 g (mg)

11 3 ft 4 in (in)

12 4 yd 2 ft (ft)

13 48 in (ft)

14 65 in (ft and in)

15 150 ft (yd and ft)

16 3 lb 4 oz (oz)

17 12 st 3 lb (lb)

18 186 lb (stones and lb)

19 2 tons 12 cwt (cwt)

20 60 oz (lb and oz)

In questions **21** to **30** write the given quantity, roughly, in terms of the unit in brackets.

21 5 kg (lb)

22 9 lb (kg)

23 50 miles (km)

24 50 km (miles)

25 10 ft (m)

26 250 g (oz)

27 2 m (in)

28 2 in (cm)

29 $\frac{1}{2}$ in (mm)

30 20 cm (in)

Find, giving your answer in the unit in brackets

31 3 m + 60 cm (m)

32 5 kg + 500 g (kg)

33 116 g + 0.04 kg + 3940 mg (g)

34 2 m + 84 cm + 142 mm (cm)

35 1.6 t − 490 kg (kg)

36 0.9 m − 426 mm (cm)

1 State whether each of the following angles is acute, obtuse or reflex.

a 93°   **b** 216°   **c** 34°   **d** 254°   **e** 310°   **f** 89°

2 Find the sizes of the marked angles.

a
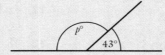

d
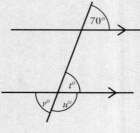

b
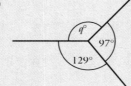

c
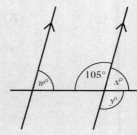

e

3 Find

a the supplement of   **i** 48°   **ii** 126°

b the third angle in a triangle in which the sizes of the other two angles are 67° and 45°

c the angles in a quadrilateral if all the angles are the same size.

4 If you stand facing west and turn clockwise through $\frac{3}{4}$ of a revolution, in which direction are you facing?

5 For each shape state

a the number of axes of symmetry

b whether or not the shape has rotational symmetry.

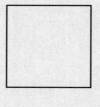

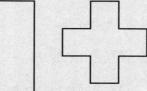

A   **B**   **C**   **D**

6 Find the sizes of the angles marked with a letter.

a

$a°$

$52°$ $43°$

b

$b°$

$32°$

$c°$

c

$e°$

$d°$ $f°$

d

$h°$

$18°$

$i°$ $g°$ $72°$

e

$j°$ $k°$

$l°$ $49°$

$64°$

f

$n°$

$m°$ $p°$

g

$s°$

$38°$ $r°$

$q°$ $57°$

h

$54°$

$v°$

$65°$ $u°$ $t°$

7 Which of these shapes are congruent?

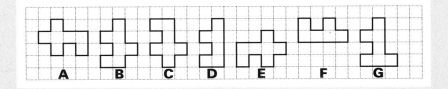

A   B   C   D   E   F   G

EXERCISE 1.5
(Summarising
data and
probability)

1 Find the range, mode, median and mean of the lengths

15 cm,   3 cm,   14 cm,   10 cm,   12 cm,   13 cm,   10 cm.

2 Find the range, mode, median and mean of the weights

66 kg,   56 kg,   54 kg,   62 kg,   59 kg,   62 kg,   61 kg.

3 One card is chosen from a pack of 52 ordinary playing cards. What is the probability that the card is

a black   **b** a diamond   **c** an ace   **d** the king of spades?

4 A dice is rolled. What is the probability that the number of spots uppermost is

a 4   **c** a prime number
b an even number   **d** a number that is both prime and odd?

5 A bag contains 3 red counters, 4 blue counters and 7 white counters. One counter is taken from the bag. What is the probability that this counter is

a blue   **b** red or white   **c** not red?

6

Mark	0	1	2	3	4	5
Frequency	1	0	4	7	10	3

The table shows the marks obtained by the pupils in Class 8G in a spelling test.

a How many pupils took the test?

b What was the modal mark?

c Find the median mark.

d Calculate the mean mark.

7 The table shows the number of goals scored by the first team in a hockey club last season.

Number of goals	0	1	2	3	4	5	6
Frequency	13	17	8	5	3	1	2

a How many games did they play?

b Find   **i** the mode
               ii the median
               iii the mean number of goals scored per match.

c What is the range?

d Illustrate the distribution with a bar chart.

REVISION EXERCISE 1.6
(Area and volume)

1 Find the perimeter and area of each shape, clearly stating the units involved.

   a A square of side 4 in.

   b A rectangle measuring 3 ft by 2 ft.

   c A square of side 200 mm.

   d A rectangle measuring 12 cm by 16 cm.

e 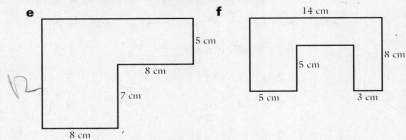 **f**

2 a The area of a square is 64 cm². How long is one side?

   b The area of a rectangle is 48 cm². If it is 8 cm long, how wide is it?

3 Find, giving your answer in the unit in brackets, the volume of a cuboid measuring

   a 3 cm by 6 cm by 8 cm (cm³)

   b 20 cm by 15 cm by 10 cm (cm³)

   c 8 cm by 4 cm by 1.4 m (cm³)

   d 1.5 m by 2 m by 1.5 m (m³)

   e 50 mm by 70 mm by 80 mm (mm³)

   f 4.5 cm by 5.4 cm by 6 cm (cm³)

4 Find the volume of a cube of side

   a 7 cm   **c** $\frac{3}{4}$ m   **e** 1.5 cm

   b 90 mm   **d** $\frac{5}{8}$ in   **f** 0.3 m

5 Express

   a 3 cm² in mm²   **d** 2 km² in m²   **g** 8000 mm³ in cm³

   b 3000 cm² in m²   **e** 12 cm³ in mm³   **h** 1.5 litres in cm³

   c 0.4 m² in cm²   **f** 0.003 m³ in cm³   **i** 3 m³ in litres

6 Give roughly

   a the number of pints in 10 litres

   b the approximate equivalent in litres of $2\frac{1}{2}$ gallons.

REVISION EXERCISE 1.7 (Algebra)

1 Simplify

a $3a - 2a + 5a$

b $4b + 6b - 5b$

c $3b - 6 + 4b - 7$

d $3c + 8c - 5c + 2c$

e $5c - 8 - 3c - 4$

f $9a + 3b - 4b + 3a$

g $2a - 4 + 5a - 7 + 3a$

h $5x - 7y - 2x + 4y$

i $3x + 4y - 4x + 5y$

j $8b - 2c - 6c - 3b$

k $5x - 10y - 4x + 12y$

l $3x - 5y + 5x + 2y$

Solve the equations.

2 $x + 8 = 14$

3 $x - 4 = 16$

4 $6y = 36$

5 $3x - 5 = 7$

6 $7 + 4x = 23$

7 $x - 0.8 = 1$

8 $5x - 7 = 2x + 8$

9 $15 - 4x = 3$

10 $3a = 2$

11 $9x - 11 - 3x - 13 = 0$

REVISION EXERCISE 1.8 (Directed numbers and formulas)

1 Put $<$ or $>$ between the numbers in each of the following pairs.

a 8   5   **b** -7 5   **c** 6 -5 **d** -4 -6

2 Find

a $-3 - 4 + 9$

b $8 - 4 - 7$

c $(+3) - (+4) - (+5)$

d $3 - (-7)$

e $-6 - (+2)$

f $-5 - (+3) - (-6)$

g $7 + (3 - 5)$

h $(8 - 12) - (7 - 11)$

i $-10 - (+5)$

j $(7 - 2) - (9 - 12)$

3 Find   **a** $(-7) \times 2$ **c** $(-42) \div 6$ **e** $(-12) \div 4$

b $(-8) \div 2$ **d** -8×3 **f** $(-25) \times 3$

4 Given that $P = 4a$, find P when   **a** $a = 3$ **b** $a = 8$

5 Use the formula $p = \dfrac{48}{v}$ to find p when   **a** $v = 1$ **b** $v = 8$

6 Use the formula $C = 5k - 3$ to find C when

a $k = 3$ **b** $k = 10$

7 Given that $P = 2q - r$ find P when

a $q = 2, r = -3$ **b** $q = -4, r = 7$ **c** $q = -2, r = -8$

WORKING WITH NUMBERS

We sometimes have to work with very large numbers and very small numbers. For example, Sian had to research some facts about the Solar System.

Here are a few extracts from information that Sian found in a science dictionary.

> **Earth** The planet that orbits the Sun between the planets Venus and Mars at a mean distance from the Sun of 149 600 000 km. It has a mass of about 5.976×10^{24} kg.
>
> **Sun** The star at the centre of the Solar System. The Sun has a mass of 1.9×10^{30} kg.
>
> **Venus** A planet having its orbit between Mercury and Earth. Its mean distance from the Sun is 108×10^{6} km and its mean diameter is 12 100 km.
>
> In a vacuum light travels 1 km in $3.335\,609 \times 10^{-6}$ seconds.

- If Sian wants to make sense of this information, she needs to know what the numbers such as 1.9×10^{30} and $3.335\,609 \times 10^{-6}$ mean.
- If she wants to find the difference between the masses of Earth and Venus, she needs to be able to work with such numbers.

EXERCISE 1A Discuss what you need to know and be able to do to answer these questions.

1 Why do you think the distance of Venus from the Sun is written as 108×10^{6} km whereas the distance of Earth from the Sun is written as 149 600 000 km?

2 The time it takes light to travel 1 km is given as $3.335\,609 \times 10^{-6}$ seconds. Is this a large or a small number?

3 Which planet is further from the sun, Earth or Venus?

4 How long does it take for light from the Sun to reach Venus?

Discussion of the problems in the last exercise show that there are occasions when we need to understand and work with numbers written in the form 1.9×10^{30}.

Firstly we will remind ourselves of the meaning of 10^{30}.

POSITIVE
INDICES

We have seen, in Book 7A, that 3^2 means 3×3
and that $2 \times 2 \times 2$ can be written as 2^3.
So 10^{30} means thirty 10s multiplied together – this is a very large number!

The superscript (the small upper number) is called the *index* or *power*.
The lower number is called the *base*.
It follows that 2 can be written as 2^1, although we do not normally do this.
It also follows that, if a is any number, then

$$a^1 = a$$

EXERCISE 1B

Find the value of 2^5.

$2^5 = 2 \times 2 \times 2 \times 2 \times 2$
$\quad = 32$

Remember $2 \times 2 \times 2 \times 2 \times 2 = 4 \times 2 \times 2 \times 2$
$= 8 \times 2 \times 2$
$= 16 \times 2$
$= 32$

Find the value of

1 4^1 **3** 10^3 **5** 10^1 **7** 1.2^2

2 5^3 **4** 2^7 **6** 10^6 **8** 1.1^2

Find the value of 3.6×10^2.

$3.6 \times 10^2 = 3.6 \times 100$
$\quad\quad\quad = 360$

Remember that numbers can be multiplied in
any order, so $3.6 \times 10^2 = 3.6 \times 10 \times 10$
$= 3.6 \times 100$
and to multiply by 100, we move the
decimal point 2 places to the right and fill in
gaps with zeros.

Find the value of

These are
extra questions

9 7.2×10^3 **12** 3.82×10^3 **15** 4.63×10^1

10 8.93×10^2 **13** 2.75×10^1 **16** 5.032×10^2

11 6.5×10^4 **14** 5.37×10^5 **17** 7.09×10^2

18 Write these numbers as powers of 10.

     a 100   **b** 1000   **c** 1 000 000   **d** 10 000 000 000

19 a Use the information at the beginning of this chapter to write down,
as an ordinary number, the distance of Venus from the Sun.

     b Now answer question **3** in **Exercise 1A**.

20 Why do you think that the mass of the Sun is written in the form 1.9×10^{30} kg and not as an ordinary number?

These are 'challenge' questions

21 The mass of the Earth is 5.98×10^{24} kg. How many tonnes is this?

22 A satellite is orbiting Earth at constant speed so that it travels 2.8×10^4 kilometres each hour. How far does it travel in one year, that is, in 8.76×10^3 hours?

MULTIPLYING NUMBERS WRITTEN IN INDEX FORM

If you did the last question in the exercise above, you had to multiply 2.8×10^4 by 8.76×10^3. The calculation is easier if we can find $10^4 \times 10^3$ without having to write down all the zeros.

We can write $2^2 \times 2^3$ as a single number in index form because both bases are 2,

i.e.
$$2^2 \times 2^3 = (2 \times 2) \times (2 \times 2 \times 2)$$
$$= 2 \times 2 \times 2 \times 2 \times 2$$
$$= 2^5$$
$$\therefore \qquad 2^2 \times 2^3 = 2^{2+3} = 2^5$$

But we cannot do the same with $2^2 \times 5^3$ because the numbers multiplied together are not all 2s (nor are they all 5s),
i.e. the bases of the two numbers are *not* the same.

> We can multiply together different powers of the *same base* by adding the indices

but we cannot multiply together powers of different bases in this way.

EXERCISE 1C

> Write $a^3 \times a^4$ as a single expression in index form.
>
> $a^3 \times a^4 = a^{3+4}$
> $\qquad = a^7$

Write as a single expression in index form

1 $3^5 \times 3^2$	**5** $10^4 \times 10^3$	**9** $10^9 \times 10^8$
2 $7^5 \times 7^3$	**6** $b^3 \times b^2$	**10** $p^6 \times p^8$
3 $9^2 \times 9^8$	**7** $5^4 \times 5^4$	**11** $4^7 \times 4^9$
4 $2^2 \times 2^7$	**8** $12^4 \times 12^5$	**12** $r^5 \times r^3$

DIVIDING
NUMBERS
WRITTEN IN
INDEX FORM

If we want to write $2^5 \div 2^2$ as a single number in index form then

$$2^5 \div 2^2 = \frac{2^5}{2^2} = \frac{2 \times 2 \times 2 \times 2 \times 2}{2 \times 2} = 2^3$$

i.e.

$$\frac{2^5}{2^2} = 2^{5-2} = 2^3$$

> We can divide different powers of the *same base*
> by subtracting the indices

EXERCISE 1D

> Write $a^7 \div a^3$ as a single expression in index form.
>
> $a^7 \div a^3 = a^{7-3} = a^4$

Write as a single expression in index form

These are
extra questions

1 $4^4 \div 4^2$ **4** $10^8 \div 10^3$ **7** $6^{12} \div 6^7$

2 $7^9 \div 7^3$ **5** $q^9 \div q^5$ **8** $b^7 \div b^5$

3 $5^6 \div 5^5$ **6** $15^8 \div 15^4$ **9** $p^4 \div p^3$

These are
'challenge'
questions

10 $6^4 \times 6^7$ **13** $c^6 \div c^3$ **16** $4^2 \times 4^3 \div 4^4$

11 $3^9 \div 3^6$ **14** $a^9 \times a^3$ **17** $3^6 \div 3^2 \times 3^4$

12 $2^8 \div 2^7$ **15** $2^2 \times 2^4 \times 2^3$ **18** $b^2 \times b^3 \times b^4$

NEGATIVE
INDICES

When we use the rule for dividing numbers in index form, we may find we have a negative index.

Consider $2^3 \div 2^5$

Subtracting the indices gives $2^3 \div 2^5 = 2^{3-5} = 2^{-2}$

But, as a fraction $2^3 \div 2^5 = \frac{2 \times 2 \times 2}{2 \times 2 \times 2 \times 2 \times 2} = \frac{1}{2^2}$

Therefore 2^{-2} means $\frac{1}{2^2}$

In the same way 5^{-3} means $\frac{1}{5^3}$

and, if a and b are any two numbers, $a^{-b} = \frac{1}{a^b}$

EXERCISE 1E

Find the value of 5^{-2}.

$$5^{-2} = \frac{1}{5^2}$$

$$= \frac{1}{25}$$

Find the value of

___ These are extra questions

1 2^{-2}	**6** 4^{-2}	**11** 4^{-3}	**16** 5^{-3}
2 3^{-3}	**7** 3^{-4}	**12** 6^{-2}	**17** 10^{-2}
3 2^{-4}	**8** 5^{-1}	**13** 15^{-1}	**18** 2^{-3}
4 3^{-1}	**9** 3^{-2}	**14** 6^{-1}	**19** 10^{-1}
5 7^{-1}	**10** 4^{-1}	**15** 7^{-2}	**20** 8^{-2}

Find the value of 1.7×10^{-2}.

$$1.7 \times 10^{-2} = 1.7 \times \frac{1}{10^2}$$

$$= \frac{1.7}{100} = 0.017$$

Remember: to divide by 100 move the decimal point two places to the left.

Find the value of

21 3.4×10^{-3}	**26** 4.67×10^{-5}
22 2.6×10^{-1}	**27** 3.063×10^{-1}
23 6.2×10^{-2}	**28** 2.805×10^{-2}
24 8.21×10^{-3}	**29** 51.73×10^{-4}
25 5.38×10^{-4}	**30** 30.04×10^{-1}

31 Write these numbers as fractions.   **a** 10^{-3} **b** 10^{-5}

32 Write these numbers as powers of 10.

   a $\frac{1}{100}$ **b** $\frac{1}{1000}$ **c** $\frac{1}{1\,000\,000}$

33 Write these numbers as decimals.   **a** 10^{-4} **b** 10^{-2}

34 Write these numbers as powers of ten.

   a 0.001 **b** $0.000\,000\,000\,01$

Write $2 \div 2^3$ as a single number in index form.

$2 \div 2^3 = 2^1 \div 2^3 = 2^{-2}$

Write as a single number in index form

___ These are
extra questions

35 $5^2 \div 5^4$ **40** $10^3 \div 10^6$

36 $3 \div 3^4$ **41** $b^5 \div b^9$

37 $6^4 \div 6^7$ **42** $4^8 \div 4^3$

These are
'challenge'
questions

38 $2^5 \div 2^3$ **43** $c^5 \div c^4$

39 $a^5 \div a^7$ **44** $2^a \div 2^b$

**THE MEANING
OF a^0**

Consider $2^3 \div 2^3$

Subtracting indices gives $2^3 \div 2^3 = 2^0$

Simplifying $\frac{2^3}{2^3}$ gives $\frac{\cancel{2} \times \cancel{2} \times \cancel{2}}{\cancel{2} \times \cancel{2} \times \cancel{2}} = 1$

So 2^0 means 1

In the same way $a^3 \div a^3 = a^0$ (subtracting indices)

But $a^3 \div a^3 = \frac{a \times a \times a}{a \times a \times a} = 1$ (simplifying the fraction)

> Any number with an index of zero is equal to 1
> i.e. $a^0 = 1$

The next exercise contains a mixture of questions involving indices.

EXERCISE 1F

___ These are
extra questions

Find the value of

1 2^2

2 5^{-2}

3 4^3

4 3^{-1}

5 7^0

6 5^3

7 3^4

8 3^0

9 4^1

10 6^{-2}

11 10^{-3}

12 8^{-1}

Write as a single term in index form

13 $2^3 \times 2^4$

14 $4^6 \div 4^3$

15 $3^{-2} \times 3^4$

16 $a^4 \times a^3$

17 $a^7 \div a^3$

18 $5^4 \times 5^{-2}$

19 $3^5 \div 3^5$

20 $b^3 \div b^3$

21 $4^{-2} \times 4^6$

22 $5^3 \div 5^9$

These are
'challenge'
questions

23 $2^2 \times 2^4 \times 2^3$

24 $a^2 \times a^4 \times a^6$

25 $3^5 \times 3^2 \div 3^3$

26 $7^3 \times 7^3 \div 7^6$

27 $\dfrac{4^2 \times 4^6}{4^3}$

28 $a^3 \times a^2 \times a^5$

29 $3^2 \div 3^6 \times 3^2$

30 $b^3 \times b^{-3}$

31 $5^{-2} \times 5^{-3}$

32 $\dfrac{a^3 \times a^4}{a^7}$

33 Which of these numbers is the larger?

a 10^2 or 10^5 **b** 10^2 or 10^{-2} **c** 10^0 or 10^{-1}

34 Write these numbers in order of size with the smallest first.

$$10^4, \quad 10^{-3}, \quad 10^1, \quad 10^0, \quad 10^3, \quad 10^{-2}$$

35 Write these numbers in as short a form as possible.

a $3 \times 10^3 \times 4 \times 10^5$ **c** $(8 \times 10^9) \div (4 \times 10^5)$

b $2 \times 10^{20} \times 5 \times 10^{15}$ **d** $(6 \times 10^5) \div (3 \times 10^8)$

36 The first five terms of a sequence are 2, 4, 8, 16, 32.

a Write down the next three terms.

b Write down the tenth term in index form.

c Find an expression for the nth term.

This is
a 'challenge'
question

37 Cars and lorries can run out of control on steep hills. On some roads there are escape lanes; these are sand filled tracks designed to stop vehicles quickly. One of these is to be designed so that, when a vehicle enters it with a speed of 100 miles per hour,
after t seconds its speed is $(100 \times 3^{-t} - 1)$ miles per hour.

a What speed is expected after 1 second?

b What speed is expected after 3 seconds?

c About how long does it take to stop a vehicle?

d What is the value of $(100 \times 3^{-t} - 1)$ when $t = 6$?
Does this mean anything in relation to the speed of a vehicle?
Explain your answer.

STANDARD FORM

The mass of Earth, given at the beginning of this chapter, is

5.976×10^{24} kg,   i.e. $5\,976\,000\,000\,000\,000\,000\,000\,000$ kg

The mass of Earth is a very large number, and when written out in full it is not easy to judge its size without counting the zeros.
When a number is written 5.976×10^{24} it is in *standard form*; the advantages of using standard form are that it is easier to judge the size of the number and it is quicker to write down.

The same advantages apply to very small numbers. For example, the time it takes light to travel 1 km is

$3.335\,609 \times 10^{-6}$ seconds,   i.e, $0.000\,003\,335\,609$ seconds

> A number written in standard form is a number between
> 1 and 10 multiplied by a power of 10.

So 1.3×10^2 and 3.72×10^{-3} are in standard form,
but 13×10^5 and 0.26×10^{-3} are not in standard form because the first number is not between 1 and 10.
In previous exercises, when you were asked to find the value of expressions such as 2.5×10^{-3}, you were changing a number given in standard form into an ordinary number. Now we look at how to write an ordinary number in standard form.

CHANGING NUMBERS INTO STANDARD FORM

To write 6800 in standard form, the decimal point has to be placed between the 6 and the 8 to give a number between 1 and 10.
Counting then tells us that, to change 6.8 to 6800, we must move the decimal point three places to the right (that is, multiply by 10^3)

i.e. $$6800 = 6.8 \times 1000 = 6.8 \times 10^3$$

To give 0.019 34 in standard form, the point has to go between the 1 and the 9 to give a number between 1 and 10.
This time counting tells us that, to change 1.934 to 0.019 34, we have to move the point two places to the left (i.e. divide by 10^2)

so $$0.01934 = 1.934 \div 100 = 1.934 \times 10^{-2}$$

EXERCISE 1G

Write the following numbers in standard form.

___ These are extra questions

1 2500	**6** 39 070	**11** 26 030
2 630	**7** 4 500 000	**12** 547 000
3 15 300	**8** 530 000 000	**13** 30 600
4 260 000	**9** 40 000	**14** 4 060 000
5 9900	**10** 80 000 000 000	**15** 704

Write 0.006 043 in standard form.

$$0.006\,043 = 6.043 \times 10^{-3}$$

$0.006\,043 = 6.043 \div 1000$

Write the following numbers in standard form.

16 0.026	**21** 0.79	**26** 0.907
17 0.0048	**22** 0.0069	**27** 0.0805
18 0.053	**23** 0.000 007 5	**28** 0.088 08
19 0.000 018	**24** 0.000 000 000 4	**29** 0.000 704 4
20 0.52	**25** 0.684	**30** 0.000 000 000 073

31 88.92	**36** 84	**41** 5090
32 0.000 050 6	**37** 351	**42** 268 000
33 0.000 000 057	**38** 0.09	**43** 30.7
34 503 000 000	**39** 0.007 05	**44** 0.005 05
35 99 000 000	**40** 36	**45** 0.000 008 8

Write these numbers as ordinary numbers.

— These are
extra questions

46 3.78×10^3 **48** 5.3×10^6 **50** 6.43×10^{-8}

47 1.26×10^{-3} **49** 7.4×10^{14} **51** 4.25×10^{12}

52 Write these numbers in standard form and then place them in order of size, with the smallest first.

    a 576 000 000, 20 000 000 000, 997 000 000, 247 000, 37 500

    b 0.005 27, 0.600 05, 0.9906, 0.000 000 050 2, 0.003 005

    c 0.0705, 7.080 000, 79.3, 0.007 008 09, 560 800

53 a Without using a calculator, find 1 200 000 × 400 000 and write your answer in standard form.

    b Now use your calculator to find 1 200 000 × 400 000. Write down exactly what is showing on the display of your calculator. What do you think it means?

**SIGNIFICANT
FIGURES**

Danny measured her height as 1678 mm. She measured it again to make sure; this time the reading was 1676 mm.

- Try measuring your own height in millimetres three times. You will almost certainly find that you have three different measurements.
- Because it is not possible to measure height exactly, we need to round the measurement.

It should be possible to measure your height to the nearest 10 mm, that is, to the nearest centimetre. In Danny's case, it is reasonable to give her height as 1680 mm to the nearest 10 mm.

If Danny used a measure graduated in centimetres, her first measurement would be 167.8 cm; this is 168 cm to the nearest centimetre.

If the ruler were graduated in metres, the reading would be 1.678 m. To the nearest centimetre, this gives a rounded height of 1.68 m correct to 2 decimal places.

We could also give this measurement to the same degree of accuracy in kilometres,

i.e. 0.001 68 km correct to 5 decimal places (d.p.).

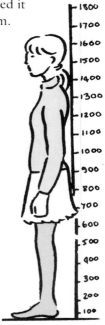

So, depending on which unit is chosen, Danny could give her height as

> 1680 mm to the nearest 10 mm,
> 168 cm to the nearest centimetre,
> 1.68 m correct to 2 d.p.,
> 0.001 68 km correct to 5 d.p.

Notice that the three figures 1, 6 and 8 occur in all four numbers and that it is the 8 that has been corrected in each case.

The figures 1, 6, and 8 are called the *significant figures* and in all four cases the number is given correct to 3 significant figures.

Using significant figures rather than decimal places has advantages. For example, if you are asked to measure your height and give the answer correct to 3 significant figures (s.f.), then you can choose any convenient unit. You do not need to be told which unit to use.

Reading from left to right, the *first significant figure* in a number is *the first non-zero figure*.
So 1 is the first significant figure in 170.6,
and 2 is the first significant figure in 0.025 09.

The *second significant figure* is the *next* figure to the right, *whether or not it is zero*, (7 in the case of 170.6 and 5 in the case of 0.025 09).

The *third significant figure* is the next figure to the right again (0 in both cases), and so on.

EXERCISE 1H

For 0.001 503 write down **a** the first significant figure

b the third significant figure

a The first s.f. is 1

b The third s.f. is 0

> The first significant figure is always the first non-zero figure, the second significant figure is the next figure to the right whether or not it is zero, and so on.

In each of the following numbers write down the significant figure specified in the bracket.

1 36.2 (1st)

2 378.5 (3rd)

3 0.0867 (2nd)

4 3.786 (3rd)

5 47.632 (2nd)

6 5.083 (3rd)

7 34.807 (4th)

8 0.076 03 (3rd)

9 54.06 (3rd)

10 5.7087 (4th)

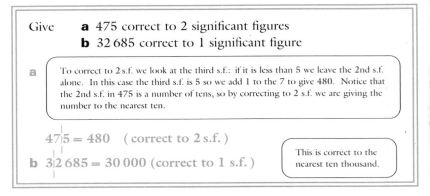

Give   **a** 475 correct to 2 significant figures
        b 32 685 correct to 1 significant figure

a   To correct to 2 s.f. we look at the third s.f.: if it is less than 5 we leave the 2nd s.f. alone. In this case the third s.f. is 5 so we add 1 to the 7 to give 480. Notice that the 2nd s.f. in 475 is a number of tens, so by correcting to 2 s.f. we are giving the number to the nearest ten.

47|5 = 480   (correct to 2 s.f.)

b 3|2 685 = 30 000 (correct to 1 s.f.)

This is correct to the nearest ten thousand.

___ **These are extra questions**

Give the following numbers correct to 1 significant figure.

11 59 727   **14** 80 755   **17** 908

12 4164   **15** 476   **18** 26

13 4 396 185   **16** 51 488   **19** 980

Give the following numbers correct to 2 significant figures.

20 4673   **23** 72 601   **26** 6992

21 57 341   **24** 50 047   **27** 476

22 59 700   **25** 9973   **28** 597

Give 0.021 94 correct to 3 significant figures.

0.0219|4 = 0.0219   (correct to 3 s.f.)

The fourth s.f. is 4 so we leave the third s.f. alone.

Give the following numbers correct to 3 significant figures.

29 0.008 463   **32** 78.49   **35** 7.5078

30 0.825 716   **33** 46.8451   **36** 369.649

31 5.8374   **34** 0.007 854 7   **37** 53.978

Give each of the following numbers correct to the number of significant figures indicated in the bracket.

38 46.931 06   (2)   **43** 4537   (1)

39 0.006 845 03 (4)   **44** 37.856 72   (3)

40 576 335   (1)   **45** 6973   (2)

41 497   (2)   **46** 0.070 865   (3)

42 7.824 38   (3)   **47** 0.067 34   (1)

> Find $50 \div 8$ correct to 2 significant figures.
>
> (To find an answer correct to 2 s.f. we first work to 3 s.f.)
>
> $$\begin{array}{r} 6.2\,|5 \\ 8\,\overline{)50.0\,|0} \end{array}$$
>
> So $\quad 50 \div 8 = 6.3$ (correct to 2 s.f.)

Without using a calculator find, correct to 2 significant figures

48 $20 \div 6$	**51** $53 \div 4$	**54** $0.7 \div 3$
49 $10 \div 6$	**52** $125 \div 9$	**55** $0.23 \div 9$
50 $25 \div 2$	**53** $143 \div 5$	**56** $0.0013 \div 3$

Depending on the circumstances, we may have to round numbers to a given number of significant figures or to a particular place value.
The next exercise contains a mixture of rounding questions. Make sure that you read each one carefully.

EXERCISE 1I

Round each number to the accuracy given in brackets.

1 1547 (nearest 10)	**7** 16.903 (nearest 10)
2 2.578 (nearest unit)	**8** 0.0527 (3 d.p.)
3 73.79 (2 s.f.)	**9** 4.0579 (3 s.f.)
4 8.896 (2 d.p.)	**10** 2022 (nearest 10)
5 2993 (nearest 100)	**11** 0.000 357 (1 s.f.)
6 55.575 (1 d.p.)	**12** 36 835 (nearest 1000)

> A building firm stated that, to the nearest 100, it built 2600 homes last year. What is the greatest number of homes that it could have built and what is the least number of homes that it could have built?
>
> The smallest number that can be rounded up to 2600 is 2550. The biggest number that can be rounded down to 2600 is 2649.
>
> So the firm built at most 2649 homes and at least 2550 homes.

13 A bag of marbles is said to contain 50 marbles to the nearest 10. What is the greatest number of marbles that could be in the bag and what is the least number?

14 To the nearest thousand, the attendance at a particular Premier Division football match was 45 000. What is the largest number of spectators who could have been there and what is the smallest number who could have attended?

15 1500 people came to the school fête. If this number is correct to the nearest hundred, give the maximum and the minimum number of people who could have come.

16 The annual accounts of Scrub plc (soap manufacturers) gave the company's profit as £3 000 000 to the nearest million. What is the smallest profit that the company could have made?

17 The chairman of A. Brick (Builders) plc says that they employ 2000 people. If this number is correct to the nearest 100, what is the least number of employees that the company can have?

18 To the nearest metre, Vicky's garden is 12 m long. Vicky ran the length of her garden. What is the shortest length she may have run?

19 To the nearest centimetre, Michael's handspan is 9 cm.

   a What is the greatest width that Michael's handspan could be?

   b By how much could Michael's handspan be shorter than 9 cm?

20 Martin was asked how far, roughly, he had to travel to school.
Being the proud owner of a new pedometer, he said 1563 metres.
What do you think a reasonable answer to this question would be and why?

This is a 'challenge' question

21 A rectangle is 1.5 cm long and 2.7 cm wide, both measurements being correct to 1 decimal place. Find the area, rounding your answer to an appropriate number of decimal places.
Explain why you have chosen that number.

22 Claire is going to travel by train from London to Birmingham. She wants to know how long the train journey will take. Explain, with reasons, which of the following answers is sensible.

A $1\frac{1}{2}$ to 2 hours **C** 1 hour 45 minutes to the nearest $\frac{1}{4}$ hour

B 1 hour 41 minutes

23 A 10 metre length of string is cut into 7 equal pieces. How long is each piece? Round your answer to an appropriate number of significant figures.

24 One side of the playground at school is 30 m long to the nearest metre. Mark ran up and down this side 20 times.

a Find the least distance he may have run

b What is the difference between the least distance he may have run and the distance he thinks he has run using the given measurement?

ROUGH ESTIMATES

If you were asked to find 1.397×62.57 you could do it by long multiplication or you could use a calculator. Whichever method you choose, it is essential first to make a rough estimate of the answer. You will then know whether or not the actual answer you get is reasonable.

One way of estimating the answer to a calculation is to write each number correct to 1 significant figure.

So $$1.397 \times 62.57 \approx 1 \times 60 = 60$$

EXERCISE 1J

> Correct each number to 1 significant figure and hence give a rough answer to **a** 9.524×0.0837 **b** $54.72 \div 0.761$
>
> **a** $9.524 \times 0.0837 \approx 10 \times 0.08 = 0.8$
>
> **b** $\dfrac{54.75}{0.761} \approx \dfrac{50}{0.8} = \dfrac{500}{8} \approx 60$ ⟨ Giving $500 \div 8$ to 1 significant figure. ⟩

Correct each number to 1 significant figure and hence give a rough answer for each calculation.

1 4.78×23.7 **4** 354.6×0.0475 **7** 0.0062×574

2 56.3×0.573 **5** 576×256 **8** $7.835 \div 6.493$

3 $0.0674 \div 5.24$ **6** $82.8 \div 146$ **9** 4736×729

10 34.7×21

11 8.63×0.523

12 $(2.37)^3$

13 $(0.175)^{-2}$

14 $(0.19)^5$

15 0.0326×12.4

16 $0.007\,24 \times 0.783$

17 $(2.95)^4$

18 $(0.109)^{-3}$

19 $(52.7)^3$

Correct each number to 1 significant figure and hence estimate
$$\frac{0.048 \times 3.275}{0.367}$$

$$\frac{0.048 \times 3.275}{0.367} \approx \frac{0.05 \times 3}{0.4} = \frac{0.15}{0.4} = \frac{1.5}{4}$$

$$= 0.4 \quad (\text{correct to 1 s.f.})$$

Correct each number to 1 significant figure and hence give a rough answer for each calculation.

20 $\dfrac{3.87 \times 5.24}{2.13}$

21 $\dfrac{0.636 \times 2.63}{5.47}$

22 $\dfrac{21.78 \times 4.278}{7.96}$

23 $\dfrac{6.38 \times 0.185}{0.628}$

24 $\dfrac{(23.7)^2}{47.834}$

25 $\dfrac{89.03 \times 0.079\,37}{5.92}$

26 $\dfrac{975 \times 0.636}{40.78}$

27 $\dfrac{8.735}{5.72 \times 5.94}$

28 $\dfrac{0.527}{6.41 \times 0.738}$

29 $\dfrac{57.8}{(0.57)^2 \times (3.94)^2}$

30 The area of this metal machine part is given by calculating $\dfrac{3.142 \times (0.2954)^2}{2.26}$ cm^2.
Estimate this area to 1 significant figure.

31 A bullet is fired into a block of wood. The depth that it penetrates into the block is found from $\dfrac{0.5 \times 0.023 \times (126.4)^2}{1567}$ metres.
Estimate this depth to 1 significant figure.

These are 'challenge' questions

32 £1576 was put into a savings account 2 years ago.
The amount now in the account is calculated using
£1576 × 1.057 × 1.5 × 1.049 × 0.8.

  a Estimate the amount in the account.

  b Explain whether you think your estimate is more or less than the actual amount.

33 **a** Estimate 0.44 × 1.49.
    Now find the true value.
    By how much does your estimate differ from the true value?

  b Repeat part **a** for 0.41 × 1.01.

  c Comment on your answers to parts **a** and **b** and suggest a way in which you could improve your estimates.

34 The number of rabbits on Chessle Island is now estimated at 24 500. It is expected that the numbers will increase to $24\,500 \times 1.25^3$ in three years.

  a Estimate the size of the rabbit population in three years.

  b How accurate do you judge your estimate to be?

USING A CALCULATOR

Remember that a scientific calculator works out multiplications and divisions before additions and subtractions.

If you key in $\boxed{2}\ \boxed{+}\ \boxed{3}\ \boxed{\times}\ \boxed{4}\ \boxed{=}$, your calculator will work

out 3×4 before adding it to 2, i.e. $2 + 3 \times 4 = 2 + 12 = 14$.
If you want to add 2 and 3 and multiply the result by 4, either use

brackets, i.e. press $\boxed{(}\ \boxed{2}\ \boxed{+}\ \boxed{3}\ \boxed{)}\ \boxed{\times}\ \boxed{4}\ \boxed{=}$,

or do it in two stages, i.e. $\boxed{2}\ \boxed{+}\ \boxed{3}\ \boxed{=}\ \boxed{\times}\ \boxed{4}\ \boxed{=}$

The $\boxed{x^y}$ key is used to work out powers of numbers,

e.g. to find 2.56^3, press $\boxed{2}\ \boxed{.}\ \boxed{5}\ \boxed{6}\ \boxed{x^y}\ \boxed{3}\ \boxed{=}$

and to enter 10^5, press $\boxed{1}\ \boxed{0}\ \boxed{x^y}\ \boxed{5}$

The $\boxed{\pm}$ key is used to change the sign of a number from positive to negative, or viceversa.

To find the value of 1.94^{-3}, we use this key to change the index to a

negative number, i.e. press $\boxed{1}\ \boxed{.}\ \boxed{9}\ \boxed{4}\ \boxed{x^y}\ \boxed{3}\ \boxed{\pm}\ \boxed{=}$

If the result of a calculation is very large or very small, the calculator will display the result in standard form. For example, if a calculator is used to find $(2370)^5$, the result is $7.477\,247 \times 10^{16}$, but the display will not show the 10; it may look like 7.477247^{16} or $7.477247 \text{ E } 16$.

Note that the instructions given here apply to many *but not all* calculators. You need to use the manual that comes with your calculator.

EXERCISE 1K

> Use your calculator to find 0.025×3.981, giving the answer correct to 3 significant figures.
>
> Estimate: $0.025 \times 3.981 \approx 0.03 \times 4 = 0.12$
> Calculator: $0.025 \times 3.981 = 0.09952 \ldots$
> $= 0.0995$ (correct to 3 s.f.)
>
> > There is no need to write down all the figures in the display
> > For an answer correct to 3 s.f., it is enough to write down the first 4 significant figures.

For each calculation, first make a rough estimate of the answer, then use your calculator to give the answer correct to 3 significant figures.

___ These are
extra questions

1 2.16×3.28 **4** 4.035×2.116 **7** 2.304×3.251

2 2.63×2.87 **5** 3.142×2.925 **8** 8.426×1.086

3 1.48×4.74 **6** 6.053×1.274 **9** 5.839×3.618

10 $9.571 \div 2.518$ **13** $4.931 \div 3.204$ **16** $384 \div 21.8$

11 $5.393 \div 3.593$ **14** $8.362 \div 5.823$ **17** $45.8 \div 143.7$

12 $7.384 \div 2.51$ **15** $23.4 \div 56.7$ **18** $537.8 \div 34.6$

19 63.8×2.701 **22** $(34.2)^3$ **25** 39.03×49.94

20 $40.3 \div 2.74$ **23** 5007×2.51 **26** $2000 \div 52.66$

21 $400 \div 35.7$ **24** $5703 \div 154.8$ **27** $(36.8)^2$

28 $0.366 - 0.37 \times 0.52$

29 $0.0526 \times 0.372 + 0.027$

30 $6.924 + 1.56 \div 0.007\,93$

31 $0.638 \times 825 - 54.3$

32 $52 \times 0.0895 - 0.489$

33 $0.0826 - 0.348 \times 0.582$

34 $24.78 \times 0.0724 + 8.25$

35 $0.00835 \times 0.617 - 0.002\,47$

36 $0.5824 + 1.054 \times 6.813$

37 $0.74 + 8.42 \div 0.56$

Use your calculator to find $(\,2.486 - 1.295\,) \times 3.057$, giving the answer corrected to 3 significant figures.

Estimate:
$(\,2.486 - 1.295\,) \times 3.057 \approx (\,2 - 1\,) \times 3 = 1 \times 3 = 3$

This can be done in one stage using the brackets buttons:

press `(` `2` `.` `4` `8` `6` `−` `1` `.` `2` `9` `5` `)`
`x` `3` `.` `0` `5` `7` `=`

It can also be done in two stages:

press `2` `.` `4` `8` `6` `−` `1` `.` `2` `9` `5` `=`
and write down the result but do not clear the display. Then

press `x` `3` `.` `0` `5` `7` .

Calculator:
$(\,2.486 - 1.295\,) \times 3.057 = 3.640\ldots$
$= 3.64$ (correct to 3 s.f.)

First make an estimate of the answer, then use your calculator to find, correct to 3 significant figures

38 $54.6 \times (\,22.05 - 8.17\,)$

39 $6.04 \div (\,1.958 - 0.872\,)$

40 $(\,0.824 + 0.057\,) \times 27.45$

41 $(\,2.798 - 21.25\,) \div 12.099$

42 $\dfrac{0.014}{1.53 - 0.9889}$

43 $32.03 \times (\,17.09 - 16.9\,)$

44 $0.51 \div (\,0.45 + 0.327\,)$

45 $(\,1.033 + 0.29\,) \times 4.47$

46 $(\,0.029 - 0.0084\,) \div 1.88$

47 $\dfrac{24.5 + 9.992}{101.7}$

48 Use a calculator to find, correct to 3 significant figures, the answers to questions **20** to **29** in **Exercise 1J**. Compare the answers with the estimates.

49 Alia had to calculate $5.87 \times (27.4 + 4.82)$ to find the area, in square metres, of a paved patio.
Find the area, giving your answer correct to 2 significant figures.

50 Tony needed to calculate $2.57 + 8.36 \times 0.19$.
He used his calculator and wrote down the result as 2.08 correct to 3 significant figures.

a Without using your calculator, show that Tony's answer is wrong.

b How did Tony get his answer, and why is he wrong?

51 Olive estimated 2.49×1.49 as roughly 2.

a Calculate 2.49×1.49 and find the difference between your answer and Olive's estimate of the answer.

b Suggest how Olive could improve the accuracy of her estimate.

These are 'challenge' questions

52 Without calculating the answer, which of these estimates is likely to be nearest to the value of $\dfrac{1.29 \times 0.59}{(1.45)^2}$ and why?

A $\dfrac{1 \times 0.6}{1}$ **B** $\dfrac{1 \times 1}{1}$ **C** $\dfrac{1 \times 0.6}{2}$

Use your calculator to find $3.057 \times 2.485 \times 10^{-5}$, giving the answer corrected to 3 significant figures.

Estimate:
$$(3.057 \times 2.485 \times 10^{-5} \approx 3 \times 2 \div 100\,000 = 6 \div 100\,000$$
$$= 0.000\,06$$

Press

[3] [.] [0] [5] [7] [x] [2] [.] [4] [8] [5] [x] [1] [0] [xʸ] [5] [±] [=]

The display shows $7.5966^{\;-05}$; this means 7.5966×10^{-5}

Calculator:
$$3.057 \times 2.485 \times 10^{-5} = 7.5966\ldots \times 10^{-5}$$
$$= 7.60 \times 10^{-5} \text{ (correct to 3 s.f.)}$$

This is reasonably close to the estimate, so it is probably correct.
Notice that the answer is written with a zero at the end. This is to show that the third significant figure is zero.

Use your calculator, giving answers correct to 3 significant figures, to find

53 $12.5 \times 5.027 \times 10^3$ **55** $2988 \times 1.05 \times 10^{-3}$

54 $0.45 \times 1.39 \times 10^{-4}$ **56** $2.5 \div (3.1 \times 10^{-5})$

These are 'challenge' questions

57 Light travels 1 km in about 3.3×10^{-6} seconds.
How long does it take light to travel 24.5 kilometres?

58 The distance travelled by light in a vacuum during one year is equal to 9.4650×10^{15} metres. (This is called a light-year.)
How far does light travel in 1 second? (1 year $=$ 365 days)

INVESTIGATIONS

1 a Everyone has two biological parents.
Going back one generation, each of your parents has two biological parents.
Copy and complete the tree – fill in the number of ancestors for five generations back. Do not fill in names!

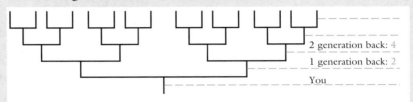

2 generation back: 4
1 generation back: 2
You

b Giving your answers as a power of 2 and as a number in standard form, how many ancestors does this table suggest you have
  i five generations back
  ii six generations back
  iii ten generations back?

c If we assume that each generation spans 25 years, how many generations are needed to go back 1000 years?

d Find the number of ancestors the table suggests that you would expect to have 6000 years back. Give your answer as a power of 2 and as a number in standard form. What other assumptions are made to get this answer?

e About 6000 years ago, according to the bible, Adam and Eve were the only people on the earth. This contradicts the answer from part **d**. Suggest some reasons for this contradiction.

2 If you read advertisements for computers, you will notice specifications such as '250 Mb hard disk', '1 Gb hard disk'.
Mb stands for megabytes and Gb stands for gigabytes.
Mega and Giga are prefixes used to describe very large numbers.
There are other prefixes used to describe very small numbers.
Find out what Mega and Giga mean.
Find out what other prefixes are used to describe very large and very small numbers and what they mean.

PROBABILITY

There are 30 sweets in a bag, 6 of which are green.
Peter takes one sweet picked at random out of this bag.
He has 6 chances in 30 of picking a green sweet so the probability that
Peter chooses a green one is $\frac{6}{30} = \frac{1}{5}$.

Sometimes we are interested in the chances of an event not happening.

- If Peter does not like green sweets, he is more likely to be interested in the probability of not picking a green one.
- The organiser of an outdoor pop concert will be concerned to choose a date when the chances of it not raining are high.

The next exercise explores the relationship between the probability that an event does happen and the probability that an event does not happen.

EXERCISE 2A

1 An ordinary pack of 52 playing cards is well shuffled and then cut.

  a What is the probability that the card showing is an ace?

  b What is the probability that the card showing is not an ace?

  c How are the answers to parts **a** and **b** related?

2 A six-sided dice is thrown. Find the relationship between the probability of throwing a six and that of not throwing a six.

3 Mr. Impresario wants to book Wembley Stadium for a pop concert. He needs to do this one year in advance and has a choice of three dates; one in April, one in June and one in September. He wants to choose the date with the highest likelihood of it not raining. Discuss what information is needed to make a choice.

PROBABILITY THAT AN EVENT DOES NOT HAPPEN

The last exercise should convince you that the probabilities of an event happening and an event not happening add up to 1. This example explains why this is always the case.

If one card is drawn at random from an ordinary pack of playing cards, the probability that it is a club is given by

$$P(\text{a club}) = \frac{13}{52} = \frac{1}{4}$$

Now there are 39 cards that are not clubs so the probability that the card is not a club is given by

$$P(\text{ not a club }) = \frac{39}{52} = \frac{3}{4}$$

i.e. $\quad\quad P(\text{ not a club }) + P(\text{ a club }) = \frac{3}{4} + \frac{1}{4} = 1$

Hence $\quad\quad\quad\quad\quad\quad P(\text{ not a club }) = 1 - P(\text{ a club })$

This relationship is true in any situation because

$$\begin{pmatrix} \text{The number of} \\ \text{ways in which} \\ \text{an event, } A, \\ \text{can } not \text{ happen} \end{pmatrix} = \begin{pmatrix} \text{The total} \\ \text{number of} \\ \text{possible} \\ \text{outcomes} \end{pmatrix} - \begin{pmatrix} \text{The number} \\ \text{of ways in} \\ \text{which } A \\ \text{can happen} \end{pmatrix}$$

i.e. $\quad\quad P(\text{ } A \text{ does not happen }) = 1 - P(\text{ } A \text{ does happen })$

'A does not happen' is shortened to $\bar{A}$, where $\bar{A}$ is read as 'not A'.

Therefore $\quad\quad\quad\quad\quad\quad P(\bar{A}) = 1 - P(A)$

For example, if the probability that I will pass my driving test is **0.7**, then the probability that I will fail (i.e. not pass) is $1 - 0.7 = 0.3$.

EXERCISE 2B

One number is chosen at random from the whole numbers 1 to 50. What is the probability that it is not a multiple of 7?

The multiples of 7 are 7, 14, 21, 28, 35, 42, 49.

$$P(\text{ multiple of 7 }) = \frac{\text{number of choices that are multiples of 7}}{\text{total number of possible choices}}$$

$$= \frac{7}{50}$$

$$\therefore \ P(\text{ not a multiple of 7 }) = 1 - \frac{7}{50}$$

$$= \frac{43}{50}$$

We could find this probability by listing the numbers from 1 to 50 that are not multiples of 7. But it is easier to list and then count the numbers that are multiples of 7.

1 The probability that Ann will have to wait more than five minutes for the next bus is **0.4**. What is the probability that she will not have to wait more than five minutes?

2 Barry is not very good at getting
his first serve in when playing tennis.
His long-term average is one in ten.
What is the probability that Barry's
next first serve

a goes in

b does not go in?

3 A number is chosen at random from the first 20 positive whole
numbers. What is the probability that it is not a prime number?

4 A card is drawn at random from an ordinary pack of playing cards.
What is the probability that it is not a two?

5 One letter is chosen at random from the letters of the alphabet.
What is the probability that it is not a vowel?

6 A box of 60 coloured crayons contains a mixture of colours, 10 of
which are red. If one crayon is removed at random, what is the
probability that it is not red?

7 A number is chosen at random from the first 10 whole numbers.
What is the probability that it is not exactly divisible by 3?

8 One letter is chosen at random from the letters of the word
ALPHABET
What is the probability that it is not a vowel?

9 In a raffle, 500 tickets are sold. If you buy 20 tickets, what is the
probability that you will not win first prize?

10 If you throw an ordinary six-sided dice, what is the probability that
you will not get a score of 5 or more?

11 There are 200 packets hidden in a lucky dip. Five packets contain
£1 and the rest contain 1 p. What is the probability that you will not
draw out a packet containing £1?

12 When an ordinary pack of playing cards is cut, what is the
probability that the card showing is not a picture card? (The
picture cards are the jacks, queens and kings.)

13 A letter is chosen at random from the letters of the word

SUCCESSION

What is the probability that the letter is

a N   **b** S   **c** a vowel   **d** not S?

14 A card is drawn at random from an ordinary pack of playing cards. What is the probability that it is

a an ace   **c** not a club

b a spade   **d** not a seven or an eight?

15 A bag contains a set of snooker balls (i.e. 15 red and 1 each of the following colours: white, yellow, green, brown, blue, pink and black). What is the probability that one ball removed at random is

a red   **c** black

b not red   **d** not red or white?

16 There are 60 cars in the station car-park. Of the cars, 22 were made in Britain, 24 were made in Japan and the rest were made in Europe but not in Britain. What is the probability that the first car to leave was made in

a Japan   **c** Europe but not Britain

b not Britain   **d** America?

17 One number is chosen at random from the whole numbers 1 to 100 inclusive.

What is the probability that the number

a is not a multiple of ten

b is not prime

c has two digits?

18 A train operator states that, over the last year, 3 out of every 10 trains that ran arrived early.

a Use this information to estimate the probability that, when you next travel on one of these trains, it will not arrive early.

b Jane said 'This means that there is a 7 in 10 chance that a train will arrive late.' Is she right? Give reasons for your answer.

c Say why the answer to part **a** can only be an estimate.

19 Royal Standard Company deals in car insurance. The company report for one year gave the number of accident claims for cars driven by its policy holders and included these details:

Make of car	Ford	Vauxhall	Jaguar
Claims	148	127	13

a Penny looked at these figures and said 'Jaguars must be safe cars to drive'. Eddy set out to prove that Penny could not use these figures to justify her statement.
What other information does Eddy need to be able to do this?

b What information do you need to decide whether one make of car is safer than another make?

POSSIBILITY SPACE FOR TWO EVENTS

When we throw two dice there are two events to consider: the way the first dice lands and the way the second dice lands.
When two events are involved it is not always easy to ensure that all the possible outcomes are taken into account.

For example, before parting with her money at this stall, Sue wants to know what her chances are of winning a can. She knows that there is only one way that the dice can land showing two fives. To work out the probability of winning Sue needs to find how many other ways there are in which the dice can land. She should try listing the possible outcomes: 1,1 1,2 1,3 2,4 ... but can she be sure that she hasn't missed any?

Situations involving two events need an organised approach to ensure that all the possible outcomes are listed. We will start by considering a simple case.

Suppose a 2 p coin and a 10 p coin are tossed together. One possibility is that the 2 p coin will land head up and that the 10 p coin will also land head up.

If we use H for a head on the 2 p coin and *H* for a head on the 10 p coin, we can write this possibility more briefly as the ordered pair (H, *H*).

To list all the possibilities, we use a table.

The possibilities for the 10 p coin are written across the top and the possibilities for the 2 p coin are written down the side. They could also go the other way; it does not matter which coin is placed along the top.

	10 p coin	
	H	T
2 p coin H		
T		

When both coins are tossed we can see all the combinations of heads and tails that are possible and then fill in the table.

	10 p coin	
	H	T
2 p coin H	(H, H)	(H, T)
T	(T, H)	(T, T)

A display of all the possibilities is called a *possibility space*.

EXERCISE 2C

1 Two bags each contain 3 white counters and 2 black counters. One counter is removed at random from each bag. Copy and complete the following possibility space for the possible combinations of the two counters.

	1st bag				
	○	○	○	●	●
○	(○,○)	(○,○)	(○,○)	(○,●)	
○					
2nd bag ○					
●					(●,●)
●	(●,○)				

2 An ordinary six-sided dice is tossed and a 10 p coin is tossed. Copy and complete the following possibility space.

	Dice					
	1	2	3	4	5	6
10 p coin H		(H, 2)				
T				(T, 4)		

3 One bag contains 2 red counters, 1 yellow counter and 1 blue counter. Another bag contains 2 yellow counters, 1 red counter and 1 blue counter. One counter is taken at random from each bag. Copy and complete the following possibility space.

		1st bag			
		R	R	Y	B
	R		(R, R)		
2nd bag	Y				(Y, B)
	Y				
	B	(B, R)			

4

A top like the one in the diagram is spun twice. Copy and complete the possibility space.

		1st spin		
		1	2	3
	1			
2nd spin	2			
	3			

When there are several entries in a possibility space it can take a long time to fill in the ordered pairs. To save time we use a cross in place of each ordered pair. We can see which ordered pair a particular cross represents by looking at the edges of the table.

Two ordinary six-sided dice are tossed. Draw up a possibility space showing all the possible combinations in which the dice may land. Use the possibility space to find the probability that a total score of at least 10 is obtained.

		1st dice					
		1	2	3	4	5	6
	1	×	×	×	×	×	×
	2	×	×	×	×	×	×
2nd dice	3	×	×	×	×	×	×
	4	×	×	×	×	×	⊗
	5	×	×	×	×	⊗	⊗
	6	×	×	×	⊗	⊗	⊗

We have ringed the entries giving a score of 10 or more. There are 36 entries (i.e. possibilities) in the table and 6 of them give a score of 10 or more.

$$P(\text{ score of at least } 10) = \frac{6}{36} = \frac{1}{6}$$

5 Copy the possibility space in the example above and use it to find the probability of getting a score of

    a 4 or less   **b** 9   **c** a double

    d Find the chance that Sue will win a can on the stall shown on page 42.

6 Use the possibility space for question **1** to find the probability that the two counters removed

    a are both black   **b** contain at least one black.

7 Use the possibility space for question **3** to find the probability that the two counters removed are

    a both blue   **c** one blue and one red

    b both red   **d** such that at least one is red.

8 A 5 p coin and a 1 p coin are tossed together. Make your own possibility space for the combinations in which they can land. Find the probability of getting two heads.

9 One bag of coins contains three 10 p coins and two 50 p coins. Another bag contains one 10 p coin and one 50 p coin. One coin is removed at random from each bag. Make a possibility space and use it to find the probability that a 50 p coin is taken from each bag.

10 One bookshelf contains two story books and three text books. The next shelf holds three story books and one text book. Draw a possibility space showing the various ways in which you could pick up a pair of books, one from each shelf. Use this to find the probability that

    a both books are story books   **b** both are text books.

11 The four aces and the four kings are removed from an ordinary pack of playing cards. One card is taken from the set of four aces and one card is taken from the set of four kings. Make a possibility space for the possible combinations of two cards and use it to find the probability that the two cards

    a are both black   **c** include at least one black card

    b are both spades   **d** are both of the same suit.

12 A six-sided dice has two of its faces blank and the other faces are numbered 1, 3, 4 and 6. This dice is tossed with an ordinary six-sided dice (faces numbered 1, 2, 3, 4, 5, 6). Make a possibility space for the ways in which the two dice can land and use it to find the probability of getting a total score of

    a 6   **b** 10   **c** 1   **d** at least 6.

THE NUMBER OF TIMES AN EVENT IS LIKELY TO HAPPEN

We sometimes need to estimate how often an event might happen. For example, if we are organising the stall shown on page 42, we will need to estimate how many wins to expect; we can then decide how many cans to provide.

We know that, on one turn, the probability of winning a can is $\frac{1}{36}$,

i.e.   about 1 in every 36 turns will result in a win.

Estimating that there will be about 500 turns, then we can see that there are likely to be about $500 \div 36$ wins, that is, about 14 wins, so we need about 14 cans.

> If p is the probability that an event happens once, when the situation is repeated n times we expect the event to happen about $p \times n$ times.

For example, if we toss a coin 20 times,
we expect to get about $\frac{1}{2} \times 20$ heads.

EXERCISE 2D

1 a What is the probability of getting a head on one toss of a coin?

   b How many heads would you expect to get if you tossed the coin 100 times?

   c Will you get this number of heads? Give a reason for your answer.

   d Explain what your reaction would be if you tossed a particular coin 100 times and got 10 heads.

2 On a sailing course for beginners, past experience shows that about 1 in 4 pupils capsize on their first solo sailing trip.

   a There are 20 pupils out sailing solo for the first time. How many are expected to capsize?

   b Support boats are needed to help those who capsize. If one boat can cope with only one capsize at a time, how many boats should be provided?

3 Jamestown airport has 500 flights leaving each day. On average, the departure of 1 plane in 25 is delayed. How many delayed flights are expected on one day?

4 Two ordinary six-sided dice are thrown 360 times. How many double sixes are there likely to be?

5 The probability that the same symbol
shows in each of the three windows
on this slot machine is $\frac{1}{64}$.

When this happens, the machine pays
out £10, otherwise it pays out nothing.

a How many times is the machine
likely to pay out in 100 turns?

b Each turn costs 20 p. How much is it likely to cost for 100 turns,
taking into account likely winnings?

**PRACTICAL
WORK**

1 This is an experiment to find out if you can see into the future!
You need to work in pairs and you need one coin.
One of you is the tosser and recorder and the other is the guesser.

a The guesser predicts whether the coin will land head up or tail
up. The tosser then tosses the coin. If the guesser has no
psychic powers,

i what is the probability that he/she guesses the actual outcome?

ii when this experiment is repeated 100 times, about how many
times do you expect the guesser to predict the actual outcome?

b Now perform the experiment described at least 100 times and
record each result as right or wrong as appropriate.
Use an observation sheet in the form of a tally chart.

c Compare what you expected to happen with what did happen,
using appropriate diagrams as illustrations. Comment on the
likelihood of the guesser being able to predict which way the
coin will land.

d How could you make your results more reliable?

e Suggest other experiments that you could perform to test
whether someone can see into the future.

2 Ten (very desirable) posters have been
donated to your class. You have to
use these posters for a fund-raising
activity to raise money for charity.
Plan a suitable activity and write a
report.
Give all your ideas, including
any that you reject. Include the
advantages and the disadvantages of
each scheme that you consider.

TRY YOUR
LUCK!
WIN A
POSTER

MULTIPLICATION AND DIVISION OF FRACTIONS

Fractions are part of everyday language; the following sentences could be overheard anywhere.

'I can get a third off the train fare with a young person's rail card.'

'My slice is only $\frac{1}{8}$ of the cake. If I give you half of it, you'll have more than I have.'

'Here is $6\frac{1}{2}$ in of wire. Cut it into $\frac{3}{8}$ in lengths – you should get at least 20 pieces out of it.'

- If you don't know what fractions mean, you cannot make much sense of these statements.
- The last two sentences contain statements which could be wrong.

In order to check whether these are correct, we need to be able to calculate with fractional quantities.

MULTIPLYING FRACTIONS

A particular fraction describes the size of part of a quantity.

For example, $\frac{3}{4}$ of £10 means 3 out of 4 equal-sized portions of £10.

Finding equivalent fractions, and adding and subtracting fractions, can be revised using the Summary and Exercises at the beginning of this book.

However none of these processes will help us to find, for example, the fraction of a cake that half of one-eighth of it is, but we can use a cake diagram to illustrate what $\frac{1}{2}$ of $\frac{1}{8}$ means.

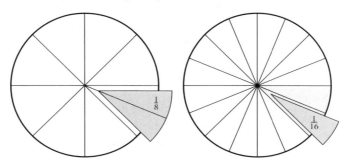

Now we can see that $\frac{1}{2}$ of $\frac{1}{8}$ of the cake is equal to $\frac{1}{16}$ of the cake.

The next step is to interpret the meaning of $\frac{1}{2}$ of $\frac{1}{8}$ in mathematical symbols;

when we have to find   '3 of a quantity',
we do so by calculating   '3 × the quantity'.
In the same way to find   '$\frac{1}{2}$ of a quantity',
we calculate   '$\frac{1}{2}$ × the quantity'.

Hence $\frac{1}{2}$ of $\frac{1}{8}$ means calculate $\frac{1}{2} \times \frac{1}{8}$

$\left(\text{and conversely, } \frac{1}{2} \times \frac{1}{8} \text{ means } \frac{1}{2} \text{ of } \frac{1}{8}.\right)$

From the cake diagram, we know that $\frac{1}{2} \times \frac{1}{8} = \frac{1}{16}$

EXERCISE 3A

Draw cake diagrams to find

1 $\frac{1}{2} \times \frac{1}{4}$ **3** $\frac{1}{2} \times \frac{3}{4}$ **5** $\frac{1}{3} \times \frac{2}{5}$

2 $\frac{1}{3} \times \frac{1}{2}$ **4** $\frac{2}{3} \times \frac{1}{3}$ **6** $\frac{1}{4} \times \frac{1}{3}$

7 Use a copy of the diagram to find $\frac{3}{4} \times \frac{2}{3}$

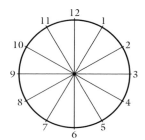

8 Use a copy of the diagram to find $\frac{2}{5} \times \frac{3}{10}$

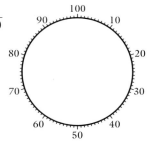

9 **a** For questions **1** to **6** find a relationship between the numerators in the given fractions and the numerator in the result.

   b Repeat part **a** for the denominators.

**THE RULE FOR
MULTIPLYING
FRACTIONS**

You may have discovered this rule in question **9** of the exercise on the previous page:

> To multiply fractions find the product of the numbers in the numerators and the product of the numbers in the denominators.

The 'product of two numbers' means multiply them together.

For example, $\dfrac{2}{3} \times \dfrac{4}{7} = \dfrac{2 \times 4}{3 \times 7} = \dfrac{8}{21}$

EXERCISE 3B

Calculate

1 $\dfrac{1}{2} \times \dfrac{3}{4}$ **5** $\dfrac{3}{4} \times \dfrac{1}{5}$ **9** $\dfrac{1}{3} \times \dfrac{2}{5}$

2 $\dfrac{2}{3} \times \dfrac{1}{5}$ **6** $\dfrac{2}{5} \times \dfrac{3}{5}$ **10** $\dfrac{3}{5} \times \dfrac{3}{4}$

3 $\dfrac{2}{7} \times \dfrac{3}{5}$ **7** $\dfrac{1}{2} \times \dfrac{1}{2}$ **11** $\dfrac{2}{3} \times \dfrac{2}{3}$

4 $\dfrac{1}{9} \times \dfrac{1}{2}$ **8** $\left(\dfrac{1}{3}\right)^2$ **12** $\left(\dfrac{2}{5}\right)^2$

Calculate $\dfrac{1}{3} \times \dfrac{3}{4}$

$$\dfrac{1}{3} \times \dfrac{3}{4} = \dfrac{1}{{}_1\cancel{3}} \times \dfrac{\cancel{3}^{\,1}}{4} = \dfrac{1 \times 1}{1 \times 4}$$

> The numbers can sometimes be simplified by cancelling common factors before finding the product.

$$= \dfrac{1}{4}$$

The diagram shows that

$\dfrac{1}{3}$ of $\dfrac{3}{4} = \dfrac{1}{4}$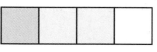

Calculate

13 $\dfrac{3}{4} \times \dfrac{4}{7}$ **15** $\dfrac{2}{3} \times \dfrac{3}{5}$ **17** $\dfrac{4}{7} \times \dfrac{5}{12}$

14 $\dfrac{2}{3} \times \dfrac{7}{8}$ **16** $\dfrac{2}{5} \times \dfrac{3}{4}$ **18** $\dfrac{2}{9} \times \dfrac{3}{7}$

Sometimes there is more than one common factor that can be cancelled.

Calculate $\dfrac{4}{25} \times \dfrac{15}{16}$

$$\dfrac{\cancel{4}^{1}}{_{5}\cancel{25}} \times \dfrac{\cancel{15}^{3}}{_{4}\cancel{16}} = \dfrac{1 \times 3}{5 \times 4}$$

4 is a common factor, so is 5.

$$= \dfrac{3}{20}$$

Calculate

19 $\dfrac{7}{8} \times \dfrac{4}{21}$ **23** $\dfrac{7}{9} \times \dfrac{3}{21}$ **27** $\dfrac{4}{15} \times \dfrac{25}{64}$

20 $\dfrac{3}{4} \times \dfrac{16}{21}$ **24** $\dfrac{3}{4} \times \dfrac{14}{15}$ **28** $\dfrac{2}{3} \times \dfrac{33}{40}$

21 $\dfrac{21}{22} \times \dfrac{11}{27}$ **25** $\dfrac{4}{5} \times \dfrac{15}{16}$ **29** $\dfrac{3}{7} \times \dfrac{28}{33}$

22 $\dfrac{8}{9} \times \dfrac{33}{44}$ **26** $\dfrac{10}{11} \times \dfrac{33}{35}$ **30** $\dfrac{48}{55} \times \dfrac{5}{12}$

Calculate $\dfrac{3}{5} \times \dfrac{15}{16} \times \dfrac{4}{7}$

$$\dfrac{3}{_{1}\cancel{5}} \times \dfrac{\cancel{15}^{3}}{_{4}\cancel{16}} \times \dfrac{\cancel{4}^{1}}{7} = \dfrac{3 \times 3 \times 1}{1 \times 4 \times 7}$$

Remember, multiply two numbers and then multiply the result by the third number.

$$= \dfrac{9}{28}$$

Calculate

31 $\dfrac{3}{7} \times \dfrac{5}{9} \times \dfrac{14}{15}$ **35** $\dfrac{3}{10} \times \dfrac{5}{9} \times \dfrac{6}{7}$ **39** $\dfrac{6}{5} \times \dfrac{4}{3} \times \dfrac{10}{4}$

32 $\dfrac{11}{21} \times \dfrac{30}{31} \times \dfrac{7}{55}$ **36** $\dfrac{5}{7} \times \dfrac{3}{8} \times \dfrac{21}{30}$ **40** $\dfrac{9}{8} \times \dfrac{1}{3} \times \dfrac{4}{27}$

33 $\dfrac{15}{16} \times \dfrac{8}{9} \times \dfrac{4}{5}$ **37** $\dfrac{1}{2} \times \dfrac{7}{12} \times \dfrac{18}{35}$ **41** $\dfrac{7}{16} \times \dfrac{9}{11} \times \dfrac{8}{21}$

34 $\dfrac{5}{6} \times \dfrac{8}{25} \times \dfrac{3}{4}$ **38** $\dfrac{7}{11} \times \dfrac{8}{9} \times \dfrac{33}{28}$ **42** $\dfrac{5}{14} \times \dfrac{21}{25} \times \dfrac{5}{9}$

43 A cook added $\frac{1}{2}$ cup of water to a stew. The cup holds $\frac{1}{10}$ litre. What fraction of a litre is added to the stew?

44 There are 24 balls in a bag and 9 of them are black. Two thirds of the black balls have white spots on them.

    a What fraction of the balls in the bag are black?

    b What fraction of the balls in the bag are black with white spots on them?

45 Three-quarters of the pupils in a class got a grade B in a chemistry exam. Two-fifths of those who got grade B in chemistry also got grade B in biology. What fraction of the class got Bs in both exams?

46 Three-quarters of the area of some reclaimed land is designated for housing and the rest is to be used for industrial development. One-fifth of the area set aside for housing is left as open space.

    a What fraction of the area of reclaimed land is to be used for industrial purposes?

    b What fraction of the area of reclaimed land is to be left as open space?

47 Place these fractions in order of size with the smallest first.

    a $\frac{2}{3}$ of $\frac{1}{5}$, $\frac{1}{2} \times \frac{2}{5}$

    b $\frac{3}{4} \times \frac{1}{3}$, $\frac{1}{12}$ of eight-ninths.

48 State, with reasons, whether two-thirds of one-fifth is the same as one-third of two-fifths. Without working them out, state with reasons, whether twelve-seventeenths of thirteen-nineteenths is the same as thirteen-seventeenths of twelve-nineteenths.

49 Simplify   **a** $\frac{a}{2} \times \frac{a}{3}$ **b** $\frac{2x}{5} \times \frac{x}{4}$ **c** $\frac{a}{3} \times \frac{b}{4}$

50 This calculation has a mistake in it. Describe the mistake.

$$\frac{1}{3} \times \frac{3}{5} \times \frac{25}{27} = \frac{1}{5} \times \frac{25}{27} = \frac{2}{27}$$

MULTIPLYING
WHOLE
NUMBERS AND
MIXED NUMBERS

To find the area of a rectangular rug measuring $3\frac{1}{2}$ feet by $4\frac{1}{8}$ feet, we have to calculate $3\frac{1}{2} \times 4\frac{1}{8}$.

Mixed numbers cannot be multiplied together unless they are first changed into improper fractions.

So we change $3\frac{1}{2}$ into $\frac{7}{2}$ and we change $4\frac{1}{8}$ into $\frac{33}{8}$, then

$$3\frac{1}{2} \times 4\frac{1}{8} = \frac{7}{2} \times \frac{33}{8}$$

$$= \frac{231}{16} = 14\frac{7}{16}$$

EXERCISE 3C

Calculate $2\frac{1}{3} \times 1\frac{1}{5}$

$2\frac{1}{3} \times 1\frac{1}{5} = \frac{7}{3} \times \frac{6}{5}$

> 3 is a common factor of the top and bottom so it can be cancelled.

$$= \frac{14}{5} = 2\frac{4}{5}$$

Calculate

1 $1\frac{1}{2} \times \frac{2}{5}$

2 $2\frac{1}{2} \times \frac{4}{5}$

3 $3\frac{1}{4} \times \frac{3}{13}$

4 $4\frac{2}{3} \times \frac{9}{14}$

5 $2\frac{1}{5} \times \frac{5}{22}$

6 $1\frac{1}{4} \times \frac{2}{5}$

7 $2\frac{1}{3} \times \frac{3}{8}$

8 $\frac{10}{11} \times 2\frac{1}{5}$

9 $3\frac{1}{2} \times \frac{2}{3}$

10 $4\frac{1}{4} \times \frac{4}{21}$

11 $5\frac{1}{4} \times 2\frac{2}{3}$

12 $3\frac{5}{7} \times 1\frac{1}{13}$

13 $8\frac{1}{3} \times 3\frac{3}{5}$

14 $2\frac{1}{10} \times 7\frac{6}{7}$

15 $6\frac{3}{10} \times 1\frac{4}{21}$

16 $4\frac{2}{7} \times 2\frac{1}{10}$

17 $6\frac{1}{4} \times 1\frac{3}{5}$

18 $5\frac{1}{2} \times 1\frac{9}{11}$

19 $8\frac{3}{4} \times 2\frac{2}{7}$

20 $16\frac{1}{2} \times 3\frac{7}{11}$

We can use the rule for multiplying fractions when we have to find the product of a fraction and a whole number; we treat the whole number as a fraction,

e.g.   we write 2 as $\frac{2}{1}$

Calculate $6 \times 7\frac{1}{3}$

$6 \times 7\frac{1}{3} = \frac{\cancel{6}^{2}}{1} \times \frac{22}{\cancel{3}_{,1}}$ | Cancelling top and bottom by 3. |

$\qquad\qquad = \frac{44}{1} = 44$

Calculate

21 $5 \times 4\frac{3}{5}$ **25** $18 \times 6\frac{1}{9}$ **29** $5\frac{5}{7} \times 21$

22 $2\frac{1}{7} \times 14$ **26** $4 \times 3\frac{3}{8}$ **30** $3 \times 6\frac{1}{9}$

23 $3\frac{1}{8} \times 4$ **27** $3\frac{3}{5} \times 10$ **31** $1\frac{3}{4} \times 8$

24 $4\frac{1}{6} \times 9$ **28** $2\frac{5}{6} \times 3$ **32** $28 \times 1\frac{4}{7}$

Find $\frac{2}{3}$ of £27.

| Remember, the word 'of' means 'multiply by', so $\frac{2}{3}$ of £27 means $\frac{2}{3} \times £27$. |

$\frac{2}{3}$ of £27 $= £\frac{2}{3} \times 27$

$\qquad\qquad = £\frac{2}{\cancel{3}_{,1}} \times \frac{\cancel{27}^{9}}{1} = £18$

Find

33 $\frac{1}{3}$ of 12   **39** $\frac{1}{6}$ of 42

34 $\frac{2}{5}$ of 45   **40** $\frac{3}{7}$ of 42

35 $\frac{3}{4}$ of £84   **41** $\frac{1}{3}$ of £18

36 $\frac{7}{10}$ of 50 litres   **42** $\frac{2}{5}$ of 25 gallons

37 $\frac{3}{5}$ of £1   **43** $\frac{2}{3}$ of 25 millilitres

38 $\frac{3}{8}$ of 64 miles   **44** $\frac{2}{5}$ of 100 milligrams

45 A bag of flour weights $1\frac{1}{2}$ kg. What is the weight of 20 bags?

46 Three and a half sacks of sand are added to a cement mix. Each sack of sand weighs 26 kg. What weight of sand is added to the cement?

47 The area of a garden is 40 m^2. Two fifths of the area is used for growing soft fruit. How many square metres are used for growing soft fruit?

48 A paving stone measures $\frac{3}{4}$ yd by $\frac{3}{8}$ yd. What is the area of this paving stone?

49 If $\frac{19}{20}$ of my body weight is water and I weigh $52\frac{1}{2}$ kg, how much of my weight is water?

50 A cuboid measures $4\frac{3}{8}$ inches by 5 inches by 3 inches.
Find its volume in cubic inches.

51 Which is heavier, $\frac{7}{8}$ lb or $\frac{1}{2}$ kg?

52 This is an extract from an American mail order catalogue.

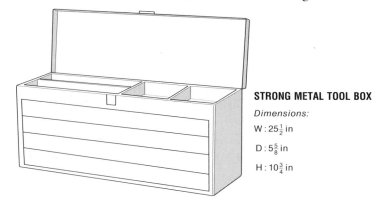

STRONG METAL TOOL BOX

Dimensions:

W : $25\frac{1}{2}$ in

D : $5\frac{5}{8}$ in

H : $10\frac{3}{4}$ in

What is the volume of this tool box?

53 Which of these boxes has the largest capacity?
Which has the smallest capacity? (They are *not* drawn to scale.)

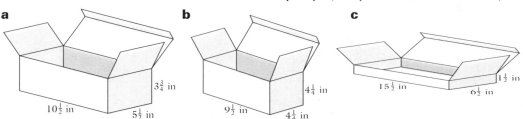

a $3\frac{3}{4}$ in $10\frac{1}{2}$ in $5\frac{1}{2}$ in

b $4\frac{1}{4}$ in $9\frac{1}{2}$ in $4\frac{1}{4}$ in

c $15\frac{1}{2}$ in $6\frac{1}{2}$ in $1\frac{1}{2}$ in

**DIVIDING BY
A FRACTION**

When we want to find how many 6s there are in 84, we calculate $84 \div 6$. In the same way, to find the number of $\frac{1}{2}$s in 4, we calculate $4 \div \frac{1}{2}$.

The diagram shows that there are 8 halves in 4, that is, $4 \div \frac{1}{2} = 8$; but we know that $4 \times 2 = 8$,

so
$$\frac{4}{1} \div \frac{1}{2} = \frac{4}{1} \times \frac{2}{1}$$

Now consider the number of $\frac{3}{8}$s in $1\frac{1}{2}$, that is, the value of $1\frac{1}{2} \div \frac{3}{8}$.

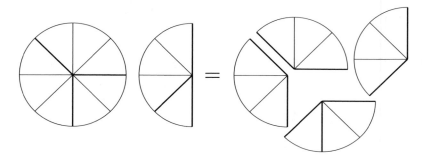

From the diagram we see that $1\frac{1}{2}$ cakes can be divided into 4 pieces, each of which is $\frac{3}{8}$ of a cake, so $1\frac{1}{2} \div \frac{3}{8} = 4$,

i.e. $\qquad \frac{3}{2} \div \frac{3}{8} = 4, \qquad$ but $\qquad \frac{\cancel{3}}{\cancel{2}} \times \frac{\cancel{8}}{\cancel{3}} = 4$

These two examples demonstrate the rule that

> to divide by a fraction
> we turn that fraction upside down
> and multiply by it.

EXERCISE 3D

a Find the value of $36 \div \frac{3}{4}$ $\qquad$ **b** Divide $\frac{7}{16}$ by $\frac{5}{8}$

a $\dfrac{36}{1} \div \dfrac{3}{4} = \dfrac{36^{12}}{1} \times \dfrac{4}{\cancel{3}}$ $\qquad$ **b** $\dfrac{7}{16} \div \dfrac{5}{8} = \dfrac{7}{{}_{2}\cancel{16}} \times \dfrac{\cancel{8}^{1}}{5}$

$\qquad\qquad = \dfrac{48}{1}$ $\qquad\qquad\qquad = \dfrac{7}{10}$

$\qquad\qquad = 48$

1 How many $\frac{1}{2}$ s are there in 7?

2 How many $\frac{1}{4}$ s are there in 5?

3 How many times does $\frac{1}{7}$ go into 3?

4 How many $\frac{3}{5}$ s are there in 9?

5 How many times does $\frac{2}{3}$ go into 8?

Find

6 $8 \div \frac{4}{5}$

7 $18 \div \frac{6}{7}$

8 $40 \div \frac{8}{9}$

9 $72 \div \frac{8}{11}$

10 $28 \div \frac{14}{15}$

11 $15 \div \frac{5}{6}$

12 $14 \div \frac{7}{8}$

13 $35 \div \frac{5}{7}$

14 $44 \div \frac{4}{9}$

15 $27 \div \frac{9}{13}$

16 $36 \div \frac{4}{7}$

17 $34 \div \frac{17}{19}$

18 $\frac{21}{32} \div \frac{7}{8}$

19 $\frac{9}{25} \div \frac{3}{10}$

20 $\frac{3}{56} \div \frac{9}{14}$

21 $\frac{21}{22} \div \frac{7}{11}$

22 $\frac{8}{75} \div \frac{4}{15}$

23 $\frac{35}{42} \div \frac{5}{6}$

24 $\frac{28}{27} \div \frac{4}{9}$

25 $\frac{22}{45} \div \frac{11}{15}$

26 $\frac{15}{26} \div \frac{5}{13}$

27 $\frac{49}{50} \div \frac{7}{10}$

28 $\frac{8}{21} \div \frac{4}{7}$

29 $\frac{9}{26} \div \frac{12}{13}$

Find $3\frac{1}{8} \div 8\frac{3}{4}$

$$3\frac{1}{8} \div 8\frac{3}{4} = \frac{25}{8} \div \frac{35}{4}$$

$$= \frac{25^5}{_2 8} \times \frac{4^1}{_7 35}$$

$$= \frac{5}{14}$$

When mixed numbers are involved, they must first be changed into improper fractions,

i.e. $3\frac{1}{8}$ must be changed to $\frac{25}{8}$

and $8\frac{3}{4}$ must be changed to $\frac{35}{4}$.

Check: $\frac{5}{14} \times 8\frac{3}{4} = \frac{5}{14} \times \frac{35^5}{4} = \frac{25}{8} = 3\frac{1}{8}$

Calculate

30 $5\frac{4}{9} \div \frac{14}{27}$

31 $3\frac{1}{8} \div 3\frac{3}{4}$

32 $7\frac{1}{5} \div 1\frac{7}{20}$

34 $5\frac{5}{8} \div 6\frac{1}{4}$

33 $4\frac{2}{7} \div \frac{9}{14}$

35 $6\frac{4}{9} \div 1\frac{1}{3}$

36 Divide $8\frac{1}{4}$ by $1\frac{3}{8}$

38 Divide $5\frac{1}{4}$ by $2\frac{11}{12}$

37 Divide $6\frac{2}{3}$ by $2\frac{4}{9}$

39 Divide $7\frac{1}{7}$ by $1\frac{11}{14}$

Calculate

40 $10\frac{2}{3} \div 1\frac{7}{9}$

42 $9\frac{3}{4} \div 1\frac{5}{8}$

41 $8\frac{4}{5} \div 3\frac{3}{10}$

43 $12\frac{1}{2} \div 8\frac{3}{4}$

44 Divide $11\frac{1}{4}$ by $\frac{15}{16}$

46 Divide $10\frac{5}{6}$ by $3\frac{1}{4}$

45 Divide $9\frac{1}{7}$ by $1\frac{11}{21}$

47 Divide $8\frac{2}{3}$ by $5\frac{7}{9}$

48 The filling for one sandwich needs $1\frac{1}{2}$ ounces of cheese. How many sandwiches can be filled using 12 ounces of cheese?

49 'Here is $6\frac{1}{2}$ in of wire. Cut it into $\frac{3}{8}$ in lengths – you should get at least 20 pieces out of it.'
This assertion appeared at the start of this chapter. Is it correct?
Give a reason for your answer.

50 If $1\frac{1}{4}$ lb of apples cost 80 p, what is the cost of 1 lb of these apples?

51 One 2 p coin is $\frac{1}{8}$ in thick.
A pile of 2 p coins is $1\frac{1}{2}$ in high.

    a How many coins are there in the pile?

    b What is the value of the coins in the pile?

52 Hassan read 30 pages of a book in $\frac{3}{4}$ hour.

    a What fraction of an hour did it take him, on average, to read one page?

    b How many minutes did it take him to read one page?

53 Which of these statements are true?

A $7 \div 3 = \frac{7}{3}$ **B** $\frac{2}{3} \div 2 = \frac{1}{3}$ **C** $\frac{4}{7} \div \frac{1}{2} = \frac{2}{7}$ **D** $\frac{5}{3} = 3 \div 5$

54 At Bertorelli's café, cakes are divided into eighths and each slice sold for £1.20. At Angelino's café they divide cakes of the same size into twelfths and sell each slice for 90 p.

a How many of Angelino's slices are equivalent to one of Bertorelli's slices?

b Which cake is more expensive?

55 This gift box has a volume of 49 cubic inches. It is $3\frac{1}{2}$ inches wide by $3\frac{1}{2}$ inches deep. How high is it?

56 This stack of $100 bills is $25\frac{1}{2}$ mm high and is worth $10 000. What is the thickness of one bill as a fraction of a millimetre?

57 Simplify **a** $\frac{x}{2} \div \frac{1}{2}$ **b** $\frac{x}{3} \div \frac{3}{2}$

MIXED OPERATIONS WITH FRACTIONS

When a calculation involves brackets and a mixture of multiplication, division, addition and subtraction, remember the order in which these operations are performed:

> work out the calculation in brackets first,
> then do any multiplication and division
> before doing any addition and subtraction.

Calculate $4\frac{1}{3} \times 1\frac{1}{8} \div 2\frac{1}{4}$

$$4\frac{1}{3} \times 1\frac{1}{8} \div 2\frac{1}{4} = \frac{13}{3} \times \frac{9}{8} \div \frac{9}{4}$$

$$= \frac{13}{3} \times \frac{9}{28} \times \frac{4}{9}$$

$$= \frac{13}{6} = 2\frac{1}{6}$$

> Remember that the sign before a number tells you what to do with just that number.
> So we divide by $\frac{9}{4}$, i.e. multiply by 4 over 9.

Find

1 $\frac{5}{8} \times 1\frac{1}{2} \div \frac{15}{16}$

2 $2\frac{3}{4} \times \frac{5}{6} \div \frac{11}{12}$

3 $\frac{2}{3} \times 1\frac{1}{5} \div \frac{12}{25}$

4 $\frac{2}{5} \times \frac{9}{10} \div \frac{27}{40}$

5 $\frac{3}{4} \times 2\frac{1}{3} \div \frac{21}{32}$

6 $3\frac{2}{5} \times \frac{4}{5} \div \frac{8}{15}$

7 $\frac{3}{5} \times \frac{9}{11} \div \frac{18}{55}$

8 $\frac{1}{4} \times \frac{11}{12} \div \frac{22}{27}$

9 $\frac{3}{7} \times \frac{2}{5} \div \frac{8}{21}$

Calculate $\frac{2}{5} - \frac{1}{2} \times \frac{3}{5}$

> Multiplication must be done first, so we start by working out $\frac{1}{2} \times \frac{3}{5}$.

$$\frac{2}{5} - \frac{1}{2} \times \frac{3}{5} = \frac{2}{5} - \left(\frac{1}{2} \times \frac{3}{5}\right)$$

$$= \frac{2}{5} - \frac{3}{10}$$

$$= \frac{4}{10} - \frac{3}{10} = \frac{1}{10}$$

> Remember that fractions can only be added or subtracted when they have the same denominator.

Calculate

10 $\frac{1}{2} + \frac{1}{4} \times \frac{2}{5}$

11 $\frac{2}{3} \times \frac{1}{2} + \frac{1}{4}$

12 $\frac{4}{5} - \frac{3}{10} \div \frac{1}{2}$

13 $\frac{2}{7} \div \frac{2}{3} - \frac{3}{14}$

14 $\frac{4}{5} + \frac{3}{10} \times \frac{2}{9}$

15 $\frac{3}{4} \div \frac{1}{2} + \frac{1}{8}$

16 $\frac{1}{7} + \frac{5}{8} \div \frac{3}{4}$

17 $\frac{5}{6} \times \frac{3}{10} - \frac{3}{16}$

18 $\frac{3}{7} - \frac{1}{4} \times \frac{8}{21}$

Remember, work out the calculation in the bracket first.

19 $\left(\frac{4}{9} - \frac{1}{3}\right) \times \frac{6}{7}$

24 $\frac{3}{8} \div \left(\frac{2}{3} + \frac{1}{4}\right)$

20 $\frac{3}{5} \times \left(\frac{2}{3} + \frac{1}{2}\right)$

25 $\left(\frac{4}{7} + \frac{1}{3}\right) \div 3\frac{4}{5}$

21 $\frac{7}{8} \div \left(\frac{3}{4} + \frac{2}{3}\right)$

26 $\frac{5}{9} \times \left(\frac{2}{3} - \frac{1}{6}\right)$

22 $\left(\frac{3}{10} + \frac{2}{5}\right) \div \frac{7}{15}$

27 $\left(\frac{6}{11} - \frac{1}{2}\right) \div \frac{3}{4}$

23 $\left(\frac{5}{11} - \frac{1}{3}\right) \times \frac{3}{8}$

28 $\frac{9}{10} \div \left(\frac{1}{6} + \frac{2}{3}\right)$

29 State whether each of the following statements is true or false.

a $\frac{1}{2} \times \frac{2}{3} + \frac{1}{3} = \frac{1}{3} + \frac{1}{3}$ 　　**b** $\frac{3}{4} - \frac{1}{2} \times \frac{2}{3} = \frac{1}{4} \times \frac{2}{3}$

30 Find $\frac{2}{3}$ of $\frac{3}{8}$ added to $1\frac{1}{2}$.

31 A pharmacist counts 58 tablets and puts them into a bottle. Each tablet weights $\frac{1}{4}$ gram and the empty bottle weighs $112\frac{1}{2}$ grams. What is the weight of the full bottle?

32 Nina usually walks to school in the morning and back home again in the afternoon. The distance from her home to school is $1\frac{1}{4}$ km. One week she was given a lift home from school on two afternoons and was given a lift halfway to school one morning. How far did she walk to and from school that week (5 days)?

33 Place $+$, $-$, $\times$, or $\div$ in the space to make these calculations correct.

a $\frac{2}{3} \quad \frac{4}{5} = \frac{5}{6}$ 　　**b** $\frac{3}{4} \quad \frac{8}{9} = \frac{2}{3}$ 　　**c** $\frac{4}{5} \quad \frac{2}{3} = 1\frac{7}{15}$

34 Place $<$ or $>$ between each pair of calculations.

a $5 \times \frac{1}{2} \quad 5 \times 2$ 　　**b** $3\frac{1}{4} \times \frac{3}{5} \quad 3\frac{1}{4} \div \frac{3}{5}$

35 The formula $v = u + at$ is used in physics.
Find the value of v when $u = \frac{1}{2}$, $a = \frac{3}{4}$ and $t = 1\frac{1}{2}$.

36 This formula is also used in physics: $s = \frac{1}{v} + \frac{1}{u}$.
Find the value of s when $v = 2\frac{3}{4}$ and $u = 1\frac{3}{8}$.

37 The first four terms of a sequence are $\frac{x}{2}$, $\frac{x^2}{4}$, $\frac{x^3}{8}$, $\frac{x^4}{16}$.

a Write down the next two terms.

b Write down the rule for generating the sequence.

1 Find   **a** $4\frac{1}{2} \times 3\frac{1}{3}$ **b** $3\frac{2}{5} \div \frac{3}{10}$

2 Find   **a** $\frac{8}{9} + \frac{21}{27}$ **b** $2\frac{1}{3} + \frac{4}{9} + 1\frac{5}{6}$

3 Put $>$ or $<$ between the following pairs of numbers
  a $\frac{4}{7}$ $\frac{5}{8}$ **b** $\frac{11}{9}$ $1\frac{3}{10}$

4 Calculate   **a** $5\frac{1}{4} - 1\frac{2}{3} \div \frac{2}{5}$ **b** $3\frac{3}{8} \times \left(8\frac{1}{2} - 5\frac{5}{6}\right)$

5 Arrange in ascending order: $\frac{7}{15}$, $\frac{1}{3}$, $\frac{2}{5}$.

6 What is $1\frac{1}{2}$ subtracted from $\frac{2}{3}$ of $5\frac{1}{4}$?

7 Find   **a** $4\frac{1}{2} \times 3\frac{2}{3} - 10\frac{1}{4}$ **b** $3\frac{1}{2} \div \left(2\frac{1}{8} - \frac{3}{4}\right)$

8 What is $1\frac{2}{3}$ of 1 minute 15 seconds (in seconds)?

9 Fill in the missing numbers   **a** $\frac{4}{5} = \frac{}{30}$ **b** $\frac{2}{7} = \frac{6}{}$

10 Express as mixed numbers   **a** $\frac{25}{8}$ **b** $\frac{49}{9}$ **c** $\frac{37}{6}$

11 A man can paint a door in 1 hour 15 minutes. How many similar doors can he paint in $7\frac{1}{2}$ hours?

1 Investigate what happens to a number when you multiply it by a fraction that is less than 1.
Does it matter whether the number itself is more or less than 1 ?
What happens to a number when it is divided by a fraction less than 1 ?

2 a These are called continued fractions. Write down the next three patterns.

$$\frac{1}{1+1}, \quad \frac{1}{1+\dfrac{1}{1+1}} \quad \frac{1}{1+\dfrac{1}{1+\dfrac{1}{1+1}}}$$

  b Evaluate each of these fractions, giving your answers as
    i a fraction in as simple a form as possible
    ii as a decimal correct to 5 decimal places.
  c Investigate what happens as the pattern continues.

FRACTIONS AND PERCENTAGES

Percentages and fractions are used a great deal in everyday life.
Kate earns £260 for a 5-day working week but is on strike because her company's offer of a 4.5% pay increase is considered unsatisfactory.
Her union want her to remain on strike until they secure a 5% increase.
Kate would like to know

- how much more she will get each week if she accepts the company offer
- how much more she will get if she holds out for the figure the union is seeking
- how much she loses each day she is on strike
- how many days she will have to work to make up her lost pay.

They remain on strike for the higher figure and eventually settle for 4.8% after being on strike for 16 working days.

- How long does she have to work to make up the lost money?

Studying this chapter should enable you to answer these and similar questions.

You can use Revision Exercise 1.2 on page 9 to revise the basic work on fractions, decimals and percentages covered in Book 7A.

EXPRESSING ONE QUANTITY AS A FRACTION OF ANOTHER

Many quantities can be divided into equal parts.
For instance, there are seven days in a week, so one day is $\frac{1}{7}$ of a week and 5 days is $\frac{5}{7}$ of a week.

Similarly, 30 minutes can be expressed as a fraction of $2\frac{1}{2}$ hours.

There are 150 minutes in $2\frac{1}{2}$ hours.

so 30 minutes is $\frac{30}{150} = \frac{1}{5}$ of $2\frac{1}{2}$ hours.

> To express one quantity as a fraction of another,
> first express both quantities in the same unit
> and then put the first quantity over the second quantity.

Express the first quantity as a fraction of the second quantity.

a 15 cm, 75 cm   **b** 680 g, 2 kg

a 15 cm as a fraction of 75 cm is

$$\frac{\overset{1}{\cancel{15}}}{\underset{5}{\cancel{75}}} = \frac{1}{5}$$

b 2 kg = 2 × 1000 g
          = 2000 g

> Both quantities must be in the same unit so we change the larger unit to the smaller unit.

Then 680 g as a fraction of 2 kg is

$$\frac{\overset{17}{\cancel{680}}}{\underset{50}{\cancel{2000}}} = \frac{17}{50}$$

In questions **1** to **14** express the first quantity as a fraction of the second quantity.

1 15 minutes, 45 minutes

2 35 seconds, 120 seconds

3 70 cm, 200 cm

4 £2, £5

5 35 g, 80 g

6 20 minutes, 2 hours

7 500 m, 3 km

8 25 cm², 400 cm²

9 4 in, 3 ft

10 4 pints, 2 gallons

11 55 p, £2.25

12 5.6 cm, 12 cm

13 800 g, 2.25 kg

14 40 mm², 4 cm²

15 Sally's journey to school costs 55 p on one bus and 35 p on another bus.

    a Find the total cost of her journey

    b What fraction of the total cost arises from each bus?

16 Amos has a 25 acre field. He plants $15\frac{1}{2}$ acres with wheat. What fraction of his field does Amos plant with wheat?

17 Simon gets £2.40 a week pocket money. If he spends £1.35, what fraction of his pocket money is left?

18 When she left home, Carol's school bag weighed $3\frac{1}{4}$ kg. When she got to school, she took out her maths text book. Her school bag then weighed $2\frac{1}{2}$ kg. What was the weight of her maths book as a fraction of the weight of her bag when she left home?

19 This pumpkin weighed $5\frac{1}{2}$ kg before it was carved.
It now weighs $1\frac{1}{2}$ kg.
What fraction of its weight has been removed?

FINDING A
FRACTION OF A
QUANTITY

In Book 7A, to find two-thirds of a quantity we first found one-third, by dividing the quantity by 3, and then multiplied the result by 2 to get $\frac{2}{3}$ of the quantity. Now that you are more familiar with multiplying and dividing fractions we can combine the working, and use a simpler method.

We can find $\frac{2}{3}$ of 15 by calculating $\frac{2}{3} \times 15$

Note that   ' of ' can be replaced by ' $\times$ '

EXERCISE 4B

Find three-fifths of 95 metres.

$$\frac{3}{{}_{1}5} \times \frac{95^{19}}{1} = 57$$

$\frac{3}{5}$ of 95 metres is 57 metres.

Find

1 $\frac{3}{5}$ of 20 metres

2 $\frac{5}{9}$ of 45 dollars

3 $\frac{9}{10}$ of 50 litres

4 $\frac{3}{8}$ of 88 miles

5 $\frac{7}{16}$ of 48 gallons

6 $\frac{1}{4}$ of £2

7 $\frac{2}{9}$ of 36 pence

8 $\frac{3}{10}$ of £1

9 $\frac{2}{7}$ of 42 pence

10 $\frac{4}{5}$ of 1 year (365 days)

Find $\dfrac{4}{9}$ of $5\dfrac{5}{8}$

$\dfrac{4}{9}$ of $5\dfrac{5}{8} = \dfrac{4}{9} \times 5\dfrac{5}{8}$

$\qquad = \dfrac{\cancel{4}^{1}}{_{1}9} \times \dfrac{\cancel{45}^{5}}{_{2}\cancel{8}}$

$\qquad = \dfrac{5}{2}$

$\qquad = 2\dfrac{1}{2}$ $\qquad$ Check: $2\dfrac{1}{2} \div 5\dfrac{5}{8} = \dfrac{\cancel{5}''}{2} \times \dfrac{\cancel{8}^{4}}{\cancel{45}} = \dfrac{4}{9}$

Find

11 $\dfrac{4}{7}$ of $4\dfrac{2}{3}$ $\qquad\qquad\qquad$ **14** $\dfrac{5}{12}$ of $1\dfrac{3}{5}$

12 $\dfrac{2}{3}$ of $2\dfrac{1}{4}$ $\qquad\qquad\qquad$ **15** $\dfrac{3}{11}$ of $4\dfrac{5}{7}$

13 $\dfrac{3}{5}$ of $2\dfrac{1}{7}$ $\qquad\qquad\qquad$ **16** $\dfrac{5}{8}$ of $5\dfrac{3}{5}$

17 Martine was given £12. She spent $\dfrac{7}{12}$ of it on a tape and $\dfrac{3}{8}$ of it on a book.
How much did she spend on $\quad$ **a** the tape $\quad$ **b** the book?

18 Fiona bought a 1 kg bag of dried fruit. She used $\dfrac{2}{5}$ of the fruit in one cake and $\dfrac{3}{10}$ of it in a second cake.
How many grams of fruit did she use for

a the first cake $\qquad$ **b** the second cake?

19 Due to illness Sally was away from work for the whole of April. She spent $\dfrac{1}{6}$ of the month in hospital, $\dfrac{2}{5}$ of it in a convalescent home and the remainder of the month at home.
How many days was she

a in hospital $\qquad$ **b** in a convalescent home $\qquad$ **c** at home?

20 When the 288 members of a club voted for a new president, $\dfrac{7}{12}$ of the members voted for Harvey, $\dfrac{5}{12}$ of those who did not vote for Harvey voted for Metson, and all those who did not vote either for Harvey or for Metson voted for Critchley. How many members voted for each candidate?

FINDING THE
WHOLE OF A
QUANTITY

Sometimes we know what a fraction of a quantity is, and from this we can find the quantity itself.

For example, if $\frac{2}{3}$ of the pupils in a class are girls and we also know that there are 18 girls in the class

then $\quad \frac{2}{3}$ of the pupils in the class $= 18$ pupils

so $\quad \frac{1}{3}$ of the pupils in the class $= 9$ pupils

Therefore there are $\quad 9 \div \frac{1}{3} = \frac{9}{1} \times \frac{3}{1}$

$$= 27 \text{ pupils in the class altogether}$$

In the following exercise remember that to add and subtract fractions, they must first be expressed as equivalent fractions with the same denominator.

EXERCISE 4C

Anne ate $\frac{1}{4}$ of the contents of a box of chocolates and Brian ate $\frac{1}{3}$. There were 10 chocolates left. How many chocolates were there originally?

Fraction of the contents eaten $= \dfrac{1}{4} + \dfrac{1}{3}$

$$= \dfrac{3}{12} + \dfrac{4}{12} = \dfrac{7}{12}$$

Fraction left $= 1 - \dfrac{7}{12} = \dfrac{5}{12}$

But $\dfrac{5}{12}$ of the contents of the box $= 10$ chocolates

so $\dfrac{1}{12}$ of the contents is $10 \div 5 = 2$ chocolates

Original number of chocolates $= 2 \times 12 = 24$

> Another way to find the original number is to let x be this number.
> Then $\qquad \frac{5}{12} \times x = 10$
> $\therefore \qquad\qquad 5x = 120 \text{ so } x = 24$

1 After spending $\frac{5}{6}$ of my pocket money I have 80 p remaining.

   a What fraction of my pocket money is left?

   b How much pocket money do I receive?

2 A man poured $\frac{5}{8}$ of a full can of oil into his car engine and had 3 litres left over.

   a What fraction of the contents is left in the can?

   b How much did he use?

3 John and Fred go into business together, John putting up $\frac{7}{12}$ of the capital and Fred putting up £10 400 which is the remainder of the capital.

a What fraction of the capital did Fred contribute?

b How much did John contribute?

At a local election with three candidates, Mr Brown received $\frac{3}{10}$ of the votes cast, Mrs Collard $\frac{2}{5}$ and Miss Durley $\frac{3}{13}$.

If there were 243 spoilt papers, how many people voted and how many votes did the winner receive?

Total fraction of votes cast $= \dfrac{3}{10} + \dfrac{2}{5} + \dfrac{3}{13}$

$$= \frac{39}{130} + \frac{52}{130} + \frac{30}{130} = \frac{121}{130}$$

Fraction of votes not cast for these candidates

$$= 1 - \frac{121}{130} = \frac{130}{130} - \frac{121}{130} = \frac{9}{130}$$

As $\dfrac{9}{130}$ of the total $= 243$, $\quad \dfrac{1}{130}$ of the total $= \dfrac{243}{9} = 27$

$\therefore$ Total $= 27 \times 130 = 3510$

Since $\frac{2}{5} \left(\frac{52}{130} \right)$ is larger than both $\frac{3}{10} \left(\frac{39}{130} \right)$ and $\frac{3}{13} \left(\frac{30}{130} \right)$. Mrs Collard received the largest fraction of the votes.

Number of votes for Mrs Collard

$= \dfrac{2}{5}$ of 3510

> Remember that $\frac{2}{5}$ of 3510 means $\frac{2}{5} \times 3510$.

$= \dfrac{2}{\cancel{5}} \times \dfrac{\cancel{3510}^{702}}{1} = 1404$

$\therefore$ 3510 people voted and the winner received 1404 votes.

---

4 Anne, Betty and Cheryl decide to open a hairdressing salon. Anne contributes $\frac{7}{20}$ of the total capital, Betty $\frac{3}{10}$ and Cheryl the remaining £2814.

a What fraction of the capital does Cheryl provide?

b How much is the capital?

c How much do Anne and Betty each contribute?

5 In a book containing four short stories, the first is $\frac{1}{6}$ of the whole, the second $\frac{1}{8}$, the third has 126 pages and the fourth is $\frac{1}{3}$ of the whole.

   a What fraction of the book is the third story?

   b How many pages are there in the book?

6 A load of coal was divided between four brothers. The first received $\frac{1}{3}$ of it, the second $\frac{2}{5}$, the third $\frac{1}{7}$ and the fourth 273 kg. Calculate the original mass of coal and the amount received by the first brother.

7 A retailer buys a quantity of apples. On Thursday $\frac{2}{7}$ were sold, on Friday $\frac{3}{8}$ and on Saturday $\frac{1}{5}$. If 156 apples remain unsold, how many were sold on Friday?

The remaining questions are mixed problems involving fractions.

8 The product of two numbers is 8. If one of the numbers is $3\frac{1}{3}$, find the other.

9 The product of two numbers is 21. If one of the numbers is $2\frac{4}{7}$, find the other.

10 A tennis ball bounces to $\frac{2}{3}$ of the height from which it is dropped and then to $\frac{2}{3}$ of the height to which it rises after each subsequent bounce.

   If it is dropped from a height of 6 m, how high will it rise after it has struck the ground for the second time?

11 When the smaller of two fractions is divided by the larger, the result is $\frac{7}{9}$. If the smaller fraction is $2\frac{2}{3}$, find the larger.

12 If it takes $3\frac{1}{3}$ minutes to fill $\frac{3}{8}$ of a water storage tank, how long will it take to fill it completely?

13 When the larger of two fractions is divided by the smaller, the result is $1\frac{7}{18}$. If the smaller fraction is $2\frac{2}{5}$, find the larger.

14 In the tenth year of a school $\frac{3}{5}$ of the pupils study geography and $\frac{1}{3}$ of those who do not study geography study history.

   If 32 members of Year 10 study neither history nor geography, how many pupils are there in Year 10?

In everyday life it is commonplace to come across harder fractions and percentages than those we have considered so far. For example,

- Last month the rate of inflation rose by 0.02% to give an annual figure of **4.83%**.
- Buildapower offer a discount of $3\frac{3}{4}$% for bills paid immediately in cash and $2\frac{1}{2}$% for accounts settled within 3 days.

'Per cent' means 'out of 100'.

For example, $2\frac{1}{2}$% means $2\frac{1}{2}$ out of 100', that is, $2\frac{1}{2} \div 100$.

Now $2\frac{1}{2} \div 100 = \frac{5}{2} \div \frac{100}{1} = \frac{\cancel{5}^{1}}{2} \times \frac{1}{\cancel{100}_{20}} = \frac{1}{40}$

To express a percentage as a fraction divide the percentage by 100.

Express as fractions in their lowest terms

1 60%	**3** 75%	**5** 45%	**7** 42%
2 38%	**4** 96%	**6** 24%	**8** 68%

Express $137\frac{1}{2}$%, as a fraction in its lowest terms.

> A percentage can be expressed as a fraction by dividing by 100.

$137\frac{1}{2}$% $= \dfrac{275}{2}$%

$= \dfrac{275}{2} \div \dfrac{100}{1}$

$= \dfrac{\cancel{275}^{11}}{2} \times \dfrac{1}{\cancel{100}_{4}}$

$= \dfrac{11}{8} = 1\dfrac{3}{8}$

Express as fractions in their lowest terms

9 $33\frac{1}{3}$%	**12** $66\frac{2}{3}$%	**15** $37\frac{1}{2}$%	**18** $87\frac{1}{2}$%
10 $12\frac{1}{2}$%	**13** $62\frac{1}{2}$%	**16** $5\frac{1}{3}$%	**19** $6\frac{1}{4}$%
11 $2\frac{1}{2}$%	**14** 125%	**17** $17\frac{1}{2}$%	**20** 150%

Express **a** $6\frac{1}{2}\%$ **b** $27\frac{1}{3}\%$ as decimals.

a First express $6\frac{1}{2}\%$ as 6.5%.

$6\frac{1}{2}\% = \dfrac{6.5}{100} = 0.65$

b $27\frac{1}{3}\% = \dfrac{27.33...}{100} = 0.273$ (correct to 3 s.f.)

Reminder: work to 4 s.f. if answers are required correct to 3 s.f.

Express the following percentages as decimals, giving your answers correct to 3 significant figures where necessary.

21 $5\frac{1}{2}\%$ **24** $62\frac{1}{4}\%$ **27** 120%   **30** 180%

22 145%   **25** 350%   **28** 231%   **31** $5\frac{1}{3}\%$

23 $58\frac{1}{3}\%$ **26** $48\frac{2}{3}\%$ **29** $85\frac{2}{3}\%$ **32** $54\frac{1}{7}\%$

EXPRESSING FRACTIONS AS PERCENTAGES

Suppose that $\frac{4}{5}$ of the pupils in a school, that is, 4 out of 5, have been away for a holiday.

All of a quantity is 100% of that quantity

so $\quad \frac{4}{5}$ of all the pupils $= \frac{4}{5} \times 100\% = 80\%$

We can check this by expressing $\frac{4}{5}$ as an equivalent fraction

$\frac{4}{5} = \frac{80}{100} = 80\%$

> A fraction can be converted into a percentage
> by multiplying that fraction by 100%.

This does not alter its value, since 100% is 1.

It follows that a decimal can be expressed as a percentage by multiplying the decimal by 100.

> Express **a** $\frac{7}{20}$ **b** $\frac{41}{60}$ as percentages,
>
> giving your answer correct to 1 decimal place where necessary.
> Do not use a calculator.
>
> > To convert a fraction to a percentage, multiply by 100%.
>
> **a** $\frac{7}{20} = \frac{7}{\cancel{20}} \times \frac{\cancel{100}^{5}\%}{1} = 35\%$
>
> **b** $\frac{41}{60} = \frac{41}{\cancel{60}} \times \frac{\cancel{100}^{10}\%}{1} = 68.3\%$ (correct to 1 d.p.) $\quad \begin{array}{r} 68.33 \\ \hline 6)410.00 \end{array}$

Express the following fractions as percentages, giving your answers correct to 1 decimal place where necessary. Do not use a calculator.

1 $\frac{1}{2}$ **6** $\frac{1}{4}$ **11** $\frac{3}{4}$ **16** $\frac{3}{5}$

2 $\frac{7}{10}$ **7** $\frac{3}{20}$ **12** $\frac{9}{20}$ **17** $\frac{7}{20}$

3 $\frac{13}{20}$ **8** $\frac{4}{25}$ **13** $\frac{7}{5}$ **18** $\frac{31}{25}$

4 $\frac{1}{3}$ **9** $\frac{3}{8}$ **14** $\frac{5}{8}$ **19** $\frac{7}{8}$

5 $\frac{21}{40}$ **10** $\frac{23}{60}$ **15** $\frac{8}{3}$ **20** $\frac{8}{5}$

> Express **a** 1.24 **b** 0.045 as percentages.
>
> > To convert a decimal to a percentage, multiply by 100%.
>
> **a** $1.24 = 1.24 \times 100\% = 124\%$
>
> **b** $0.045 = 0.045 \times 100\% = 4.5\%$

Express the following decimals as percentages. Do not use a calculator.

21 0.22   **24** 0.04   **27** 0.74   **30** 0.16

22 1.72   **25** 2.64   **28** 3.41   **31** 6.35

23 0.0625   **26** 0.845   **29** 0.075   **32** 0.1825

Express each fraction, first as a decimal, then as a percentage.
Give answers correct to 2 decimal places. Use a calculator where you find it necessary.

33 $\dfrac{2}{7}$ **36** $\dfrac{5}{11}$ **39** $\dfrac{4}{9}$ **42** $\dfrac{5}{13}$

34 $\dfrac{9}{22}$ **37** $\dfrac{34}{57}$ **40** $\dfrac{14}{37}$ **43** $\dfrac{16}{93}$

35 $\dfrac{34}{9}$ **38** $\dfrac{67}{13}$ **41** $\dfrac{345}{17}$ **44** $\dfrac{56}{5}$

INTERCHANGING FRACTIONS, DECIMALS AND PERCENTAGES

Now we are in a position to change one form for expressing part of a whole into another form, i.e. we can interchange fractions, decimals and percentages.

EXERCISE 4F

Express
a $5\dfrac{1}{2}\%$ as   **i** a decimal   **ii** a fraction in its lowest terms
b 0.88 as   **i** a percentage   **ii** a fraction in its lowest terms
c $\dfrac{9}{16}$ as   **i** a decimal   **ii** a percentage.

a   **i** $5\dfrac{1}{2}\% = 5.5\% = 5.5 \div 100 = 0.055$

   ii $5\dfrac{1}{2}\% = \dfrac{11\%}{2} = \dfrac{11}{2} \div \dfrac{100}{1} = \dfrac{11}{2} \times \dfrac{1}{100} = \dfrac{11}{200}$

b   **i** $0.88 = 0.88 \times 100\% = 88\%$

   ii $0.88 = \dfrac{88}{100} = \dfrac{22}{25}$

c   **i** $\dfrac{9}{16} = 0.5625$ $\boxed{\dfrac{9}{16} = 9 \div 16 = 0.5625}$

   ii $\dfrac{9}{16} = 0.5625 \times 100\% = 56.25\%$ $\boxed{\text{Use the result from part }\textbf{i.}}$

Copy and complete the following tables.

	Fraction	Percentage	Decimal
1	$\dfrac{13}{20}$		
2	$\dfrac{17}{40}$		

	Fraction	Percentage	Decimal
3		$12\dfrac{1}{2}\%$	
4		225%	

	Fraction	Percentage	Decimal			Fraction	Percentage	Decimal
5			0.875	**13**				1.4
6			1.05	**14**	$2\frac{3}{4}$		
7		$3\frac{3}{4}\%$		**15**			115%	
8	$\frac{17}{50}$			**16**	$17\frac{1}{2}$		
9		$\frac{1}{2}\%$		**17**				1.36
10			0.035	**18**			345%	
11	$1\frac{1}{2}$			**19**				0.002
12		120%		**20**			5.6%	

USING ONE PERCENTAGE TO FIND ANOTHER

Suppose that in the town of Doxton, 55 families in every 100 own a car. Every family either owns a car or does not own a car, so if we are given one percentage we can deduce the other. It follows that in Doxton, 45 in every 100 families do not own a car.

EXERCISE 4G

> If 66% of homes have a telephone, what percentage do not?
>
> All homes either have or do not have a telephone.
> If 66% have a telephone, then (100 − 66)% do not,
> i.e.   34% do not.

1 In a box of oranges, 8% are bad. What percentage are good?

2 Twelve per cent of the cars that come to an MOT testing station fail to pass first time. What percentage pass first time?

3 A hockey team won 62% of their matches and drew 26% of them. What percentage did they lose?

4 Deductions from a youth's wage were: income tax 18%, other deductions 14%. What percentage did he keep?

5 In an election, 8% of the electorate failed to vote, 40% of the electorate voted for Mrs Long, 32% for Mr Singhe and the remainder voted for Miss Berry. If there were only three candidates, what percentage voted for Miss Berry?

6 In a book, 98% of the pages contain text, diagrams or both. If 88% of the pages contain text and 32% contain diagrams, what percentage contain

    a neither text nor diagrams   **c** only text

    b only diagrams   **d** both text and diagrams?

EXPRESSING ONE QUANTITY AS A PERCENTAGE OF ANOTHER

John has a tray holding 20 eggs, 6 of which are cracked. If we wish to find 6 as a percentage of 20, we express 6 as a fraction of 20 and then change the fraction to a percentage by multiplying by 100%

so 6 as a percentage of 20 is $\dfrac{6}{_{,}20} \times \dfrac{\overset{5}{\cancel{100}}\%}{1} = 30\%$

> To express one quantity as a percentage of another, we divide the first quantity by the second and multiply this fraction by 100%.

EXERCISE 4H

> Express 20 cm as a percentage of 3 m. Give your answer as
>
> **a** a mixed number   **b** a decimal, correct to 3 significant figures.
>
> **a** (First express 3 m in centimetres to give both quantities in the same unit.)
>
> $$3\,m = 300\,cm$$
>
> The first quantity as a percentage of the second quantity is
>
> $$\dfrac{20}{300} \times \dfrac{100\%}{1} = 6\dfrac{2}{3}\%$$
>
> **b** $\dfrac{2}{3} = 2 \div 3 = 0.666\ldots$
>
> $$\therefore \quad 6\dfrac{2}{3}\% = 6.67\% \ (\text{correct to 3 s.f.})$$

Express the first quantity as a percentage of the second. When necessary give each answer as   **a** a mixed number

                          b a decimal, correct to 3 significant figures.

1 3, 12	**3** 60 cm, 4 m	**5** 24 cm, 40 cm
2 30 cm, 50 cm	**4** 15, 20	**6** 4 in, 12 in

7 5, 50

14 40, 20

8 35 m, 56 m

15 600 m, 2000 m

9 50 m, 5 m

16 $3\frac{1}{2}$ yd, 7 yd

10 20 m², 80 m²

17 200 mm², 800 mm²

11 75 cm², 200 cm²

18 25 cm², 125 cm²

12 50 m², 15 m²

19 4 litres, 10 litres

13 3.6 t, 5 t

20 33.6 g, 80 g

In questions **21** to **34** first express both quantities in the same unit.

21 1200 g, 3 kg

24 900 g, 2.5 kg

22 45 p, £1.35

25 28 cm, 1.2 m

23 98 mm, 2.45 m

26 74 p, £1.11

27 46 cm², 1 m²

31 100 cm³, 1 litre

28 10 cm², 200 mm²

32 25 000 cm³, 1 m³

29 39 ft², 60 yd²

33 6 pints, 3 gallons

30 0.1 m², 25 000 mm²

34 0.01 m³, 125 000 cm³

There are 350 pupils in Year 8 and 287 of them study geography. What percentage of Year 8 study geography?

> We want the number of pupils studying geography as a percentage of the number of pupils in Year 8, i.e. 287 as a percentage of 350.

Percentage of 350 pupils studying geography $= \dfrac{287}{350} \times \dfrac{100\%}{1}$

$= 82\%$

---

35 There are 60 boys in the Year 9, 24 of whom study chemistry. What percentage of ninth-year boys study chemistry?

36 In a history test, Pauline scored 28 out of a possible 40. What was her percentage mark?

37 Out of 20 cars tested in one day by an MOT testing station, 4 of them failed. What percentage of the cars tested failed?

38 There are 60 photographs in a book, 12 of which are coloured. What percentage of the photographs are coloured?

39

Forty-two of the 60 choristers in a choir wear spectacles. What percentage of the 60 choristers do not wear spectacles?

40 Each week a boy saves £3 of the £12 he earns. What percentage of his earnings does he spend?

41 A secretary takes 56 letters to the post office for posting; 14 are first class and the remainder are second class. What percentage of the 56 letters go second class?

42 Judy obtained 80 marks out of a possible 120 in her end-of-term maths examination. What was her percentage mark?

43 Jane's gross wage is £120 per week, but her take-home pay is only £78. What percentage is this of her gross wage?

FINDING A PERCENTAGE OF A QUANTITY

Kate earns £260 a week. If Kate wants to find out the money value of a 4.5% pay rise, she needs to find 4.5% of £260.

Now **4.5% = 0.045**

so $\quad$ 4.5% of £260 $= \frac{4.5}{100} \times £260$

$$= £0.045 \times 260$$

$$= £11.70$$

> To find a percentage of a quantity, change the percentage to a decimal and multiply by the quantity.

EXERCISE 4I

> Find the value of·   **a** 12% of 450   **b** 7.3% of 3.75 m.
> Give answers that are not exact correct to 3 significant figures.
>
> **a** 12% of 450 $= \frac{12}{100} \times 450 = 0.12 \times 450 = 54$
>
> **b** 7.3% of 3.75 m $= \frac{7.3}{100} \times 375$
> $$= 0.073 \times 375 \text{ cm}$$
> $$= 27.375 \text{ cm}$$
> $$= 27.4 \text{ cm (correct to 3 s.f.)}$$

Giving your answers correct to 3 significant figures where necessary, find the value of

1 40% of 120

2 12% of 800 g

3 74% of 75 cm

4 8% of £2

5 63% of 4 m

6 96% of 15 m^2

7 6% of 24 m

8 77% of 4 kg

9 86% of 1150 g

10 55% of 8.6 m

11 17% of 2 km

12 30% of £250

13 45% of 740

14 66% of 300 cm^2

15 5.5% of £560

16 9.75% of 80 m

17 15.8% of 5 litres

18 2.75% of £44

19 3.625% of 360 g

20 7.25% of £35.30

21 4.5% of 45 m^2

22 67% of 316 cm^2

23 44.4% of 8.5 kg

24 2.5% of 6 km

25 If 8% of a crowd of 24 500 at a football match were females, how many females attended?

26 By the end of the month a distributor had sold 74% of the 1250 washing machines he had in stock at the beginning of the month. How many washing machines had he sold?

> If 54% of the 1800 pupils in a school are boys, how many girls are there in the school?
>
> $$\text{Number of boys} = \frac{54}{100} \times 1800$$
> $$= 972$$
> $$\text{Number of girls} = 1800 - 972$$
> $$= 828$$

27 There are 80 houses in my street and 65% of them have a telephone. How many houses

    a have a telephone   **b** do not have a telephone?

28 In my class there are 30 pupils and 40% of them have a bicycle. How many pupils

    a have a bicycle   **b** do not have a bicycle?

29 Yesterday, of the 240 flights leaving London Airport, 15% were bound for North America. How many of these flights

    a flew to North America   **b** did not fly to North America?

30 In a particular year, 64% of the 16 000 Jewish immigrants entering Israel came from Eastern Europe. How many of the immigrants did not come from Eastern Europe?

31 There are 120 shops in the High Street, 35% of which sell food. How many High Street shops do not sell food?

32 Alex earns £18 000 a year. Deductions amount to 22.4% of his salary. Find

    a his total deductions   **b** his earnings after deductions.

33 Gary bought a car for £16 000. After 5 years it was worth 28% of the purchase price. Find its value after 5 years.

34 Sim's insurance premium for house contents is 0.9% of the value of the goods insured. How much must he pay if he estimates the value of the contents of his home to be £22 500?

35 Lena invests £450 in the Blackweir Building Society for one year. The building society pays interest of 5.66% on the sum invested each year. How much interest will she receive?

36 Bianca's season ticket for the train costs £345.60 a quarter. Fares are to rise by 5.6% of their present cost. How much more will her season ticket cost?

37 Last year the amount I paid for insurance was £520. This year my insurance premium will increase by 12%. Find the increase.

38 A mathematics book has 320 pages, 40% of which are on algebra, 25% on geometry and the remainder on arithmetic. How many pages of arithmetic are there?

39 Last week my gross pay was £296.53. Next week I should get a rise of 4.3% of this amount.

a By how much can I expect my gross pay to increase?

b What should my gross pay be next week?

Find the value of

40 a $22\frac{1}{2}\%$ of $40\,\text{m}^2$ **b** $5\frac{1}{4}\%$ of $56\,\text{mm}$

41 a $7\frac{1}{2}\%$ of $80\,\text{g}$ **b** $37\frac{1}{2}\%$ of $48\,\text{cm}$

42 a $62\frac{1}{2}\%$ of $8\,\text{km}$ **b** $3\frac{1}{4}\%$ of $64\,\text{kg}$

43 a $74\frac{1}{2}\%$ of $200\,\text{cm}^2$ **b** $87\frac{1}{2}\%$ of $16\,\text{mm}$

PERCENTAGE INCREASE

My telephone bill is to be increased by 8% from the first quarter of the year to the second quarter. It amounted to £64.50 for the first quarter. From this information I can find the value of the bill for the second quarter.

If £64.50 is increased by 8%, the increase is 8% of £64.50,

i.e. $\qquad\qquad$ $£0.08 \times 64.50 = £5.16$

The bill for the second quarter is therefore

$$£64.50 + £5.16 = £69.66$$

The same result is obtained if we take the original sum to be 100%. The increased amount is (100 + 8)%, or 1.08 of the original sum,

i.e.   the bill for the second quarter is $£1.08 \times 64.50 = £69.66$

The quantity 1.08 is called the multiplying factor and to increase a quantity by 12%, the multiplying factor would be 1.12.

PERCENTAGE DECREASE

Similarly if we wish to decrease a quantity by 8%, the decreased amount is (100 − 8)%, or 92%, i.e. 0.92 of the original sum.

If we wish to decrease a quantity by 15%, the new quantity is 85% of the original quantity, and the multiplying factor is 0.85.

> If a number is increased by 40%, what percentage is the new number of the original number?
>
> The new number is (100 + 40)%,
> i.e. 140% of the original number.

If a number is increased by the given percentage, what percentage is the new number of the original number?

1 50%	**4** 60%	**7** 48%	**10** $12\frac{1}{2}$%
2 25%	**5** 75%	**8** 300%	**11** 57%
3 20%	**6** 35%	**9** 175%	**12** 15%

> What multiplying factor increases a number by 44%?
>
> The multiplying factor is (100 + 44)% = 1.44

Give the multiplying factor which increases a number by

13 30%	**14** 80%	**15** 65%	**16** 130%

> If a number is decreased by 65%, what percentage is the new number of the original number?
>
> The new number is (100 − 65)%,
> i.e. 35% of the original number.

If a number is decreased by the given percentage what percentage is the new number of the original number?

17 50%	**20** 85%	**23** 4%	**26** $33\frac{1}{3}$%
18 25%	**21** 35%	**24** 66%	**27** 53%
19 70%	**22** 42%	**25** $62\frac{1}{2}$%	**28** 10%

> What multiplying factor decreases a number by 30%?
>
> The multiplying factor is $(100 - 30)\% = 0.7$

What multiplying factor decreases a number by

29 40%   **30** 75%   **31** 34%   **32** 12%

PERCENTAGE INCREASE AND DECREASE

In the last exercise we found what a given quantity must be multiplied by to increase or decrease the quantity by a particular percentage. In the next exercise we apply that knowledge.

EXERCISE 4K

> Increase 180 by 30%.
>
> The new value is 130% of the old
> i.e. the new value is $1.30 \times 180 = 234$

Increase

1 100 by 40%   **4** 550 by $36\frac{1}{2}$%   **7** 64 by $62\frac{1}{2}$%

2 200 by 85%   **5** 1600 by 73%   **8** 111 by 66.7%

3 340 by 45%   **6** 745 by 14%   **9** 145 by 120%

10 The rate of inflation for the last 12 months was 2.5%. (This means that, on average, prices are 2.5% higher than they were this time last year.) Mrs Brown's pension of £120 per week is to be increased in line with inflation. What is her new weekly pension?

11 A boy's weight increased by 15% between his fifteenth and sixteenth birthdays. If he weighed 55 kg on his fifteenth birthday, what did he weigh on his sixteenth birthday?

12 The water rates due on my house this year are 8% more than they were last year. Last year I paid £210. What must I pay this year?

13 There are 80 teachers in a school. It is anticipated that the number of staff next year will increase by 5%. How many staff should there be next year?

14 Pierre is 20% taller now than he was 2 years ago. If he was 150 cm tall then, how tall is he now?

15 A factory employs 220 workers. Next year this number will increase by 15%. How many extra workers will be taken on?

16 The government increased the price of petrol by $\frac{1}{5}$ and the following year increased the new price by a further $\frac{1}{5}$. By what fraction has the original price been increased to give the final price?

Decrease 250 by 70%.

The new value is 30% of the original value
i.e. the new value is $0.30 \times 250 = 75$

Decrease

17 100 by 30%

22 3450 by 4%

18 200 by 15%

23 93 by $33\frac{1}{3}\%$

19 350 by 46%

24 273 by $66\frac{2}{3}\%$

20 750 by 13%

25 208 by $87\frac{1}{2}\%$

21 3400 by 28%

26 248 by $37\frac{1}{2}\%$

27 Miss Kendall earns £120 per week from which income tax is deducted at 30%. Find how much she actually gets.
(This is called her *net* pay.)

28 In a certain week a factory worker earns £150 from which income tax is deducted at 12%. Find his net income after tax,
i.e. how much he actually gets.

29 Mr Hall earns £1000 per month. If income tax is deducted at 20%, find his net pay after tax.

30 The number of children attending Croydly village school is 8% fewer this year than last year. If 450 attended last year, how many are attending this year?

31 The marked price of a man's suit is £125. In a sale the marked price is reduced by 12%. Find the sale price.

The remaining questions are a mixture of percentage increase and percentage decrease.

32 A bathroom suite is marked at £650 to which is added value added tax (VAT) at $17\frac{1}{2}\%$ of the marked price to give the selling price. How much does the suite actually cost the customer?

33 A CD costs £7 plus value added tax at 20% of the cost price. How much has to be paid for the CD?

34 In a sale all prices are reduced by 10%. What is the sale price of an article marked **a** £40 **b** £85?

35 Last year in Blytham there were 75 reported cases of measles. This year the number of reported cases has dropped by 16%. How many cases have been reported this year?

36 Due to predators the population of oyster-catchers on Stoka island is decreasing each year by 15% of the number of oyster-catchers on the island at the beginning of that year. The present population is estimated to be 12 000 pairs. What is the estimated population in

a a year's time **b** 2 year's time **c** 3 year's time?

(Give each answer correct to the nearest 100, but use the uncorrected result to work with.)

37

Wendy wants to buy a CD player. She shops around and comes to the conclusion that any one of the four listed above would suit her.

a By just thinking about the figures given above and, without doing any calculations, which shop do you think sells the cheapest CD player?

b Calculate the asking price in each shop. In which shop can she buy the cheapest CD player? Does your answer agree with the thoughts you had in part **a**?

c How much more than the cheapest does the dearest CD player cost?

38 Mrs Murray has been allocated 250 tickets at £15 each for a performance of *The Mikado*. Because of the large number of tickets she bought she has been given a discount of 15%.

    a How much would all the tickets cost without the discount?

    b How much do these tickets cost Mrs Murray?

    c The tickets are divided between staff and pupils so that 8% go to staff and the remainder to pupils. How many tickets are allocated
      i to staff   **ii** to pupils?

    d $\frac{7}{10}$ of the pupils that go are girls. Of the pupils going to the performance, how many are   **i** girls   **ii** boys?

Mrs Murray booked enough 57-seater coaches to transport everybody to the theatre.

    e How many coaches did she book?

    f How many spare seats are there?

    g If each coach costs £120 + VAT at $17\frac{1}{2}$%, find the total cost of the coaches.

    h The cost of the coaches is divided equally among all those going. How much does each pupil pay?

    i How much should Mrs Murray collect from each person going on the trip? What percentage of what she collects is for the theatre ticket?

39 A car is valued at £8000. It depreciates by 20% in the first year and thereafter each year by 15% of its value at the beginning of that year. Find its value

    a after 1 year   **b** after 2 years   **c** after 3 years.

40 Mr Connah weighed 115 kg when he decided to go on a diet. He lost 10% of his weight in the first month and a further 8% of his original weight in the second month. How much did he weigh after 2 months of dieting?

41 In any year the value of a motorcycle depreciates by 10% of its value at the beginning of that year. What is its value after two years if the purchase price was £1800?

42 When John Short increases the speed at which he motors from an average of 40 mph to 50 mph, the number of miles travelled per gallon decreases by 25%. If he travels 36 miles on each gallon when his average speed is 40 mph, how many miles per gallon can he expect at an average speed of 50 mph?

43 When petrol was 50 p per litre I used 700 litres in a year. The price rose by 12% so I reduced my yearly consumption by 12%. Find

a the new price of a litre of petrol

b my reduced annual petrol consumption

c how much more (or less) my petrol bill is for the year.

44 Answer Kate's questions at the beginning of the chapter.

MIXED EXERCISE

EXERCISE 4L

1 Express $\frac{4}{25}$ **a** as a percentage **b** as a decimal.

2 Express 0.45 **a** as a percentage
   b as a fraction in its lowest terms.

3 Express 85% **a** as a decimal **b** as a fraction in its lowest terms.

4 a Express 6 mm as a percentage of 3 cm.

   b What multiplying factor would increase a quantity by 45% ?

5 Find

a $\frac{3}{5}$ of 27 m **c** 35% of 120 m^2

b $\frac{7}{12}$ of 186 cm^3 **d** $37\frac{1}{2}$% of 68 kg

6 If a number is increased by 25%, what percentage is the new number of the original number ?

7 A sculptor has a piece of wood weighing 6600 g when she begins work on a commission. During her work she cuts away 55% of it. What is the weight of the completed sculpture ?

8 a Increase 56 cm by 75%. **b** Decrease 1200 kg by 20%.

9 The annual cost of insuring the contents of a house is 0.3% of the value of the contents.
   How much will it cost to insure contents valued at £18 500 ?

10 Mrs Jones has marked $\frac{3}{5}$ of a batch of examination scripts. She still has 28 scripts left to mark. How many scripts are there altogether ?

11 In a cinema audience of 275, $\frac{2}{5}$ of the audience are women, 24% are men, **0.2** of the audience are girls and the remainder boys.

    a What fraction of the audience is men?

    b What percentage of the audience is boys?

    c How many boys are there in the audience?

    d How many of the audience are female?

INVESTIGATIONS

1 This is a group activity.
Collect references to percentages in newspapers and magazines.
Investigate what each reference means.

For example,

    a what is meant by the sentence 'The annual rate of inflation has fallen from 3.5% last month to 3.46% this month'? Does this mean that prices are rising, falling or standing still?

    b everyone in the company is to get a rise of 4.5%. This means that senior management will get an annual increase of £1350 while shop floor workers will get £500. How can this be true if everybody gets the same rise? Can you explain?

2 Investigate the different percentage rates charged for different types of insurance cover. For example, what are the percentage rates to insure the house or flat where you live, the contents of your home, your parents' motorcycle or car, or your luggage when you go on holiday? What reasons can you give for all these rates being different?

3 Building societies offer lots of different types of accounts. Get an up-to-date leaflet from one building society that gives details of all its different accounts and the rate of interest offered on each. Which account do you think you would use, and why, if

    a you are saving to go on holiday next year

    b a relative has given you £1000 and you want to keep it until you leave school?

Think of other situations in which you have money that you do not want to spend immediately. Which account would you save it in, and why?

RATIO

PARTS OF A WHOLE

Orange squash is made by diluting a concentrate with water. The concentrate and water are parts of the diluted drink and we need to know how much of each to use. This information can be given in many ways, for example,
'Add 720 ml of water to 240 ml of concentrate.'
'Dilute so that one quarter of the drink is concentrate.'
'Add one part by volume of concentrate to three parts of water.'
These instructions all result in the same mixture but some are more easily understood than others.

- The first instruction is probably more detailed than necessary. Also some arithmetic is needed if a different quantity of drink is wanted, and it is not immediately possible to compare the proportions of concentrate and water in the diluted juice.
- The second instruction is not very clear and needs some thought to work out how much water should be added. It can be useful if a particular quantity of made-up juice is required.
- The third instruction is clear, allows for different quantities of drink to be made up and gives immediate information about the quantities of concentrate and water in the diluted drink.
  This diagram sums up that information very clearly.

EXERCISE 5A

1 'Compo' is made by mixing cement and sand. The weights of sand and cement required can be given as 'Three parts sand to one part cement', or 'Three-quarters of the mixture is sand', or 'Mix 300 kg of sand with 100 kg of cement.'
Discuss which of these instructions is most helpful if you want to

a make up some mix using the sand you have in stock

b end up with 100 kg of mix

c buy enough sand and cement to make about 3000 kg mix, but you are not sure exactly how much mix you will need.

2 School uniform shirts are usually made from material that contains a mixture of cotton and polyester. Discuss the ways in which the proportions of cotton and polyester can be described on the label.

3 'Green Fingers' potting compost is made by mixing one part of sand with two parts of peat.
Discuss these statements (they are not all necessarily true).

a To make 3 litres of compost you need 1 litre of sand and 2 litres of peat.

b Half this bag of compost is sand.

c To make 3 kg of compost you need 1 kg of sand and 2 kg of peat.

d The quantity of sand is half the quantity of peat.

RATIO

The last exercise shows that there are various ways of comparing the sizes of related quantities, and that giving the relative sizes rather than the actual sizes is often more useful. Question **3** shows that, unless the nature of the quantities being compared is given, comparisons of size are not much use.

When we say that orange squash is made by adding one part by volume of concentrate to three parts by volume of water, we are *not* giving the size of the quantities to be mixed; we *are* giving information about the relative sizes of the quantities; this is called the *ratio* of the sizes, that is, the volume of concentrate to the volume of water is in the ratio 1 to 3, or, more briefly as

volume of concentrate : volume of water $= 1 : 3$

SIMPLIFYING
RATIOS

The first instructions for making orange squash was

'Add 720 ml of water to 240 ml of concentrate.'

This can be simplified to give the ratio of the two quantities,

i.e. the ratio of concentrate to water is 240 ml : 720 ml,

which can be simplified, by dividing both numbers by 240, to

1 ml : 3 ml.

We can simplify this further by omitting the units, *provided that it is made clear that the ratio is of volumes,* for example, by writing

volume of concentrate : volume of water $= 1 : 3$

The advantage of not giving units in a ratio is that any convenient unit for measuring the quantities can be used, provided that it is the *same for both quantities.*

Simplify the ratio 1 m : 2 cm

Before we can leave out the units, they must both be the same, so we use 1 m = 100 cm

$1 \text{ m} : 2 \text{ cm} = 100 \text{ cm} : 2 \text{ cm}$

Ratio of the lengths $= 100 : 2$

Divide both numbers by 2.

$= 50 : 1$

Simplify the ratios.

1 2 cm : 8 cm

2 32 p : 96 p

3 45 g : 1 kg

4 £4 : 75 p

5 48 p : £2.88

6 10 m : 50 cm

7 5 ml : 20 ml

8 £1.50 : 75 p

9 2 kg : 500 g

Simplify the ratio 2 m : 0.5 m

Ratio of lengths $= 2 : 0.5$

Multiplying by 2 gives whole numbers, and these are easier to work with than fractions or decimals.

$= 4 : 1$

Simplify the ratios.

10 $2 \text{ m} : 1\frac{1}{2}\text{ m}$

11 £1 : £1.50

12 $2\frac{1}{2}\text{ km} : 3 \text{ km}$

13 0.7 mm : 1.4 mm

14 $2\frac{1}{2}\text{ km} : 1\frac{1}{2}\text{ km}$

15 5.1 m : 25.5 cm

16 $1\frac{1}{2}\text{ in} : 1\frac{1}{2}\text{ in}$

17 2 kg : 1.4 kg

18 4.5 m : 36 cm

We can use ratios to compare more than two quantities.

Find the ratio of the lengths of the sides of this triangle.

Ratio of the lengths
$= 2.7 \text{ cm} : 1.8 \text{ cm} : 1.5 \text{ cm}$
$= 2.7 : 1.8 : 1.5$
$= 27 : 18 : 15$
$= 9 : 6 : 5$

2.7 cm

1.8 cm

1.5 cm

Simplify the ratios.

19 2 cm : 4 cm : 6 cm

20 1 m : 20 cm : 50 mm

21 2 kg : 3 kg : $1\frac{1}{2}$ kg

22 7 mg : 56 mg : 21 mg

23 25 g : 100 g : 75 g

24 6 apples : 2 oranges : 2 bananas

A family has 12 pets of which 6 are cats or kittens, 2 are dogs and the rest are birds. Find the ratio of the numbers of

a birds to dogs   **b** birds to pets.

There are 4 birds.

a Number of birds : number of dogs = 4 : 2

= 2 : 1

b Number of birds : number of pets = 4 : 12

= 1 : 3

In each question give your answer in its simplest form.

25 A couple have 6 grandsons and 4 granddaughters. Find

a the ratio of the number of grandsons to that of granddaughters
b the ratio of the number of granddaughters to that of grandchildren.

26 Square A has sides 6 cm long and square B has sides 8 cm long. Find the ratio of

a the length of the side of square A to the length of the side of square B

b the area of square A to the area of square B.

27 Tom walks 2 km to school in 40 minutes and John cycles 5 km to school in 15 minutes. Find the ratio of

a Tom's distance to John's distance

b Tom's time to John's time.

28 Mary has 18 sweets and Jane has 12. As Mary has 6 sweets more than Jane she tries to even things out by giving Jane 6 sweets. What is the ratio of the number of sweets Mary has to the number Jane has

a at the start   **b** at the end?

29 Rectangle A has length 12 cm and width 6 cm while rectangle B has length 8 cm and width 5 cm. Find the ratio of

a the length of A to the length of B

b the area of A to the area of B

c the perimeter of A to the perimeter of B

d the size of an angle of A to the size of an angle of B.

30 A triangle has sides of lengths 3.2 cm, 4.8 cm and 3.6 cm. Find the ratio of the lengths of the sides to one another.

31 Two angles of a triangle are 54° and 72°. Find the ratio of the size of the third angle to the sum of the first two.

32 For a school fête, Mrs Jones and Mrs Brown make marmalade in 1 lb jars. Mrs Jones makes 5 jars of lemon marmalade and 3 jars of orange. Mrs Brown makes 7 jars of lemon marmalade and 5 of grapefruit. Find the ratio of the numbers of jars of

a lemon to orange to grapefruit

b Mrs Jones' to Mrs Brown's marmalade

c Mrs Jones' lemon to orange.

33 A recipe for the base of a cheesecake needs 25 g of butter and 150 g of crushed biscuits. Find the ratio of the weight of butter to the weight of biscuits.

34 Amir made a model plane from a kit. The model is 40 cm long and the actual plane is 35 m long.

a Find the ratio of the length of the model to the length of the actual plane.

b Explain whether the answer is the same for the ratio of the length of the actual plane to the length of the model.

35 An enlargement of a photograph is 6.5 cm wide. The original photograph is 3.5 cm wide.

a What is the ratio of the width of the enlarged photograph to the width of the original?

b The enlarging machine is set by entering this ratio on its control panel. Explain what you think would happen if the ratio was entered in the wrong order.

36 If $p : q = 2 : 3$, find the ratio $6p : 2q$

RATIOS AS FRACTIONS

Vivian made a model of his father's car. The model is 5 cm long and the car is 5 m long. The ratio of the length of the model to the length of the car is 5 cm : 5m, i.e.

length of model : length of car $= 5 : 500$

$$= 1 : 100$$

We can also compare these lengths using fractions,

$$\frac{\text{length of model}}{\text{length of car}} = \frac{5\,\text{cm}}{5\,\text{m}} = \frac{5}{500}$$

$$= \frac{1}{100},$$

i.e.   the length of the model is $\frac{1}{100}$ of the length of the car.

FINDING UNKNOWN QUANTITIES

Sometimes the unknown quantity is obvious.
For example, orange squash is made by mixing concentrate and water in the ratio 1 : 3. To find the volume of water needed for 36 ml of concentrate, we know that

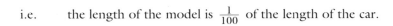

$$36\,\text{ml} : \text{volume of water} = 1 : 3$$

Now it is obvious that the volume of water needed is

$$3 \times 36\,\text{ml} = 108\,\text{ml}.$$

When the unknown quantity is not obvious, using ratios in fraction form is easier. We can then form an equation and solve it.
For example, the weights of icing sugar and butter are combined in the ratio 4 : 3 to make butter icing.
To find the weight of icing sugar needed to mix with 40 g of butter, we start by letting x g be the weight of icing sugar,

then $x : 40 = 4 : 3$

Writing these ratios as fractions gives the equation $\dfrac{x}{40} = \dfrac{4}{3}$

To find x, we have to make the left-hand side 40 times larger, i.e. we need to multiply both sides by 40:

$$\frac{40}{1} \times \frac{x}{40} = \frac{40}{1} \times \frac{4}{3}$$

giving $x = \dfrac{160}{3} = 53.3\ldots$

therefore

the weight of icing sugar needed is 53 grams (to the nearest gram).

EXERCISE 5C

Find the value of x when $2 : x = 3 : 5$

> If we write the ratio as given, x will be on the bottom of the fraction, i.e. $\dfrac{2}{x} = \dfrac{3}{5}$.
>
> It is easier to solve an equation when x is on top of a fraction; we can achieve this with ratios by changing their order. Make sure that the order of all ratios involved is changed in the same way.

If $2 : x = 3 : 5$, then $x : 2 = 5 : 3$

therefore $$\frac{x}{2} = \frac{5}{3}$$

$$\frac{2}{1} \times \frac{x}{2} = \frac{2}{1} \times \frac{5}{3}$$

$$x = \frac{10}{3} = 3\tfrac{1}{3}$$

If the value of x is obvious, write it down, otherwise form an equation to find the value of x.

1 $2 : 5 = 4 : x$ **3** $x : 9 = 3 : 5$

2 $x : 6 = 11 : 18$ **4** $x : 7 = 3 : 4$

5 $1 : 4 = 12 : x$ **9** $4 : x = 3 : 5$

6 $x : 5 = 4 : 3$ **10** $3 : 5 = x : 6$

7 $x : 4 = 1 : 3$ **11** $7 : 3 = 3 : x$

8 $4 : x = 1 : 3$ **12** $3 : x = 2 : 5$

13 $x : 1.2 = 2 : 3$ **16** $x : 4.2 = 4 : 3$

14 $1.5 : x = 2 : 3$ **17** $2.5 : x = 5 : 2$

15 $4 : 5 = x : 1\tfrac{1}{2}$ **18** $3 : 4 = x : 2\tfrac{1}{2}$

19 The ratio of the amount of money in David's pocket to that in Indira's pocket is $10 : 9$. Indira has 27 p. How much has David got?

20 Two lengths are in the ratio $3 : 7$. The second length is 42 cm. Find the first length.

21 If the ratio in question **20** was $7 : 3$, what would the first length be?

22 In a rectangle, the ratio of length to width is 9 : 4. The length is 24 cm. Find the width.

23 The ratio of the perimeter of a triangle to its shortest side is 10 : 3. The perimeter is 35 cm. What is the length of the shortest side?

24 A length, originally 6 cm, is increased so that the ratio of the new length to the old length is 9 : 2. What is the new length?

25 A class is making a model of the school building and the ratio of the lengths of the model to the lengths of the actual building is 1 : 20. The gym is 6 m high. How high, in centimetres, should the model of the gym be?

26 The ratio of lengths of a model boat to those of the actual boat is 3 : 50. Find the length of the actual boat if the model is 72 cm long.

27 A business woman has been advised that the ratio of the cost of a computer to the cost of the software she needs is about 3 : 2.
If she buys a computer costing about £1125, roughly how much will she need to spend on software?

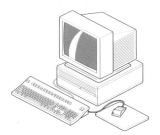

28 A lemon drink is made by mixing freshly squeezed lemon juice, sugar and water in the ratio 2 : 2 : 5 by volume.

a How many tablespoons of sugar are needed for 6 tablespoons of lemon juice?

b How many millilitres of water are needed for 25 ml of lemon juice?

29 A mix for concrete is made from cement, sand and aggregate by weight in the ratio 2 : 3 : 6.

a How much aggregate should be mixed with 10 kg of sand?

b How much cement is needed for 300 kg of aggregate?

30 A photograph is enlarged so that the ratio of the lengths of the sides of the enlargement to the lengths of the sides of the original is 5 : 2.

a The original photo measures 6 cm by 10 cm. What does the enlargement measure?

b Are the areas of the enlargement and original in the ratio 5 : 2? Explain your answer.

DIVISION IN A
GIVEN RATIO

We sometimes know the size of the final quantity made from two or more quantities but not the size of the constituent quantities. However if we know the ratio of the contributing quantities, we can find their sizes.

Sarah made up 2 litres of orange squash. Owen asked how much concentrate she used. Sarah couldn't remember, but she did know that the ratio of concentrate to water was 1 : 3

From this Sarah worked out that the concentrate was $\frac{1}{4}$ of the made-up volume, so the volume of concentrate used was $\frac{1}{4}$ of 2 litres, that is, $\frac{1}{2}$ litre.

EXERCISE 5D

Gordon has 21 kg of carpet cleaning powder with which to clean these carpets. How should he divide the powder so that each carpet gets the same quantity of powder per square metre?

27 square metres

36 square metres

> Gordon needs to divide the powder into two parts so that their weights are in the same ratio as the areas of the carpets.

The ratio of the areas is $36\,m^2 : 27\,m^2 = 4 : 3$

> Therefore the area of the larger carpet is $\frac{4}{7}$ of the total area to be cleaned, and the area of the smaller carpet is $\frac{3}{7}$ of the total area to be cleaned. So we should divide 21 kg into two parts, one of which is $\frac{4}{7}$ of 21 kg and the other is $\frac{3}{7}$ of 21 kg.

The larger carpet needs $\frac{4}{7} \times 21\,kg = 12\,kg$

The smaller carpet needs $(21 - 12)\,kg = 9\,kg$

Check: $\frac{3}{7} \times 21 = 9$

1 Divide 80 p into two parts in the ratio 3 : 2.

2 Divide 32 cm into two parts in the ratio 3 : 5.

3 Divide £45 into two shares in the ratio 4 : 5.

4 Dick and Tom share the contents of a bag of peanuts between them in the ratio 3 : 5. If there are 40 peanuts, how many do they each get?

5 Mary is 10 years old and Eleanor is 15 years old. Divide 75 p between them in the ratio of their ages.

6 In a class of 30 pupils the ratio of the number of boys to the number of girls is 7 : 8. How many girls are there?

7 Divide £20 into two parts in the ratio 1 : 7.

8 In a garden the ratio of the area of lawn to the area of flower-bed is 12 : 5. If the total area is 357 m^2, find the area of

   a the lawn   **b** the flower-bed.

9 In a bowl containing oranges and apples, the ratio of the numbers of oranges to apples is 4 : 3. If there are 28 fruits altogether, how many apples are there?

10 A bowl of punch is made by mixing wine and fruit juice in the ratio 2 : 7.

   a How much wine is there in 3 litres of this punch?

   b How much fruit juice is needed to make 12 litres of punch?

11 The fibres used to make the material for some school uniform shirts are cotton and polyester in the ratio 3 : 2 by weight.

   a How much polyester is there in a shirt that weighs 120 g?

   b Explain whether your answer to part **a** is exact or an estimate.

12 The ratio of the length of a rectangular photograph to its width is 5 : 3.

   a Find the length and width of the photograph if the perimeter is 72 cm.

   b Find the length and width of the photograph if its area is 60 cm^2.

Divide 6 m of string into three lengths in the ratio 3 : 7 : 2

> As the 6 m length of string is divided into three lengths in the ratio 3 : 7 : 2, one length is
> $$\frac{3}{3+7+2}, \text{ i.e. } \frac{3}{12}, \text{ of } 6\,\text{m}.$$
> The other lengths are also each a number of twelfths of 6 m.

First length $= \dfrac{3}{12} \times 6\,\text{m} = \dfrac{3}{12} \times 600\,\text{cm} = 150\,\text{cm}$

Second length $= \dfrac{7}{12} \times 600\,\text{cm} = 350\,\text{cm}$

Third length $= \dfrac{2}{12} \times 600\,\text{cm} = 100\,\text{cm}$

Check: 150 cm + 350 cm + 100 cm = 600 cm = 6 m

---

13   Divide £26 among three people so that their shares are in the ratio 4 : 5 : 4.

14   The perimeter of a triangle is 24 cm and the lengths of the sides are in the ratio 3 : 4 : 5. Find the lengths of the three sides.

15   In a garden, the ratio of the areas of lawn to beds to paths is $3 : 1 : \frac{1}{2}$. Find the three areas if the total area is $63\,\text{m}^2$.

16   A 25 litre bag of compost is made by mixing sand, peat and fertilizer by volume in the ratio 4 : 5 : 1. What volume of each ingredient is used ?

17   The instructions for mixing paint to give a particular shade of green are: use colours 127, 139, 250 in the ratio 2 : 1 : 7.
How much of colour 127 is needed to give 20 litres of mixed paint ?

MAP RATIO (OR REPRESENTATIVE FRACTION)

The Map Ratio of a map is the ratio of a length on the map to the length it represents on the ground. This ratio or fraction is given on most maps in addition to the scale. It is sometimes called the Representative Fraction of the map, or RF for short.

If two villages are 6 km apart and on the map this distance is represented by 6 cm, then the ratio is

$$6\,\text{cm} : 6\,\text{km} = 6\,\text{cm} : 600\,000\,\text{cm}$$
$$= 1 : 100\,000$$

so the map ratio is   1 : 100 000

and the RF is $\dfrac{1}{100\,000}$

Any length on the ground is 100 000 times the corresponding length on the map.

Find the map ratio of a map if 12 km is represented by 1.2 cm on the map.

Map ratio $= 1.2\,\text{cm} : 12\,\text{km}$
$\qquad\quad = 1.2\,\text{cm} : 1\,200\,000\,\text{cm}$
$\qquad\quad = 12 : 12\,000\,000$
$\qquad\quad = 1 : 1\,000\,000$

Multiplying both numbers by 10, then dividing both numbers by 12.

Find the map ratio of the maps in the following questions.

1 2 cm on the map represents 1 km.

2 The scale of the map is 1 cm to 5 km.

3 10 km is represented by 10 cm on the map.

4 3.2 cm on the map represents 16 km.

5 $\frac{1}{2}$ cm on the map represents 500 m.

6 100 km is represented by 5 cm on the map.

If the map ratio is 1 : 5000 and the distance between two points on the map is 12 cm, find the actual distance between the two points.

Let the actual distance be x cm.

Then $\qquad 12 : x = 1 : 5000$

or $\qquad x : 12 = 5000 : 1$

i.e. $\qquad \dfrac{x}{12} = \dfrac{5000}{1}$

so $\qquad \dfrac{12}{1} \times \dfrac{x}{12} = \dfrac{12}{1} \times \dfrac{5000}{1}$

giving $\qquad x = 60\,000$

The actual distance is 60 000 cm, that is, 600 m.

Because map ratios are always in the form 1 : n, this type of problem can be solved very simply as follows.
As the map ratio is 1 : 5000, 1 cm on the map represents 5000 cm on the ground, therefore 12 cm on the map represents 12×5000 cm on the ground,

i.e. 60 000 cm = 600 m

7 The map ratio of a map is 1 : 50 000. The distance between A and B on the map is 6 cm. What is the true distance between A and B?

8 The map ratio of a map is 1 : 1000. A length on the map is 7 cm. What real length does this represent?

9 The map ratio of a map is 1 : 10 000. Find the actual length represented by 2 cm.

10 The map ratio of a map is 1 : 200 000. The distance between two towns is 20 km. What is this in centimetres? Find the distance on the map between the points representing the towns.

11 The map ratio of a map is 1 : 2 000 000. Find the distance on the map which represents an actual distance of 36 km.

PROPORTION

When comparing quantities, words other than ratio are sometimes used. If two varying quantities are directly proportional they are always in the same ratio.

Sometimes it is obvious that two quantities are directly proportional,

e.g.   if oranges are being sold for 20 p each,

then   2 oranges cost 2 × 20 p,

6 oranges cost 6 × 20 p,

15 oranges cost 15 × 20 p.

In each case, number of oranges : cost in pence = 1 : 20
so the cost of buying oranges is proportional to the number of oranges bought.

In cases like this you would be expected to know that the quantities are in direct proportion.

EXERCISE 5F

A stack of 250 sheets of paper is 1.5 cm thick.

a How thick is a stack of 400 sheets of the same paper?

b How many sheets are there in a stack 2.7 cm thick?

a The number of sheets in the stack is proportional to the thickness of the stack, so

$$\text{numbers of sheets} : \text{thickness (cm)} = 250 : 1.5$$

The second stack is x cm thick,

$$\therefore \qquad 400 : x = 250 : 1.5$$

i.e. $\qquad x : 400 = 1.5 : 250$

so $\qquad\qquad \dfrac{x}{400} = \dfrac{1.5}{250}$

$$\dfrac{400}{1} \times \dfrac{x}{400} = \dfrac{400}{1} \times \dfrac{1.5}{250}$$

giving $\qquad\qquad x = 2.4$

The stack is 2.4 cm thick.

b If there are y sheets in the third stack then

$$y : 2.7 = 250 : 1.5$$

so $\qquad\qquad \dfrac{y}{2.7} = \dfrac{250}{1.5}$

giving $\qquad\qquad y = 450$

The third stack has 450 sheets.

> Alternatively
>
> **a** 250 sheets are 15 mm thick, so 1 sheet is $250 \div 15$ mm thick
> therefore 400 sheets are $250 \div 15 \times 400$ mm thick,
> that is, 24 mm or 2.4 cm thick.
>
> **b** 15 mm contains 250 sheets
> so 1 mm contains $250 \div 15$ sheets
> so 27 mm contains $250 \div 15 \times 27$ sheets, that is, 450 sheets.

1 Sam covers 9 m when he walks 12 paces. How far does he travel when he walks 16 paces?

2 I can buy 24 bottles of a soft drink for £8 when buying in bulk. How many bottles can I buy at the same rate for £12?

3 If 64 seedlings are allowed 24 cm² of space, how much space should be allowed for 48 seedlings?
How many seedlings can be planted in 27 cm²?

4 A ream (500 sheets) of paper is 6 cm thick. How thick a pile would 300 sheets of this paper make?

5 At a school picnic 15 sandwiches are provided for every 8 children. How many sandwiches are needed for 56 children?

Beware: some of the quantities in the following questions are not in direct proportion. Some questions need a different method and some cannot be answered at all from the given information.

> A family with two pets spends £1.50 a week on pet food. If the family gets a third pet, how much a week will be spent on pet food?
>
> We are not told what sort of animals the pets are. Different animals eat different types and quantities of food so the amount spent is not in proportion to the number of pets.

Give reasons for every answer.

6 Two tea towels dry on a clothes line in 2 hours. How long would 5 tea towels take to dry?

7 Two bricklayers build a wall in 6 hours. How long would one bricklayer take to build the wall working at the same rate?

8 House contents insurance is charged at the rate of £3.50 per thousand pounds worth of the contents. How much is the insurance if the contents are worth £3400?

9 If the insurance paid on the contents of a house is £33.60, at the rate of £4 per thousand pounds worth, what are the house contents worth?

10 It takes Margaret 45 minutes to walk 4 km. How long would it take her to walk 5 km at the same speed? How far would she go in 1 hour?

11 It takes a gardener 45 minutes to dig a flower bed of area $7.5\,\text{m}^2$. If he digs at the same rate, how long does he take to dig $9\,\text{m}^2$?

12 Fencing costs £2.40 per 1.8 m length. How much would 7.5 m cost?

13 Mrs Brown and Mrs Jones make 4 dozen sandwiches in half an hour in Mrs Jones' small kitchen. If they had 30 friends in to help, how many sandwiches could be made in the same time?

14 A recipe for 12 scones requires 2 teaspoons of baking powder and 240 g of flour. If a larger number of scones are made, using 540 g of flour, how much baking powder is needed?

MIXED EXERCISE

EXERCISE 5G

1 Express the ratio 96 : 216 in its simplest form.

2 Simplify the ratio $\frac{1}{4} : \frac{2}{5}$.

3 Divide £100 into three parts in the ratio 10 : 13 : 2

4 Two cubes have edges of lengths 8 cm and 12 cm. Find the ratio of
  a the lengths of their edges **b** their volumes.

5 Find the missing number in the ratio $x : 18 = 11 : 24$

6 What distance does 1 cm represent on a map with map ratio 1 : 10 000?

7 If $x : y = 3 : 4$, find the ratio $4x : 3y$

8 It costs £4.50 to feed a dog for 12 days. At the same rate, how much will have to be spent to feed it for 35 days?

1 People come in all shapes and sizes but we expect the relative sizes of different parts of our bodies to be more or less the same. For example, we do not expect a person's arms to be twice as long as their legs! We might expect the ratio of arm length to leg length to be about 2 : 3.

a Gather some evidence and use it to find out if the last statement is roughly correct.

b Does the age of a person make any difference?

c Investigate the ratio of shoe size to height.

2 The standard paper sizes used in the UK are called A1, A2, A3, A4, A5, and so on.
Ordinary file paper is usually A4.
You will need some sheets of A3, A4, A5 paper for this investigation.

a Investigate the ratio of length to width of A4, A5 and A3 paper. What do you notice?

b Investigate the ratio of the areas of A4 and A5 paper.

c Describe the relationship between the paper sizes.

SUMMARY 2

INDICES

When a number is written in the form 3^4, 3 is called the base and 4 is called the index or power.

3^4 means $3 \times 3 \times 3 \times 3$.

3^{-4} means $\dfrac{1}{3^4}$

$3^0 = 1$, in fact $a^0 = 1$ whatever number a stands for.

Rules of indices

We can multiply different powers of the same base by adding the indices,

e.g. $3^4 \times 3^2 = 3^{4+2} = 3^6$

We can divide different powers of the same base by subtracting the indices,

e.g. $3^4 \div 3^2 = 3^{4-2} = 3^2$

STANDARD FORM

A number written in standard form is a number between 1 and 10 multiplied by a power of ten,

e.g. 1.2×10^5

SIGNIFICANT FIGURES

The first significant figure in a number is the first non-zero figure when reading from left to right.

The second significant figure is the next figure to the right, whether or not it is zero.

The third significant figure is the next figure to the right, whether or not it is zero, and so on.

For example, in **0.0205**, the first significant figure is 2,
the second significant figure is 0,
the third significant figure is 5.

PROBABILITY

The probability that an event A does not happen is equal to one minus the probability that it does happen,

i.e. $P(\bar{A}) = 1 - P(A)$.

If p is the probability that an event happens on one occasion, then we expect it to happen np times on n occasions, e.g. the probability that a coin lands head up on one toss is $\frac{1}{2}$, so if we toss the coin 50 times, we expect 25 heads, and if we toss the coin n times we expect $\frac{1}{2}n$ heads.

(Note that *expecting* 25 heads from 50 tosses does *not* mean we will get 25 heads.)

FRACTIONS

To multiply one fraction by another fraction, we multiply their numerators and multiply their denominators,

e.g. $\quad \dfrac{1}{2} \times \dfrac{5}{3} = \dfrac{1 \times 5}{2 \times 3} = \dfrac{5}{6}$

To divide by a fraction, we turn the fraction upside down and multiply by it,

e.g. $\quad \dfrac{1}{2} \div \dfrac{5}{3} = \dfrac{1}{2} \times \dfrac{3}{5} = \dfrac{3}{10}$

To multiply or divide with mixed numbers, first change the mixed numbers to improper fractions.

To find a fraction of a quantity, we multiply the fraction by that quantity,

e.g. $\quad \dfrac{1}{2}$ of $\dfrac{3}{4}$ means $\dfrac{1}{2} \times \dfrac{3}{4}$, and $\dfrac{3}{8}$ of £24 $= £\left(\dfrac{3}{8} \times 24\right)$

To express one quantity as a fraction of another, first make sure both quantities are in the same unit, then place the first quantity over the second,

e.g. $\quad 24\,\mathrm{p}$ as a fraction of £2 is $\dfrac{24}{200}$ $\left(= \dfrac{3}{25}\right)$

or to put it another way, $24\,\mathrm{p}$ is $\dfrac{24}{200}$ of £2.

PERCENTAGES

'Per cent' means 'out of one hundred'.
Hence a percentage can be expressed as a fraction by placing the percentage over 100,

e.g. $\quad 33\% = \dfrac{33}{100}$,

and a percentage can be expressed as a decimal by dividing the percentage by 100, that is by moving the decimal point two places to the left,

e.g. $\quad 33\% = 0.33$

Reversing the process, a fraction can be expressed as a percentage by multiplying the fraction by 100, and a decimal can also be expressed as a percentage by multiplying the decimal by 100, that is, by moving the decimal point two places to the right.
For example,

$$\dfrac{2}{5} = \dfrac{2}{5} \times 100\% = \dfrac{2}{5} \times \dfrac{100}{1}\% = 40\% \qquad \text{and} \qquad 0.325 = 32.5\%$$

To find one quantity as a percentage of another quantity, we place the first quantity over the second quantity and multiply this fraction by 100,

e.g. $\quad 24\,\mathrm{p}$ as a percentage of £2 is $\dfrac{24}{200} \times \dfrac{100}{1}\% = 12\%$,

or to put it another way, $24\,\mathrm{p}$ is 12% of £2.

To find a percentage of a quantity, change the percentage to a decimal and multiply by the quantity,

e.g. $\quad 32\%$ of £18 $= 32\% \times £18 = £0.32 \times 18 \ (= £5.76)$

PERCENTAGE INCREASE

To increase a quantity by 15%,
we find *the increase* by finding 15% of the quantity, so we multiply by 0.15.

We find *the new quantity* by finding 100% + 15%,
i.e. 115% of the original quantity, so we multiply by 1.15.

PERCENTAGE DECREASE

To decrease a quantity by 15%,
we find *the decrease* by finding 15% of the quantity,

We find *the new quantity* by finding 100% − 15%,
i.e. 85% of the original quantity, so we multiply by 0.85.

RATIO

Ratios are used to compare the relative sizes of quantities.
For example, if a model of a car is 2 cm long and the real car is 200 cm long, we say that their lengths are in the ratio 2 : 200.

Ratios can be simplified by dividing the parts of the ratio by the same number,

e.g. $2 : 200 = 1 : 100$ (dividing 2 and 200 by 2).

MAP RATIO

A map ratio is the ratio of a length on the map to the length it represents on the ground. When expressed as a fraction, it is called the *Representative Fraction*.

DIRECT PROPORTION

Two quantities that can vary in size are directly proportional when they are always in the same ratio.

REVISION EXERCISE 2.1 (Chapters 1 and 2)

Do not use a calculator for this exercise.

1 Find the value of

    a 4.72×10^2 **b** 4^3 **c** 1.2^2

2 Express as a single expression in index form

    a $2^3 \times 2^5$ **b** $10^7 \times 10^{10}$ **c** $2^{10} \div 2^8$

3 Find the value of

    a 3^{-2} **c** 5.9×10^{-2} **e** 2^{-4}

    b 12^{-1} **d** 7.23×10^4 **f** 5^{-3}

4 a Write, as a single number in index form

   i $5^3 \times 5^2$ **ii** $4^{-5} \times 4^{-2}$

b Write the following numbers in standard form

   i 420 000   **ii** 0.000 32

5 a Which is the larger

   i 10^3 or 10^6 **ii** 10^{-2} or 10^{-3} **iii** 10^0 or 10^{-2} ?

b Write as an ordinary number   **i** 2.14×10^3 **ii** 5.21×10^{-2},

6 a Round 65.594 correct to

   i the nearest unit   **ii** the nearest 10   **iii** 2 s.f.

b Correct each number to 1 significant number and hence give a rough answer to

   i 18.42×0.0826 **ii** $63.92 \div 8.124$

7 The probability that Tony will score a treble 20 with his next dart is 0.2. What is the probability that Tony will not score a treble 20 with his next dart?

8 In a sale 60 shirts are on offer at reduced prices. Of these, 15 are size 13, 20 are size 14, 12 are size 15 and the remainder are size 16. Ken chooses a shirt at random. What is the probability that the shirt he chooses is

a size 14   **c** size 15 or larger

b not size 15   **d** at least size 14 ?

9 Four kings and four queens are removed from an ordinary pack of playing cards. One card is taken from the set of kings and one card is taken from the set of queens. Make a possibility space for the possible combinations of the two cards and use it to find the probability that the two cards

a are both red   **c** are both hearts

b include at least one red card   **d** are of different suits.

10 Two ordinary six-sided dice are rolled together 360 times. About how many double sixes are there likely to be?

REVISION EXERCISE 2.2 (Chapters 3 to 5)

Do not use a calculator for this exercise.

1 a Express as mixed numbers   **i** $\frac{23}{5}$ **ii** $\frac{13}{7}$

b Place the following fractions in order of size with the smallest first.

$$\frac{3}{5}, \ \frac{3}{4}, \ \frac{4}{7}, \ \frac{2}{3}$$

2 Calculate

   a $\frac{2}{3} \times \frac{5}{12}$ **b** $8\frac{2}{3} \times 1\frac{4}{13}$ **c** $9 \div \frac{2}{3}$ **d** $32 \div 1\frac{3}{5}$

3 a Find

    i $\frac{2}{3}$ of £96   **ii** $\frac{3}{4}$ of 46 cm   **iii** $\frac{5}{9}$ of 36 kg

  b Which is the smaller $\frac{7}{12}$ of 9   or $\frac{3}{4}$ of 7?

  c Which is the larger $\frac{2}{3}$ of $\frac{4}{9}$ or $\frac{4}{7}$ of $\frac{3}{4}$?

4 a Give 45 g as a fraction of 135 g.

  b Express

    i $\frac{9}{5}$ as a percentage

    ii 0.425 as a percentage

    iii 22% as a fraction in its lowest terms.

5 a Express $\frac{7}{16}$ **i** as a decimal   **ii** as a percentage.

  b Express as a percentage   **i** 2.44   **ii** 0.055

  c If 26% of the pupils in my school wear glasses, what percentage do not?

6 a Find   **i** 80% of £7.20   **ii** 5.5% of 260 g

  b There are 65 houses in my street and 40% of them now have double glazing. How many houses

    i have double glazing   **ii** do not have double glazing?

7 a If the cost of an insurance policy is increased by 55%, what percentage is the new cost of the original cost?

  b i Increase £560 by 35%   **ii** Decrease £650 by 24%

8 a Simplify the ratio   **i** 12 m : 30 m   **ii** £3 : £4.50 : £5

  b Divide £66 into two shares in the ratio 5 : 6.

9 a If $5 : 3 = x : 6$ find x.

  b The ratio of the length of a rectangular rug to its width is 9 : 5. The length is 2.7 m. Find the width.

10 a Find the map ratio of a map on which 5 cm represents 1 km.

   b The map ratio of a map is 1 : 200 000. The distance between two villages is 12 km. What is this on the map?

REVISION EXERCISE 2.3 (Chapters 1 to 5)

1 Find the value of

a 5.97×10^3 **b** 10^5 **c** 4^{-2} **d** 5^0

2 a Write in standard form **i** $0.009\,26$ **ii** $730\,000$

  b Use a calculator to find

  i 0.047×6.421 **ii** $4703 \div 2423$ **iii** 1.2^5,

  giving your answers correct to 3 significant figures.

3 A number is chosen at random from the first twelve whole numbers. What is the probability that it is

a exactly divisible by 3 **b** not exactly divisible by 3?

4 At the Eastlee Fitness Centre it is normal for about 1 in 3 of the members of each new Keep-fit class to leave within the first three weeks of enrolling. Thirty members enrol for one class. How many members are expected to attend the class held during the fourth week?

5 Calculate

a $\frac{7}{12} \times \frac{2}{21}$ **b** $3\frac{1}{3} \times 2\frac{1}{4}$ **c** $3\frac{1}{5} \div \frac{4}{7}$ **d** $8\frac{1}{8} \div 7\frac{3}{7}$

6 a How many $\frac{2}{5}$s are there in 4?

  b How many times does $\frac{3}{4}$ go into 12?

  c It takes 6 min to fill $\frac{3}{8}$ of a storage tank. How long will it take to fill the tank completely?

7 a Express, in its lowest terms, 30 minutes as a fraction of 2 hours.

  b Express 70% as a fraction in its lowest terms.

  c Express as a percentage **i** $\frac{7}{20}$ **ii** 0.44

8 a Find, giving your answer correct to 3 significant figures where necessary

  i 60% of 850 g **ii** 43% of 5.4 kg **iii** $5\frac{1}{2}$% of 7.45 kg

  b **i** A stereo priced at £200 was reduced by 30% in a sale. What was the sale price?

  ii In 1990, 360 people lived in Lorke. By 1995 this number had increased by 30%. How many people lived in Lorke in 1995?

9 a Simplify the ratio **i** £3.50 : 75 p **ii** 24 : 36 : 42

  b If $x : 3 = 12 : 18$ find x.

10 a Divide £45 among 3 persons so that their shares are in the ratio
4 : 5 : 6.

b If the map ratio of a map is 1 : 100 000 and the distance between
two places on a map is 8 cm, find the actual distance between the
two places.

**REVISION
EXERCISE 2.4
(Chapters 1 to 5)**

1 a Express as a single expression in index form

i $a^3 \times a^5$ **ii** $p^5 \div p^2$ **iii** $a^4 \div a^7$ **iv** $b^2 \times b^3 \times b^5$

b Give the number 82.049 correct to

i 3 significant figures   **ii** 2 significant figures

2 a Write the following numbers in standard form and then place
them in order of size with the smallest first.

324 000,   99 000,   1 270 000,   8700

b Use a calculator to find

i 53.8×2.835 **ii** 9.427×3.293,

giving your answers correct to 3 significant figures.

3 A card is drawn at random from an ordinary pack of 52 playing
cards. What is the probability that the card drawn is

a a 3   **b** a red 3   **c** the 3 of hearts   **d** not a red 3 ?

4 A 10 p and a 20 p coin are tossed together. Make your own
possibility space for the combinations in which they can land.
Find the probability of getting

a one head and one tail

b two tails

c at least one head.

5 Find

a **i** $3\frac{1}{3} + 2\frac{1}{4}$ **ii** $3\frac{1}{3} - 2\frac{1}{4}$ **iii** $3\frac{1}{3} \times 2\frac{1}{4}$ **iv** $3\frac{1}{3} \div 2\frac{1}{4}$

b Which is the smaller and by how much, $\frac{5}{9}$ of $\frac{3}{4}$ or $\frac{1}{2}$ of $\frac{4}{5}$?

6 Which of the following statements are true and which are false ?

A $\frac{2}{5} + 2 = \frac{4}{5}$ **C** $\frac{2}{5} + \frac{3}{5} = 1$

B $\frac{2}{5} + \frac{2}{3} = \frac{4}{8}$ **D** $\frac{3}{5} + 2 = \frac{13}{5}$

7 a Express as a decimal   **i** $8\frac{1}{2}$ **ii** $7\frac{3}{4}$ **iii** $13\frac{2}{5}$

b Express

  i $\frac{5}{12}$ as a percentage correct to 1 decimal place.

  ii 0.76 as a percentage.

8 Tracey and Helen went out for a meal. The bill came to £28.32.
Tracey paid $\frac{7}{12}$ of it and Helen paid the remainder.

a What fraction did Helen pay?

b How much did Tracey pay?

9 a Simplify the ratio $2\,\text{kg} : 1.6\,\text{kg}$

b A piece of wood, 3.15 m long, is cut into two pieces whose
lengths are in the ratio 8 : 7. How much longer is the one piece
than the other piece?

10 a Alison's pet spaniel had a litter of 6 puppies, 4 of which are
females. Find the ratio of the number of female puppies to the
number of male puppies.

b The ratio of the base of a triangle to its height is 3 : 2.
Its base is 5 cm. Find its height.

POLYGONS

David attended pottery classes and made some tiles. He wanted to make a heat-proof mat with them, and decided that he liked this shape:

When he came to fit the tiles together he found that he could not get them to cover the mat without leaving gaps.

- If David had done a little research before he started he would have found that this shape could never be arranged so that it covered a flat surface completely without leaving any gaps.

- Shapes like these tiles are called polygons. If David had been familiar with their properties before he started it would have helped him to decide which shape of polygon to use for his tile.

In this chapter we look at polygons. By the end of the chapter you should be in a position to avoid David's mistake.

The summary on page 5 will remind you of the angle facts and of the different classifications of triangles and quadrilaterals studied in Book 7A. Some of this knowledge is used in this chapter.

POLYGONS

A polygon is a plane (flat) figure bounded by straight lines.

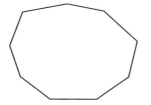

This is a nine-sided polygon.

Some polygons have names which you already know:

a three-sided polygon is a triangle

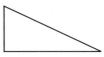

a four-sided polygon is a quadrilateral

a five-sided polygon is a pentagon

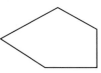

a six-sided polygon is a hexagon

an eight-sided polygon is an octagon

REGULAR POLYGONS

The polygons drawn above are not regular polygons.

A polygon is called regular when all its sides are the same length *and* all its angles are the same size. The polygons below are all regular:

Regular polygons are quite commonplace in everyday life. For example:

Paving stones

The Pentagon
in Washington

An octagonal table

EXERCISE 6A   State which of the following figures are regular polygons. Give a brief reason for your answer.

    1 Rhombus   **5** Isosceles triangle

    2 Square   **6** A triangle with one right angle

    3 Rectangle   **7** Equilateral triangle

    4 Parallelogram   **8** Circle

Make a rough sketch of each of the following polygons. (Unless you are told that a polygon is regular, you must assume that it is *not* regular.)

    9 A regular quadrilateral   **13** A regular hexagon

   10 A hexagon   **14** A pentagon

   11 A triangle   **15** A quadrilateral

   12 A regular triangle   **16** A ten-sided polygon

When the vertices of a polygon all point outwards, the polygon is *convex*.

Sometimes one or more of the vertices point inwards, in which case the polygon is *concave*.

        Convex polygon   Concave polygon

In this chapter we consider only convex polygons.

INTERIOR ANGLES   The angles enclosed by the sides of a polygon are the interior angles. For example,

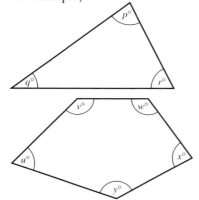

$p°$, $q°$, and $r°$ are the interior angles of the triangle,

$u°$, $v°$, $w°$, $x°$ and $y°$ are the interior angles of the pentagon.

THE EXTERIOR
ANGLES

If we produce (extend) one side of a polygon, an angle is formed outside the polygon. It is called an *exterior angle*.

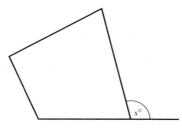

For example, $s°$ is an exterior angle of the quadrilateral.

If we produce all the sides in order we have all the exterior angles.

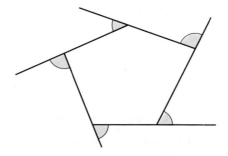

EXERCISE 6B

1 What is the sum of the interior angles of any triangle?

2 What is the sum of the interior angles of any quadrilateral?

3 In triangle ABC, find

a the size of each marked angle

b the sum of the exterior angles.

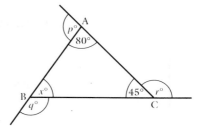

4

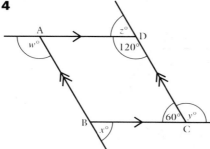

ABCD is a parallelogram. Find

a the size of each marked angle

b the sum of the exterior angles.

5 In triangle ABC, write down the value of

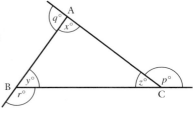

   a $x° + q°$

   b the sum of all six marked angles

   c the sum of the interior angles

   d the sum of the exterior angles.

6 ABCD is a trapezium. Find

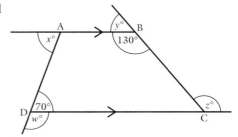

   a the size of each
      marked angle

   b the sum of the
      exterior angles.

7 Draw a pentagon. Produce the sides in order to form the five exterior angles. Measure each exterior angle and then find their sum.

8 Construct a regular hexagon of side 5 cm. (Start with a circle of radius 5 cm and then with your compasses still open to a radius of 5 cm, mark off points on the circumference in turn.) Produce each side of the hexagon in turn to form the six exterior angles.

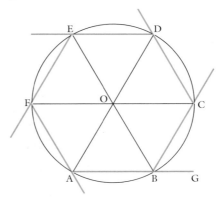

If O is the centre of the circle, join O to each vertex to form six triangles.

   a What kind of triangle is each of these triangles?

   b What is the size of each interior angle in these triangles?

   c Write down the value of $A\widehat{B}C$.

   d Write down the value of $C\widehat{B}G$.

   e Write down the value of the sum of the six exterior angles of the hexagon.

THE SUM OF THE
EXTERIOR
ANGLES OF A
POLYGON

In the last exercise, we found that the sum of the exterior angles is 360° in each case. This is true of any polygon, whatever its shape or size, as we can demonstrate.

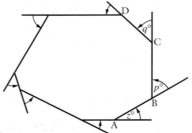

Consider walking round this polygon. Start at A and walk along AB. When you get to B you have to turn through angle $p°$ to walk along BC. When you get to C you have to turn through angle $q°$ to walk along CD, ... and so on until your return to A. If you then turn through angle $z°$ you are facing in the direction AB again. You have now turned through all the exterior angles and have made just one complete turn, i.e.

the sum of the exterior angles of a polygon is 360°.

EXERCISE 6C

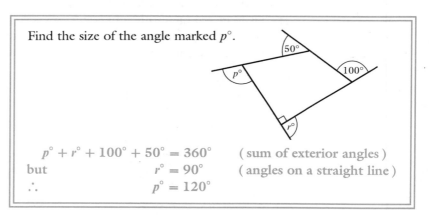

Find the size of the angle marked $p°$.

$$p° + r° + 100° + 50° = 360° \quad (\text{sum of exterior angles})$$
$$\text{but} \qquad\qquad\qquad r° = 90° \quad (\text{angles on a straight line})$$
$$\therefore \qquad\qquad\qquad p° = 120°$$

In each case find the size of the marked angle(s).

1

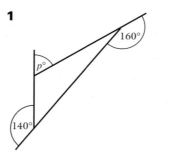

2

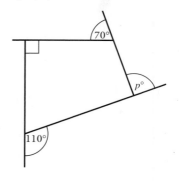

3

6

4

7

5

8

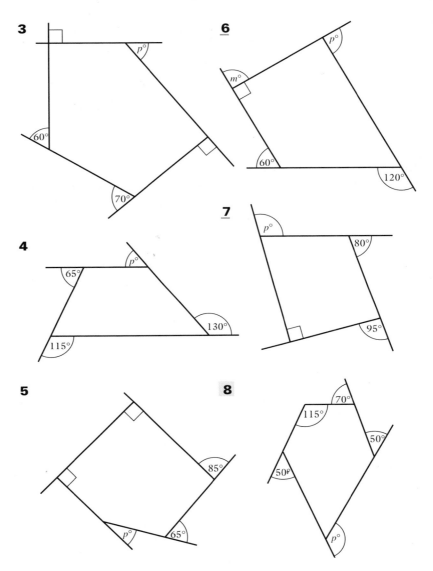

In questions **9** and **10** find the value of *x*.

9

10

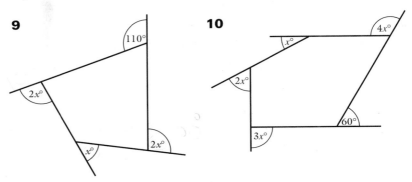

11 The exterior angles of a hexagon are $x°$, $2x°$, $3x°$, $4x°$, $3x°$ and $2x°$. Find the value of x.

12 Find the number of sides of a polygon if each exterior angle is

    a $72°$ **b** $45°$.

THE EXTERIOR ANGLE OF A REGULAR POLYGON

If a polygon is regular, all its exterior angles are the same size. We know that the sum of the exterior angles is $360°$, so the size of one exterior angle is easy to find; we just divide 360 by the number of sides of the polygon, i.e.

> in a *regular* polygon with n sides, the size of an exterior angle is
> $$\frac{360°}{n}.$$

EXERCISE 6D

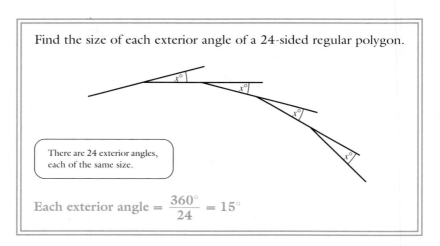

Find the size of each exterior angle of a 24-sided regular polygon.

There are 24 exterior angles, each of the same size.

$$\text{Each exterior angle} = \frac{360°}{24} = 15°$$

Find the size of each exterior angle of a regular polygon with

1 10 sides   **4** 6 sides   **7** 9 sides

2 8 sides   **5** 15 sides   **8** 16 sides

3 12 sides   **6** 18 sides   **9** 20 sides

THE SUM OF THE
INTERIOR
ANGLES OF A
POLYGON

This is an octagon.

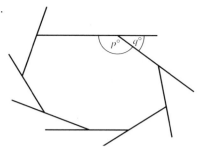

At each vertex there is an interior angle and an exterior angle and the sum of these two angles is 180° (angles on a straight line), i.e. $p° + q° = 180°$ at each one of the eight vertices.

Therefore, the sum of the interior angles and exterior angles together is

$$8 \times 180° = 1440°$$

The sum of the eight exterior angles is 360°.

Therefore, the sum of the interior angles is

$$1440° - 360° = 1080°$$

EXERCISE 6E

Find the sum of the interior angles of a 14-sided polygon.

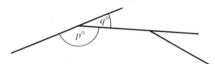

At each vertex $\qquad p° + q° = 180°$

∴ sum of interior angles and exterior angles is

$$14 \times 180° = 2520°$$

∴ sum of interior angles $= 2520° - 360°$

$$= 2160°$$

Find the sum of the interior angles of a polygon with

1 6 sides $\qquad$ **4** 4 sides $\qquad$ **7** 18 sides

2 5 sides $\qquad$ **5** 7 sides $\qquad$ **8** 9 sides

3 10 sides $\qquad$ **6** 12 sides $\qquad$ **9** 15 sides

10 List the results for questions **1** to **9** in a table.
Find an expression for the sum of the interior angles of a polygon with n sides.

**FORMULA FOR
THE SUM OF THE
INTERIOR
ANGLES**

If a polygon has n sides, then

the sum of the interior and exterior angles together is

$$n \times 180° = 180n°$$

so the sum of the interior angles only is $180n° - 360°$

which, as $180° = 2$ right angles and $360° = 4$ right angles,

can be written as $(2n - 4)$ right angles

i.e.   in a polygon with n sides, the sum of the interior angles is
$(180n - 360)°$ or $(2n - 4)$ right angles.

EXERCISE 6F

1 Find the sum of the interior angles of a polygon with

  a 20 sides   **b** 16 sides   **c** 11 sides.

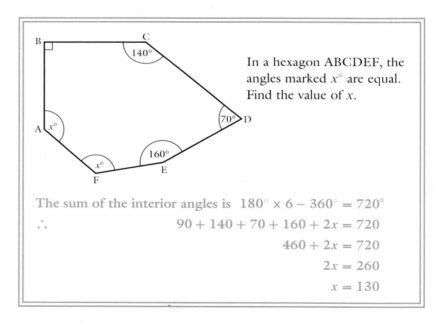

In a hexagon ABCDEF, the angles marked $x°$ are equal. Find the value of x.

The sum of the interior angles is $180° \times 6 - 360° = 720°$

$\therefore$ $90 + 140 + 70 + 160 + 2x = 720$

                                  $460 + 2x = 720$

                                        $2x = 260$

                                            $x = 130$

In each of the following questions find the size of the marked angle(s).

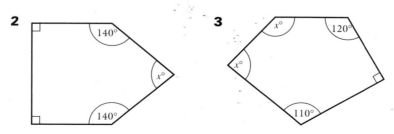

2

3

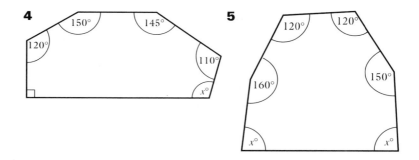

4

150° 145°
120°
110°
x°

5

120° 120°
160°
150°
x°
x°

Find the size of each interior angle of a regular nine-sided polygon.

> We can find the size of one exterior angle and then use this to find the size of an interior angle.

Sum of exterior angles = 360°
∴ each exterior angle = 360° ÷ 9 = 40°
∴ each interior angle = 180° − 40° = 140°

40°

> As the polygon is regular, all the exterior angles are equal and all the interior angles are equal.

> Alternatively we can use the formula to find the sum of the interior angles.
> Sum of interior angles = 180° × 9 − 360° = 1260°
> ∴ each interior angle = 1260° ÷ 9 = 140°

Find the size of each interior angle of

6 A regular pentagon   **8** A regular ten-sided polygon

7 A regular hexagon   **9** A regular 12-sided polygon

10 How many sides has a regular polygon if each exterior angle is
    a 20°   **b** 15°?

11 How many sides has a regular polygon if each interior angle is
    a 150°   **b** 162°?   (Find the exterior angle first.)

12 Is it possible for each exterior angle of a regular polygon to be
    a 30°   **b** 40°   **c** 50°   **d** 60°   **e** 70°   **f** 90°?
    In those cases where it is possible, give the number of sides.

13 Is it possible for each interior angle of a regular polygon to be

  a 90°   **b** 120°   **c** 180°   **d** 175°   **e** 170°   **f** 135° ?

In those cases where it is possible, give the number of sides.

14 Construct a regular pentagon with sides 5 cm long. (Calculate the size of each interior angle and use your protractor.)

15 Construct a regular octagon of side 5 cm.

ABCDE is a pentagon, in which the interior angles at A and D are each $3x°$, the interior angles at B, C and E are each $4x°$. AB and DC are produced until they meet at F. Find $B\widehat{F}C$.

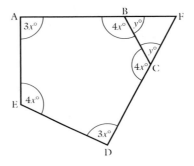

> We can find the sum of the interior angles (using the formula) and then form an equation and solve it to find x.

Sum of the interior angles of a pentagon $= 180° \times 5 - 360°$

$$= 540°$$

$\therefore$ $3x + 4x + 3x + 4x + 4x = 540$

$$18x = 540$$

$$x = 30$$

> Now that we know x, we can find $A\widehat{B}C$ and hence $C\widehat{B}F$.

$A\widehat{B}C = 120°$ and $B\widehat{C}D = 120°$

$\therefore$ $y = 60$

$\therefore$ $B\widehat{F}C = 180° - 2 \times 60°$ (angle sum of $\triangle BFC$)

$$= 60°$$

In questions **16** to **21** find the value of *x*.

16

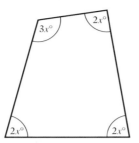

17

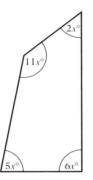

18

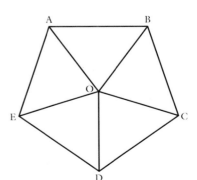

19

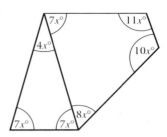

20

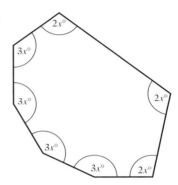

21

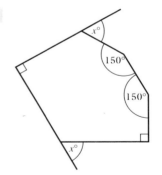

22 ABCDE is a regular pentagon.

OA = OB = OC = OD = OE.

Find the size of each angle at O.

23 ABCDEFGH is a regular octagon. O is a point in the middle of the octagon such that O is the same distance from each vertex. Find AÔB.

24 ABCDE is a regular pentagon.
AB and DC are produced until they meet at F.
Find BF̂C.

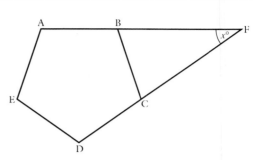

ABCDEF is a regular hexagon.
Find AD̂B.

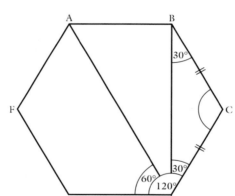

Copy the diagram and mark on it all the facts as you find them. This may help you to see what to find next.

As ABCDEF is regular, the exterior angles are all equal.

$$\text{Each exterior angle} = 360° ÷ 6$$
$$= 60°$$

∴   each interior angle $= 180° − 60°$
$$= 120°$$

△BCD is isosceles (BC = DC).

∴   CB̂D = BD̂C = 30°   (angle sum of △BCD)

AD is a line of symmetry.

∴   ED̂A = CD̂A = 60°

∴   AD̂B = 60° − 30°
$$= 30°$$

In questions **25** to **30** each polygon is regular. Give answers correct to one decimal place where necessary.

25

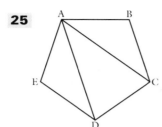

Find **a** AĈB   **b** DÂC

26

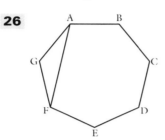

Find **a** AĜF   **b** GÂF

27

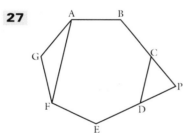

Find CP̂D

28

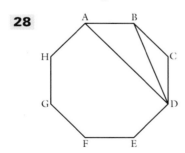

Find **a** CB̂D   **b** BD̂A

29 ABCDEFGH is a regular octagon with the side AH missing.
This shape is used as a character in a simple computer game.
Find AÊH.

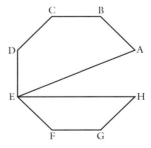

30

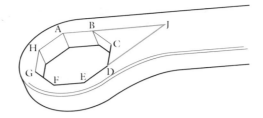

The sketch shows part of a spanner for octagonal nuts.
Find angle BĴD.

**PATTERN
MAKING WITH
REGULAR
POLYGONS**

Regular hexagons fit together without leaving gaps, to form a flat
surface. We say that they *tessellate*.

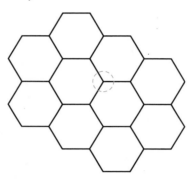

The hexagons tessellate because each interior angle of a regular hexagon
is 120°, so three vertices fit together to make 360°.

EXERCISE 6G

1 Draw a tessellation using this tile
shaped like a kite.
Trace it on card, cut it out and use
it as a template.
In how many different ways can you
position the tiles?
If the tiles can be of different colours, what is the smallest number of
different colours needed if no two tiles with touching edges are to be
the same colour.

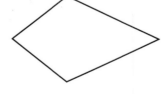

2 Draw diagrams to show that the following shapes tessellate.

a a square

b a parallelogram

c any triangle with three unequal sides

Does it follow that if a parallelogram tessellates then every triangle
must also do so? Give reasons for your answer.

3 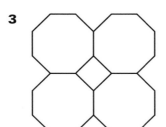 This is a pattern using regular octagons. They do not tessellate:

a Explain why they do not tessellate.

b What shape is left between the four octagons?

c Continue the pattern. (Trace one of the shapes above, cut it out and use it as a template.)

4 Apart from the hexagon, there are two other regular polygons that tessellate. Which are they, and why do they tessellate?

MIXED EXERCISE

EXERCISE 6H

1 Find the size of each marked angle.

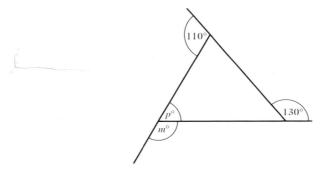

2 Find the size of each exterior angle of a regular polygon with 12 sides. What is the size of each interior angle?

3 Find the size of the angle marked $x°$.

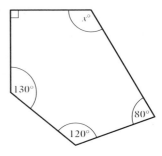

4 How many sides has a regular polygon if each interior angle is 168°?

5 Is it possible for each exterior angle in a regular polygon to be

a 45°   **b** 75°?

Explain your answers.

6 Find the size of the marked angle.

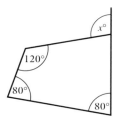

7 Which of the following statements are true and which are false?
Try to justify the statement you think is true. Illustrate any statement you think is false with a sketch.

   a Every triangle tessellates.

   b Some quadrilaterals tessellate and some do not.

   c A rhombus does not tessellate

   d Some shapes with curved edges tessellate.

   e Every regular polygon tessellates.

   f It is possible to draw a hexagon that is not regular that tessellates.

PUZZLE

This is an old Chinese puzzle, called a 'tangram'.
Draw a square of side 8 cm and divide it into five isosceles triangles, a square and a parallelogram as shown in the diagram.
As a first step join the midpoints of two adjacent sides of the square. Then draw the other lines.
Mark the regions 1 to 7 as shown.

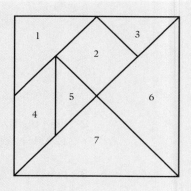

   a Arrange the two larger triangles to form a square.

   b Arrange the remaining five pieces to form an identical square to the square you arranged in part **a**.

   c Arrange all seven pieces to form a rectangle.

   d Remove one piece from the rectangle formed in part **c** and place it in a new position so that the resulting shape is a parallelogram.

   e Change the position of one piece so as to form a trapezium.

   f Using all seven pieces, can you arrange them to form
       i a regular hexagon   **ii** a hexagon that is not regular?
   If you can, show with a sketch how this can be done.

**PRACTICAL
WORK**

1 Trace this regular pentagon and
use it to cut out a template.

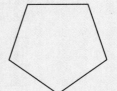

a Will pentagons tessellate?

b Use your template to
copy and continue this
pattern until you have a
complete circle of
pentagons.
What shape is left in
the middle?

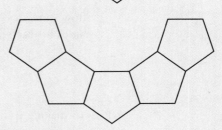

c Make up a pattern using pentagons.

2

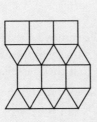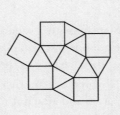

Use your template from question **1** to copy this net onto thick
paper. Cut it out and fold along the lines. Stick the edges
together using the flaps. You have made a regular dodecahedron
(a solid with 12 equal faces).

3 Regular hexagons, squares and equilateral triangles can be combined
to make interesting patterns. Some examples are given below.

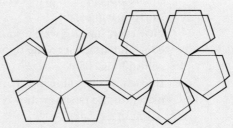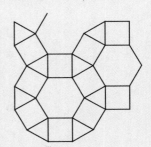

Copy these patterns and extend them. (If you make templates to
help you, make each shape of side 2 cm.)

4 Bricklayers sometimes make very interesting patterns when building a wall.

Show some designs that you have seen and try to find the different names for each pattern. (A brick is twice as long as it is wide.)

INVESTIGATIONS

Pattern making using tessellating designs can be quite fascinating.

1 The simplest shape that is used to cover a flat surface completely, without any gaps, is the square. We can see examples around us – square floor tiles, square ceramic tiles on a bathroom wall and square paving slabs.

However, starting with a square we can remove a piece from one edge and attach it to the opposite edge so that the resulting shape will still tessellate.

In each of the following changes the shaded area is removed and placed in the position shown.

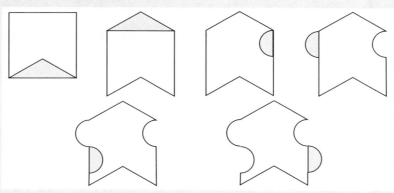

a Starting with a square tile remove a shape of your own choice from one side and place it on the opposite side. Draw about 10 of these shapes to show that they tessellate.

b Now remove another shape from the shape you finished with in part **a** and attach it to the opposite side. Draw at least 6 of these to show that they tessellate.

c Continue the process to make more interesting shapes that tessellate.

2 Starting with other shapes that tessellate, for example a rhombus, a parallelogram or a regular hexagon, repeat the instructions you followed in the first investigation.

This illustration, by M C Escher, starts with equilateral triangles and shows how they can be developed to give interesting artwork.

AREAS OF TRIANGLES AND PARALLELOGRAMS

The walls of this house need painting.

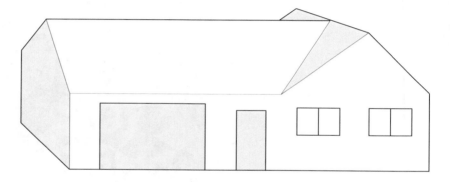

The colour of paint varies slightly from batch to batch, so it is sensible to buy all the paint needed at the same time. This means that we need a reasonably accurate estimate of the amount of paint required; to find this, we first have to work out the area that needs painting.

The shapes involved here are all rectangles except for the gable – this is triangular. Assuming that we know the measurements, finding the area of a rectangle is straightforward, but we need a way to find the area of a triangle.

- One method is to make an accurate scale drawing of the shape on squared paper and then count the squares covered. The disadvantage of doing it this way is that it is very time consuming.

- There is a formula for calculating the area of a rectangle which is quick and easy to use and it would be helpful if we had a formula that gave the area of a triangle.

1 Angela designs a patchwork quilt. This is part of her design.

To make it up she needs to buy enough material of each colour.

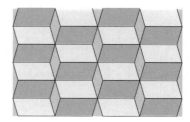

a What are the names of the shapes used in this design?

b Discuss what Angela needs to know and what she must do, in order to buy about the right amount of material.

2 A pack has to be designed to hold six of these crayons.

The ends of the crayons are equilateral triangles. Part of the design activity involves costing the packaging. Discuss what you need to know to be able to do this costing.

3 a

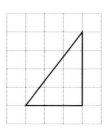

Copy this diagram on to 1 cm squared paper.
Discuss how you can find the area of this triangle using what you know about finding the area of a rectangle. Can you adapt your methods to any triangle, whatever its shape or size?

b Repeat part **a** for this parallelogram.

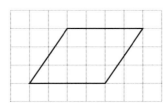

**AREA OF A
PARALLELOGRAM**

Discussion of the examples in **Exercise 7A** shows that triangles and parallelograms are shapes that appear in many everyday situations and that we need a simple way of working out their areas.

Consider this parallelogram.

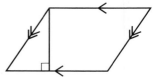

If we cut off the right-angled triangle shown above and move it to fit on the right-hand side as shown, we have transformed the parallelogram into a rectangle without changing the area.

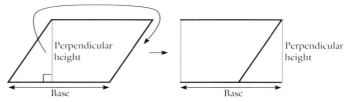

Therefore the area of the parallelogram is equal to the area of the rectangle.

Now the area of the rectangle is given by length × width which, using the labels in the diagram, is the same as base × height
The height of the rectangle is equal to the perpendicular height of the parallelogram; so

> area of parallelogram = base × perpendicular height

Provided that we know the length of the base and the perpendicular height, we can use this formula to find the area of any parallelogram.

Note that when we talk about the height of a parallelogram we mean the perpendicular height.

EXERCISE 7B

Find the area of a parallelogram of base 7 cm, height 5 cm and slant height 6 cm.

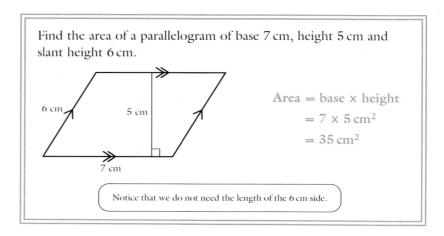

Area = base × height
= 7 × 5 cm²
= 35 cm²

Notice that we do not need the length of the 6 cm side.

Find the areas of the parallelograms.

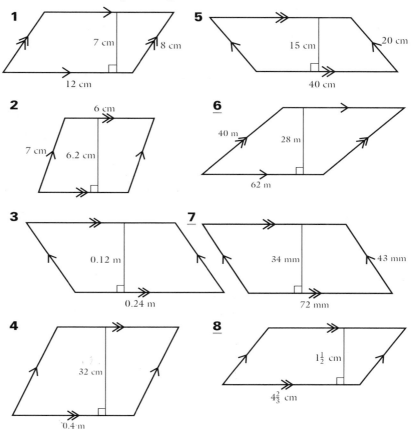

1 7 cm 8 cm 12 cm

5 15 cm 20 cm 40 cm

2 6 cm 7 cm 6.2 cm

6 40 m 28 m 62 m

3 0.12 m 0.24 m

7 34 mm 43 mm 72 mm

4 32 cm 0.4 m

8 $1\frac{1}{2}$ cm $4\frac{2}{3}$ cm

Find the area of the parallelogram.

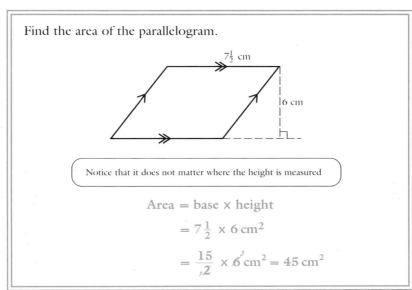

$7\frac{1}{2}$ cm

6 cm

Notice that it does not matter where the height is measured

Area = base × height

$= 7\frac{1}{2} \times 6\,\text{cm}^2$

$= \frac{15}{\cancel{2}} \times \cancel{6}\,\text{cm}^2 = 45\,\text{cm}^2$

Find the areas of the parallelograms.

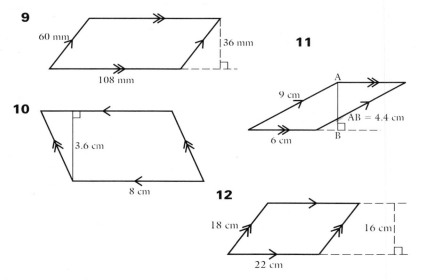

9

60 mm

36 mm

108 mm

10

3.6 cm

8 cm

11

A

9 cm

AB = 4.4 cm

6 cm

B

12

18 cm

22 cm

16 cm

In questions **13** to **18** turn the page round if necessary so that you can see which is the base and which the height.

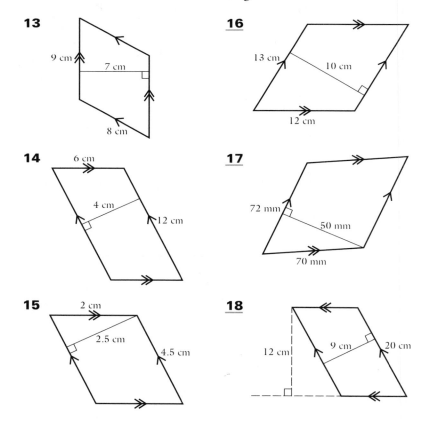

13

9 cm

7 cm

8 cm

16

13 cm

10 cm

12 cm

14

6 cm

4 cm

12 cm

17

72 mm

50 mm

70 mm

15

2 cm

2.5 cm

4.5 cm

18

12 cm

9 cm

20 cm

For each of the following questions, draw *x*- and *y*-axes for values from −4 to 4. Use 1 square to 1 unit. Draw parallelogram ABCD and find its area in square units.

19 A(−2, 0), B(2, 0), C(3, 2), D(−1, 2)

20 A(2, −2), B(4, 1), C(−1, 1), D(−3, −2)

21 A(2, 1), B(2, 4), C(−1, 2), D(−1, −1)

22 A(2, 0), B(2, 3), C(−3, 4), D(−3, 1)

23

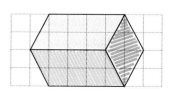

This is part of Angela's initial design work for a quilt. Angela drew it on 1 cm squared paper.

a Find the area of the green piece of the design.

b Angela calculates that she will need 250 of these green pieces to complete the quilt. Find the area of the completed quilt that will be green.

c The pieces of the patchwork need sewing together. Angela makes a seam allowance by adding 0.5 cm to the length and perpendicular height of each piece.
Find the area of each green piece that will be cut.

24 a Copy this diagram on 1 cm squared paper. You know how to find the area of a parallelogram; use this knowledge to work out the area of the shaded triangle. Write down, in terms of its base and height, how you found this area.

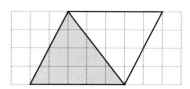

b Will the expression for the area found in part **a** work for any triangle?

25 a How many parallelograms can be drawn with these measurements?

b Do all parallelograms with base 10 cm and slant height 7 cm have the same area?

c What relationship is there between the base and the height of a parallelogram whose area is 40 cm² ?

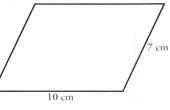

26 a How many parallelograms can be drawn with these measurements?

b Draw two *differently-shaped* parallelograms with these measurements.

c Find the area of each of your parallelograms and explain your results.

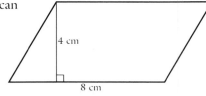

27 What can you say about the area of each of the shapes shown in the diagram? Explain your answer.

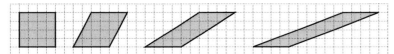

THE AREA OF A TRIANGLE

We can think of a triangle as half a parallelogram.

Area of triangle

$= \frac{1}{2} \times$ area of parallelogram

$= \frac{1}{2} \,($ base $\times$ perpendicular height $)$

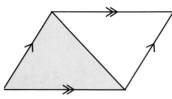

We can also see that if we enclose a triangle in a rectangle as shown below, the area of the triangle is half the area of the rectangle, i.e.

area of triangle

$= \frac{1}{2} \times$ area of enclosing rectangle

$= \frac{1}{2} \,($ base $\times$ height of triangle $)$

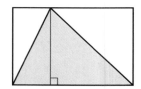

THE HEIGHT OF A
TRIANGLE

When we talk about the height of a triangle (or a parallelogram) we
mean its *perpendicular height* and *not* a slant height.

If the given triangle is drawn accurately
on squared paper, we can see that the
height of the triangle is neither
10 cm nor 7.5 cm but 6 cm.

Notice also that the foot of this
perpendicular is *not* the midpoint
of the base.

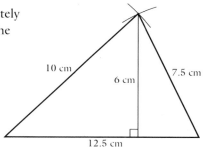

EXERCISE 7C

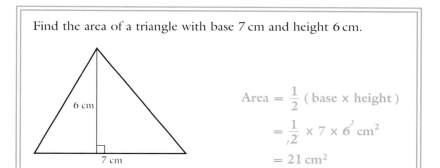

Find the area of a triangle with base 7 cm and height 6 cm.

$$\text{Area} = \frac{1}{2} \, (\, \text{base} \times \text{height} \,)$$

$$= \frac{1}{\not{2}} \times 7 \times \not{6}^{3} \; \text{cm}^2$$

$$= 21 \; \text{cm}^2$$

Find the areas of the triangles.

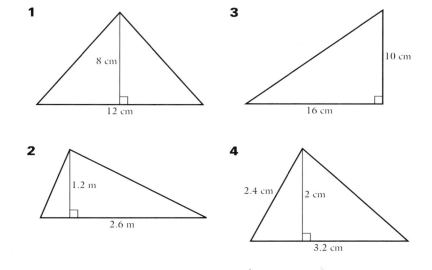

1

8 cm

12 cm

3

10 cm

16 cm

2

1.2 m

2.6 m

4

2.4 cm   2 cm

3.2 cm

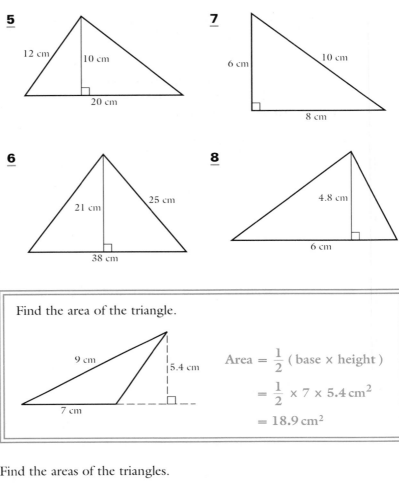

5 12 cm, 10 cm, 20 cm

7 6 cm, 10 cm, 8 cm

6 21 cm, 25 cm, 38 cm

8 4.8 cm, 6 cm

Find the area of the triangle.

9 cm, 5.4 cm, 7 cm

$$\text{Area} = \frac{1}{2} \, (\, \text{base} \times \text{height} \,)$$

$$= \frac{1}{2} \times 7 \times 5.4 \, \text{cm}^2$$

$$= 18.9 \, \text{cm}^2$$

Find the areas of the triangles.

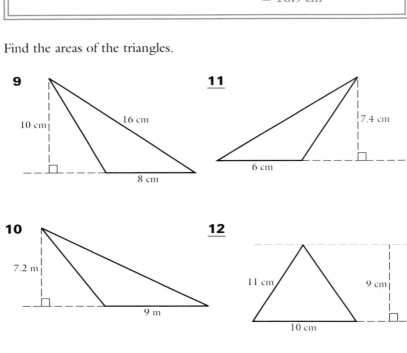

9 10 cm, 16 cm, 8 cm

11 7.4 cm, 6 cm

10 7.2 m, 9 m

12 11 cm, 9 cm, 10 cm

The base of the triangle is the side from which the perpendicular height is measured. If the base is not obvious, turn the page round and look at the triangle from a different direction.

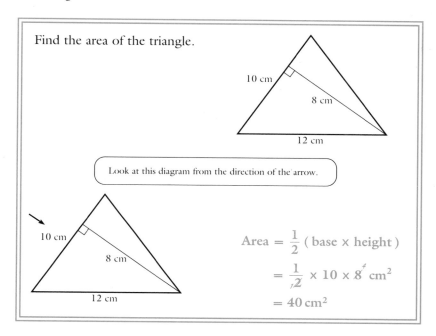

Find the area of the triangle.

Look at this diagram from the direction of the arrow.

$$Area = \frac{1}{2} \, (\, base \times height \,)$$

$$= \frac{1}{2} \times 10 \times 8 \; cm^2$$

$$= 40 \, cm^2$$

Find the areas of the triangles.

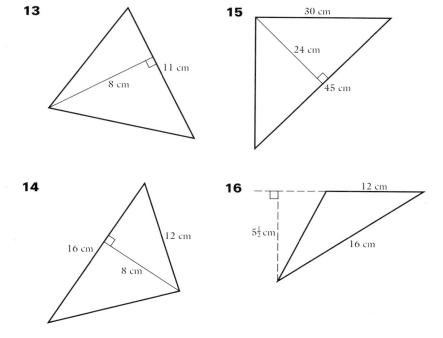

13

11 cm

8 cm

15

30 cm

24 cm

45 cm

14

12 cm

16 cm

8 cm

16

12 cm

$5\frac{1}{2}$ cm

16 cm

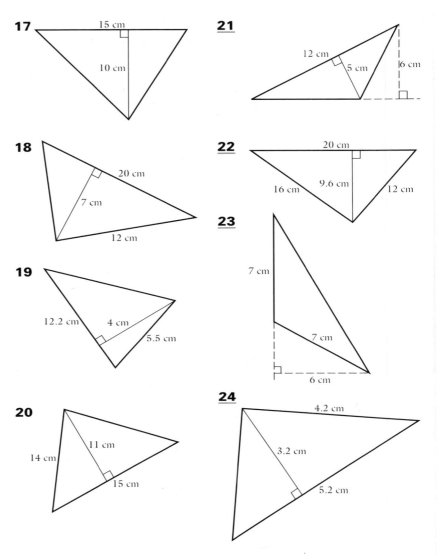

17 15 cm 10 cm

21 12 cm 5 cm 6 cm

18 20 cm 7 cm 12 cm

22 20 cm 16 cm 9.6 cm 12 cm

23 7 cm 7 cm 6 cm

19 12.2 cm 4 cm 5.5 cm

20 11 cm 14 cm 15 cm

24 4.2 cm 3.2 cm 5.2 cm

For questions **25** to **30**, use squared paper to draw axes for x and y from 0 to 6 using 1 square to 1 unit. Find the area of each triangle.

25 $\triangle ABC$ with A(1, 0), B(6, 0) and C(4, 4)

26 $\triangle PQR$ with P(0, 2), Q(6, 0) and R(6, 4)

27 $\triangle DEF$ with D(1, 1), E(1, 5) and F(6, 0)

28 $\triangle LMN$ with L(5, 0), M(0, 6) and N(5, 6)

29 $\triangle ABC$ with A(0, 5), B(5, 5) and C(4, 1)

30 $\triangle PQR$ with P(2, 1), Q(2, 6) and R(5, 3)

31 Find the area of each triangle in this diagram. The squares represent square centimetres. Explain your answers.

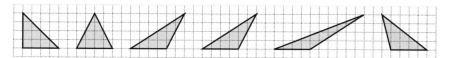

32 This company logo is made from a square and four congruent triangles. The logo is to be made from stainless steel and mounted on the front of a building. Find the area of steel needed if the inner square is made to measure 20 cm by 20 cm.

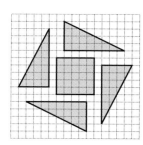

FINDING SQUARE ROOTS

Suppose we have a square tablecloth whose area is known to be $4\,\text{m}^2$. The area of a square is found by multiplying the length of a side by itself. To find the length of a side of this cloth, we have to find a number which, when it is multiplied by itself, is equal to 4.

Now $2 \times 2 = 4$, so the sides of the cloth are 2 m long.

If a square carpet has an area of $20\,\text{m}^2$, then, to find the length of a side of this carpet we have to find a number which, when it is multiplied by itself, is equal to 20. We call this number the *square root* of 20.

20 m²

The symbol $\sqrt{}$ is used to mean 'the square root of', so $\sqrt{20}$ means the square root of 20.

We know that $4 \times 4 = 16$ and that $5 \times 5 = 25$,

i.e. $4^2 = 16$ and $5^2 = 25$, so $\sqrt{16} = 4$ and $\sqrt{25} = 5$

Now we can see that $\sqrt{20}$ is a number between 4 and 5.

The $\sqrt{}$ button on a calculator can be used to find $\sqrt{20}$ correct to as many significant figures as the display will show.

Enter **2** **0** $\boxed{\sqrt{}}$; the display will show 4.472136,

so $\sqrt{20} = 4.47$ correct to 3 s.f..

Therefore, correct to 3 s.f., the length of a side of the square is 4.47 m.

EXERCISE 7D

Find the square root of each number without using a calculator.

1 9   **3** 4   **5** 49   **7** 64

2 36   **4** 8   **6** 121   **8** 144

Without using a calculator, find the value of

9 $(0.1)^2$ **13** $(0.07)^2$ **17** $(1.2)^2$ **21** 7^2

10 $(0.2)^2$ **14** $(0.06)^2$ **18** 20^2 **22** 70^2

11 $(0.01)^2$ **15** $(0.12)^2$ **19** 400^2 **23** 700^2

12 $(0.5)^2$ **16** $(1.1)^2$ **20** 40^2 **24** 7000^2

Find $\sqrt{74.5}$ correct to 3 significant figures.

$\sqrt{74.5} = 8.631\ldots$
$\qquad = 8.63$ (correct to 3 s.f.)   Press
Check: $(8.63)^2 \approx 9^2 = 81$

Use your calculator to find the square root of each number correct to 3 significant figures. Remember to check your answer by squaring it.

25 38.4   **29** 32   **33** 650   **37** 80

26 19.8   **30** 9.8   **34** 10 300   **38** 11.2

27 428   **31** 67   **35** 4 012 000   **39** 24

28 4230   **32** 5.7   **36** 8000   **40** 728

41 Find the square root of

     a 6   **b** 600   **c** 60 000   **d** 60   **e** 6000

     What do you notice?

42 Find the square root of

     a 5   **b** 0.5   **c** 0.05   **d** 5000   **e** 500

     What do you notice?

43 Without using a calculator, explain why Simon is wrong when he says 'The square root of 90 is equal to 30.'

Find the side of the square whose area is $50\,\text{m}^2$.

Length of the side $= \sqrt{50}\,\text{m}$

Length of the side $= 7.071\ldots\text{m}$

$\qquad\qquad\qquad = 7.07\,\text{m}$ (correct to 3 s.f.)

Check: $(7.07)^2 \approx 7^2 = 49$

Find the sides of the squares whose areas are given below. Give your answers correct to 3 significant figures.

44 $85\,\text{cm}^2$ **47** $32\,\text{m}^2$ **50** $749\,\text{mm}^2$

45 $120\,\text{cm}^2$ **48** $0.06\,\text{m}^2$ **51** $84\,300\,\text{km}^2$

46 $50\,\text{m}^2$ **49** $15.1\,\text{cm}^2$ **52** $0.0085\,\text{km}^2$

FINDING MISSING MEASUREMENTS

We know formulas for finding areas of rectangles, squares, parallelograms and triangles.

If A square units is the area of the figure, then

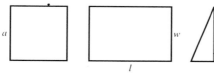
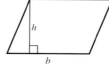

$\quad A = a^2 \qquad\qquad A = l \times w \qquad\qquad A = b \times h \qquad\qquad A = \tfrac{1}{2}(b \times h)$

If we know the area of one of these shapes, we can use the appropriate formula to find a measurement we do not know.

Suppose, for example, we are told that the area of a rectangular patio is $25\,\text{m}^2$ and that it is $10\,\text{m}$ long. If we let the width be $b\,\text{m}$, and use $A = l \times b$ with $A = 25$ and $l = 10$, we can find b, and hence the width of the patio.

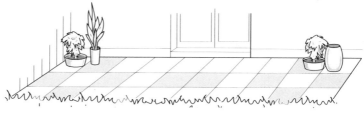

So $A = l \times b$ gives $25 = 10 \times b$

This is a simple equation that we can solve to find b,

i.e. $b = 25 \div 10 = 2.5$

Therefore the width of the patio is $2.5\,\text{m}$.

EXERCISE 7E

> The area of a rectangle is $15 \, \text{cm}^2$
> and the width is 2.5 cm.
> Find the length of the rectangle.
>
> | | |
> |-----------|--------|
> | $15 \, \text{cm}^2$ | 2.5 cm |
>
> l cm
>
> Let the length be l cm.
> Using $A = l \times b$ with $A = 15$ and $b = 2.5$ gives
> $15 = l \times 2.5$
> $l = 15 \div 2.5 = 6$
> Therefore the rectangle is 6 cm long.

The table gives the area and one measurement for various rectangles. Find the missing measurements.

	Area	Length	Width
1	$2.4 \, \text{cm}^2$	6 cm	
2	$20 \, \text{cm}^2$		4 cm
3	$36 \, \text{m}^2$		3.6 m
4	$108 \, \text{mm}^2$	27 mm	
5	$3 \, \text{cm}^2$		0.6 cm
6	$6 \, \text{m}^2$	4 m	
7	$20 \, \text{cm}^2$	16 cm	
8	$7.2 \, \text{m}^2$		2.4 m
9	$4.2 \, \text{m}^2$		0.6 m
10	$14.4 \, \text{cm}^2$		2.4 cm

> The base of a parallelogram is 24 cm
> long and the area is $120 \, \text{cm}^2$.
> Find the height of the parallelogram.
>
>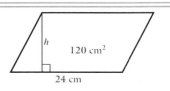
>
> Using $A = b \times h$ with $A = 120$ and $b = 24$ gives
> $120 = 24 \times h$
> Therefore $h = 120 \div 24 = 5$
> The height of the parallelogram is 5 cm.

Find the missing measurements for these parallelograms.

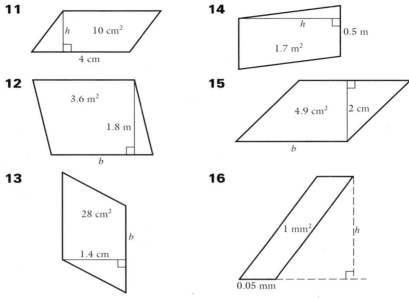

11 h $10\ \text{cm}^2$ 4 cm

14 h 0.5 m   1.7 m²

12 3.6 m²   1.8 m b

15 4.9 cm²   2 cm b

13 28 cm²   1.4 cm b

16 1 mm² h 0.05 mm

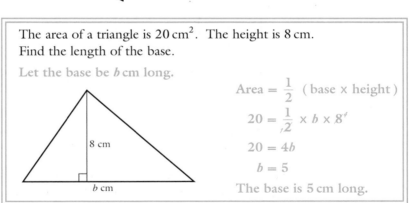

The area of a triangle is $20\ \text{cm}^2$. The height is 8 cm.
Find the length of the base.

Let the base be b cm long.

8 cm

b cm

$\text{Area} = \dfrac{1}{2}\ (\text{base} \times \text{height})$

$20 = \dfrac{1}{2} \times b \times 8$

$20 = 4b$

$b = 5$

The base is 5 cm long.

The table gives the area and one measurement for various triangles.
Find the missing measurements.

	Area	Base	Height
17	24 cm²	6 cm	
18	30 cm²		10 cm
19	48 cm²		16 cm
20	10 cm²	10 cm	
21	36 cm²	24 cm	

Find the missing measurement in each figure.

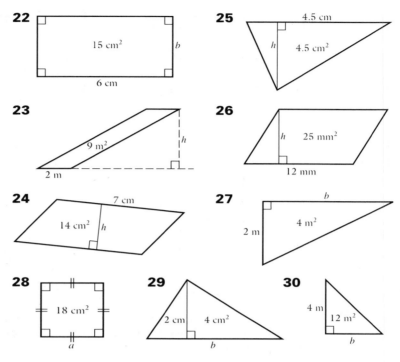

22 15 cm² 6 cm b

25 4.5 cm h 4.5 cm²

23 9 m² 2 m h

26 h 25 mm² 12 mm

24 7 cm 14 cm² h

27 b 4 m² 2 m

28 18 cm² a

29 2 cm 4 cm² b

30 4 m 12 m² b

31 Jason wants to buy enough edging to go round a square flower bed whose area is $5\,\text{m}^2$.

  a Find the distance round the edge of this flower bed.

  b Is it sensible for Jason to buy exactly the length found in part **a**? Give reasons for your answer.

32 A rectangular building plot is advertised as having an area of $300\,\text{m}^2$ and a road frontage of $12\,\text{m}$. How deep is the plot?

33 This illustration is from a developer's scheme. It shows the plan of a building plot whose area is $4300\,\text{m}^2$.

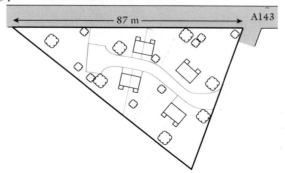

87 m — A143

How far is the back corner of the plot from the main road?

34 This silver necklace is made from sheet silver. Each pendant is either a whole rhombus whose diagonals are 2 cm and 4 cm or half one of these. Find the area of sheet silver used.

FINDING THE AREA OF A COMPOUND SHAPE

A compound shape can often be divided into shapes whose areas we can find.

EXERCISE 7F

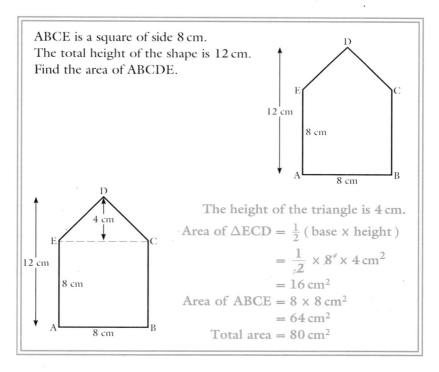

ABCE is a square of side 8 cm.
The total height of the shape is 12 cm.
Find the area of ABCDE.

The height of the triangle is 4 cm.
Area of $\triangle ECD = \frac{1}{2}$ (base × height)

$$= \frac{1}{2} \times 8 \times 4 \text{ cm}^2$$

$$= 16 \text{ cm}^2$$

Area of ABCE = 8 × 8 cm²

$$= 64 \text{ cm}^2$$

Total area = 80 cm²

Find the areas of the following shapes. Remember to draw a diagram for each question and mark in all the measurements.

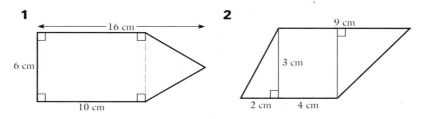

1

16 cm

6 cm

10 cm

2

9 cm

3 cm

2 cm   4 cm

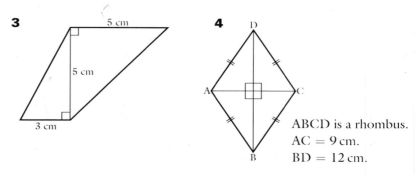

3

5 cm

5 cm

3 cm

4

D

A

C

B

ABCD is a rhombus.
AC = 9 cm.
BD = 12 cm.

For questions **5** to **17** find the areas of the figures in square centimetres.
The measurements are all in centimetres.

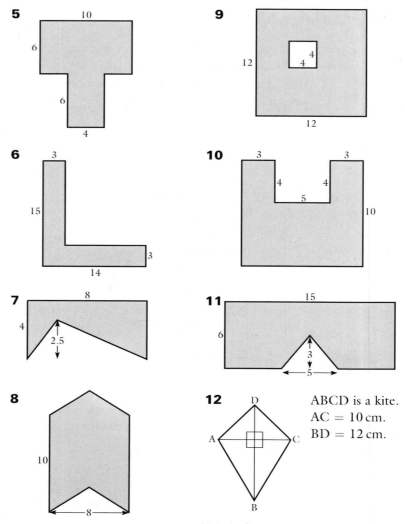

5

10

6

6

4

9

12

4

4

12

12

6

3

15

3

14

10

3

3

4

4

5

10

7

8

4

2.5

11

15

6

3

5

8

10

8

12

D

A

C

B

ABCD is a kite.
AC = 10 cm.
BD = 12 cm.

(BD is the axis of symmetry.
The diagonals cut at right angles.)

13

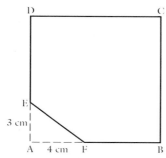

A square ABCD, of side 9 cm, has a triangle EAF cut off it.

14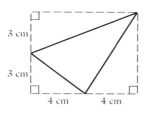

15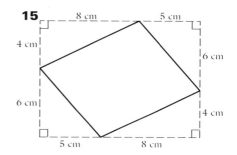

16 ABCD is a rhombus whose diagonals measure 7 cm and 11 cm.

17 ABCD is a kite whose diagonals measure 12 cm and 8 cm.
There are many kites you can draw with these measurements but their areas are all the same.
Can you explain why?

In questions **18** and **19** draw axes for x and y from -6 to $+6$, using 1 square to 1 unit.

Find the areas of the following shapes.

18 Quadrilateral ABCD with A($-2, -3$), B($3, -3$), C($0, 4$) and D($-2, 4$).

19 Quadrilateral EFGH with E($-1, 1$), F($2, -3$), G($5, 1$) and H($2, 5$).

20 A silversmith is asked to make a solid silver rectangle measuring 10 mm by 9 mm correct to the nearest millimetre.

  a Giving answers correct to 1 decimal place, what are
    i the smallest measurements that can be used
    ii the largest measurements that can be used?
  b What is the difference in the area of silver needed to make the smallest and largest possible rectangles?

21 a

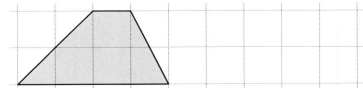

Copy this diagram on to 1 cm squared paper. Add a trapezium congruent with the one shown so that the combined shape is a parallelogram. Hence find the area of the trapezium.

b Repeat part **a** with this trapezium. Hence find the area of this trapezium in terms of a, b and h.

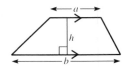

c Using your result from part **b** give, in words, an instruction for finding the area of any trapezium.

MIXED EXERCISE

EXERCISE 7G   Find the areas of the following figures.

1

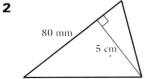

3

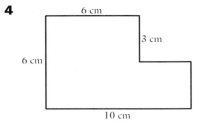

2

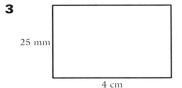

4

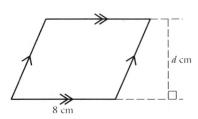

5 The area of a rectangle is 84 cm² and its width is 6 cm. Find its length.

6 The area of this parallelogram is 52 cm². Find the distance, d cm, between the parallel lines.

PRACTICAL
WORK

Design a patchwork mat measuring approximately 40 cm by 40 cm.
Base your design on a mixture of triangles, parallelograms, squares
and rectangles.
Include an estimate of the area of each colour, or texture, that you
want to use.
Here are a few ideas. You can find more ideas in other books.

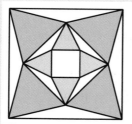 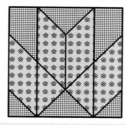 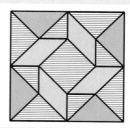

INVESTIGATION

These triangles are drawn on 1 cm grid dots.

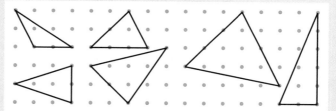

a Copy and complete this table for each triangle.

Number of dots on edge	Number of dots inside	Area (cm^2)

b Find a relationship between the number of dots on the edge, the
number of dots inside and the area of each shape. Does this
relationship hold for any triangle drawn on the grid?

c Investigate the relationship between number of dots on the edge,
the number of dots inside and the areas of rectangles and
parallelograms.

SCATTER GRAPHS

Assertions like these are commonplace.

- Young people buy more magazines than older people.
- People with large hands also have large feet.
- Not eating green vegetables makes people unhealthy.

Is there any truth in any of these statements? We can only find out by looking at some evidence.

Without evidence, or proof, any assertion is called a *hypothesis*.

In order to judge the truth of any one of the hypotheses given above, we need to start by gathering some evidence. For the first statement, we need information relating the age of a person to the number of magazines bought in, say, the last week, and we need this for several people.

EXERCISE 8A

1 Discuss what information is needed before you can assess the truth or otherwise of each of these hypotheses.

    a People with large hands also have large feet.

    b Not eating green vegetables makes people unhealthy.

    c Pupils spend more in the school canteen in the later part of a week than they do at the beginning of the week.

2 To find out if younger people do buy more magazines than older people, Mary asked several people their age and the number of magazines they had bought last week. She then arranged this information in two lists, each in order of size:

Age: 10, 12, 13, 14, 14, 15, 16, 18, 20, 35, 47, 60

Numbers of magazines: 0, 0, 1, 1, 1, 1, 2, 2, 2, 2, 3, 4.

Discuss why these lists will not help Mary with her investigation.

3 Pradesh wants to find out if there is a relationship between the ages of pupils and acceptance of wearing school uniform.
Discuss the advantages and disadvantages of each of the following method of collecting the information shown opposite.

a In two separate observation sheets, i.e.

School Year						School uniform is a good idea				
7	8	9	10	11	12	Strongly disagree	Disagree	No opinion	Agree	Strongly agree
///	/	//	//	/	///	///	////	//	/	//

b On one observation sheet, i.e.

		School uniform is a good idea				
Name	School year	Strongly disagree	Disagree	Neither agree nor disagree	Agree	Strongly agree
Jane	8			✓		
Austin	7		✓			

c Asking each individual to fill in a form, i.e.

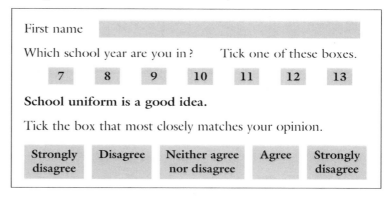

First name

Which school year are you in?   Tick one of these boxes.

  7   8   9   10   11   12   13

School uniform is a good idea.

Tick the box that most closely matches your opinion.

Strongly disagree   **Disagree**   **Neither agree nor disagree**   **Agree**   **Strongly disagree**

SCATTER GRAPHS

Each of the statements looked at in this chapter involves two pieces of information about one individual. (The technical name is *bivariate* data.) These pairs of data can be compared for several individuals.

Discussion of the examples in **Exercise 8A** shows that

- it is vital that corresponding items of information are kept together
- the simplest way of collecting two items of information about an individual is to use an observation sheet like the one in question 3, part b.

Consider this hypothesis: Tall people have larger feet than short people. This is a commonplace belief, but how true is it?
Does it mean, for example, that if my friend and I are the same height, we take the same size in shoes?
Or is there not much truth in the statement, that is, there is not much relationship between a person's height and shoe size?

We can try to find out by gathering some evidence. This sheet gives the heights (in centimetres) and the shoe sizes of 12 females in the order in which they were asked.

Person	1	2	3	4	5	6	7	8	9	10	11	12
Height (cm)	160	166	174	158	167	161	170	171	166	163	164	168
Shoe size (continental)	36	38	40	37	37	38	42	41	40	39	37	39

There is no order in either the heights or the shoe sizes and, as it is difficult to see any pattern, we will rearrange the information so that the heights are in order of size (making sure that each shoe size is kept with its corresponding height).

Height (cm)	158	160	161	163	164	166	166	167	168	170	171	174
Shoe size (continental)	37	36	38	39	37	40	38	37	39	42	41	40

Now we can see that shoe size does tend to get larger as height increases. However, the tallest person has not got the largest feet so there is not a direct relation between height and shoe size.

We get a clearer picture if we plot these points on a graph.

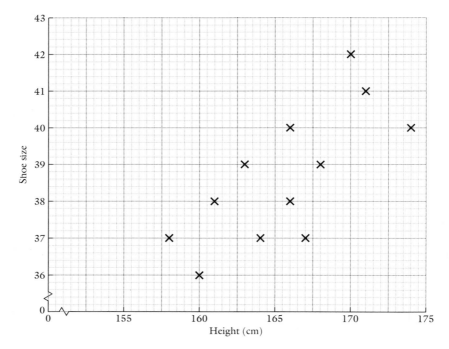

The points do not all fit on a straight line. A graph like this is called a *scatter graph*.

Now we can see that taller people tend to have larger feet but the relationship between height and shoe size is not strong enough to justify the original statement.

1 a In the example above this exercise, we state that information has been collected from 12 people, *all of whom were female*. Why is it important to say this?

b The axes for the scatter graph above have zig-zag lines near zero. Why?

2 This graph illustrates information relating the ages and prices of some second-hand cars. The cars are all the same model.

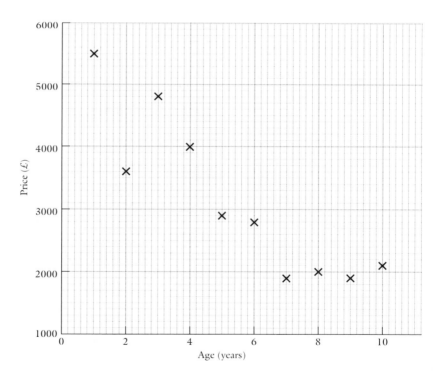

a Does this evidence support the statement 'The price of a second-hand car gets lower as the age increases'?

b Apart from age, what other factors do you think affect the price of a car?

c Is it true to say that a five-year-old car is always cheaper than a two-year-old car?

3 Lim saw this headline in a paper, 'TALL CHILDREN GET BETTER GRADES AT GCSE THAN SHORT CHILDREN.'
He did not read any further and, because he was short for his age, decided that his school career was doomed to failure.
This scatter graph shows the heights and grades obtained in end-of-year exams by a group of Year 8 pupils.

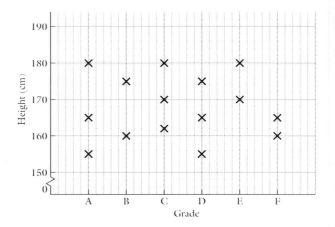

a Does the evidence here support the claim in the newspaper?

b Discuss whether end-of-year exam grades are good at predicting GCSE results.

c What does the evidence here show about the relationship between heights and results in end-of-year exams?

d Only a very small number of children are represented in this graph. We will assume that these children have been chosen randomly. Suppose you chose the children with a particular result in mind, draw sketches showing the scatter graphs it would be possible to get. (This is called *biasing* the results.)

e Discuss whether you should believe everything you read in newspapers (or books – even this one)?

Keep the graphs drawn for questions **4** to **7**: you will need them for the next exercise.

4 The table gives the French mark and the maths mark of each of 20 pupils in an end-of-term examination.

French	45	56	58	58	59	60	64	64	65	65	66	70	71	73	73	75	76	76	78	80
Maths	50	38	45	48	56	65	60	58	70	75	60	79	64	80	85	69	82	77	69	75

a Show this information on a graph; use a scale of 1 cm for 5 marks on each axis. Mark the horizontal axis from 40 to 85 for the French mark and the vertical axis from 35 to 90 for the maths mark.

b John is good at French. Is he likely to be good at maths?

5 This table shows the heights and weights of 12 people.

Height (cm)	150	152	155	158	158	160	163	165	170	175	178	180
Weight (kg)	56	62	63	64	57	62	65	66	65	70	66	67

a Show this information on a graph; use a horizontal scale of 2 cm for each 5 cm of height and mark this axis from 145 to 185. Use a vertical scale of 2 cm for each 5 kg and mark this axis from 55 to 75.

b Carlos weights 65 kg. Is he likely to be tall?

6 This table shows the number of rooms and the number of people living in each of 15 houses.

Number of rooms	3	4	4	5	5	5	6	6	6	6	7	7	7	8	8
Number of people	2	3	5	4	2	1	6	2	3	4	4	5	3	2	6

a Show this information by plotting the points on a graph; use a scale of 1 cm for one unit on each axis.

b Cheryl lives in a house with four other people. Is the house likely to have more than four rooms?

7 This table shows the number of pens and pencils and the number of books that each of 10 pupils has in a maths lesson.

Number of pens and pencils	2	3	3	5	6	6	12	15	20	25
Number of books		4	5	0	3	1	4	6	2	1	5

a Show this information by plotting the points on a graph; use a horizontal scale of 1 cm for two pens and pencils and a vertical scale of 1 cm for one book.

b Is the number of pens and pencils brought by a pupil a reliable indication of how many books that pupil has brought?

c Collect corresponding information for the pupils in your maths class and make a scatter graph from it.

If we look again at the scatter graph of height and shoe size, we see that the points are scattered about a straight line which we can draw by eye. This is called the *line of best fit*. When drawing this line, the aim is to get the points evenly distributed about the line, so that the sum of the distances from the line to points that are above it, is roughly equal to the sum of the distances from the line to points that are below it. This may mean that none of the points lies on the line.

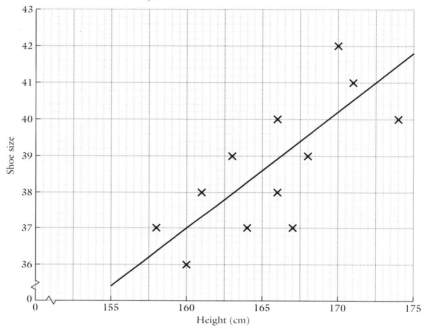

The less scatter there is about the line, the stronger is the relationship between the two quantities. We use the word *correlation* for the relationship between the two quantities.

In the diagram above, the line slopes upwards, that is, shoe size tends to increase with height. We call this *positive* correlation.

This scatter graph, from question 2 in Exercise 8B, shows that the price of cars tends to decrease as their age increases.

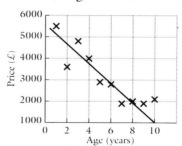

The line of best fit slopes downwards, and we say that there is *negative* correlation.

If the points are close to the line, we say that there is a strong correlation.

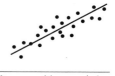

Strong positive correlation

If the points are loosely scattered about the line, we say that there is moderate correlation.

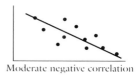

Moderate negative correlation

Sometimes the points are so scattered that there is no obvious line. Then we say that there is no correlation.

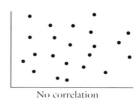

No correlation

EXERCISE 8C

1 Use the scatter graphs that you drew for questions **4** to **7** of **Exercise 8B**. For each one, draw the line of best fit, if you think there is one. Describe the correlation between the two quantities in each case.

2 Describe the correlation you would expect between these quantities. You are *not* asked to provide any evidence for your answers.

a The number of pages and the number of advertisements in a newspaper.

b The length and width of a cucumber.

c The weight of tomatoes produced by a tomato plant and its height.

d The number of miles a car will travel on one gallon of petrol and the capacity of its engine.

e The score on each dice when a red dice and a blue dice are thrown together.

f The number of games a player wins in the first set of a tennis match and the number of games the player wins in the second set of the match.

g The number of days a pupil is away from school and the number of days the pupil is late in handing in a technology project.

h The age of a pupil and that pupil's feeling about having to wear school uniform.

**COLLECTING
INFORMATION**

Up to now the information that you have been asked to collect can be measured or counted, for example, heights, weights, times, numbers of heads when three coins are tossed, numbers of brothers and sisters.
In this chapter, we have introduced quantities that are not so easily measured or counted, such as shoe size. Some quantities cannot be measured, such as exam grades, eye colour, opinions and so on.
At some stage, you will have to collect such information yourself; you will need to plan in advance, anticipate some of the problems that might arise and decide how you are going to solve them.
The next exercise contains discussion topics.

EXERCISE 8D

1 Suppose that information is to be collected about the shoe sizes of pupils in your year.

   a How many categories do you need? Should you stick to the whole number sizes?

   b Will you ask for continental sizes or English sizes?

   c If you decide on whole number sizes, what should you do about a pupil who insists that all her shoes are size $3\frac{1}{2}$?

   d What should you do about someone whose left shoe is size 3 and whose right is size 4?

   e Some people are shy about giving their shoe size. What can you do about this?

   f If you ask people to write their shoes sizes on a piece of paper anonymously, what could go wrong?

2 Information is to be collected about the eye colour of pupils in your year.

   a State the categories you would use. Why are categories needed?

   b List the problems you might meet as you collect the information.

3 Information is to be collected about the time spent by pupils on maths homework last night.

   a Pete decided to use these categories:

     Less than 15 minutes   15 to 30 minutes   30 minutes to 1 hour.

     What problems might he have with his choice of categories?

   b Jacqui used these categories:

     Less than 15 minutes   15 to 30 minutes
     30 minutes to 45 minutes   More than 45 minutes.

     i Why are these categories better than Pete's?

     ii These categories are still not ideal. How can they be improved?

   c What other problems might there be if you ask pupils how long they spent on their homework?

4 Evidence is needed about pupils' attitudes to starting the school day earlier and finishing earlier.
Discuss the problems with these attempts to get evidence.

a I like getting to school early.

Tick one of these boxes.　　　| Yes |　| No |

b The school day should start earlier and finish earlier.
Tick the box that most closely matches your opinion.

| Agree |　　| Do not mind |　　| Disagree |

c The school day should start at 8 a.m. and end at 2 p.m.
Tick the box that most closely matches your opinion.

| Strongly agree | Partly agree | No opinion | Partly disagree | Strongly disagree |

d I would like to finish the school day at 2 p.m.

Tick one of these boxes.　　　| Yes |　| No |

Discuss how you would tackle gathering the evidence.

5 Seema decided to collect information for question **4** by asking several pupils questions directly.

She asked James if he liked the idea of getting to school earlier and finishing earlier.

She then asked Cheryl if she thought that getting to school earlier and finishing earlier was a good idea.

She next approached Rafique and said, 'I don't want to have to get to school at 8 o'clock. Do you?'

Seema carried on like this, asking different questions of different pupils.

Discuss the problems caused by this approach.

QUESTIONNAIRES The information required sometimes concerns opinion on several different points. In questions **4** and **5** in the last exercise for example, you may have come to the conclusion that more than one question is needed to find out pupils' attitudes to a change in the timing of the school day. In cases like this a sheet of questions for each person might be more useful. A set of questions of this sort is called a *questionnaire*.

EXERCISE 8E **1** Copy and complete this questionnaire.
(Notice the different types of question and forms of answer.)

---

1. How tall are you ?   cm

2. Do you consider yourself to be

Tall   Average   Small

(Underline your answer.)

3. Do you like being the height you are ?

Underline your answer.

Love it   Like it   Don't mind   Dislike it   Hate it

4. I want to grow taller   0   1   2   3

(0 means 'not at all', 3 means 'very much')

Ring the number that represents your answer.

5. I am male/female. (Cross out the unwanted word.)

---

2 In the questionnaire above,

a why is item 5 needed?

b question 3 could have had an answer in a different form, such as:

2   1   0   −1   −2

Ring the number which represents your liking.

What is the problem when the question is put in this form or in the form used for item 4?

3 There are several things wrong with the wording of the following questions and choice of responses. List them, giving reasons.

a Do you like mathematics?   0   1   2   3   4

b What colour is your hair?

c How many people are there in your family?   . . . .

d Don't you agree that it is wrong to lie?

4 Two different groups conducted a survey on attitudes to turning the Town Centre into pedestrian access only.

Group 1 asked these questions:

Do cars cause pollution?

Is crossing the roads dangerous because of the traffic?

Should the Town Centre be made pedestrian only?

Group 2 asked these questions:

Would you find it difficult to get to the shops if the Town Centre was pedestrian only?

Can you manage to carry your shopping to outside the Town Centre?

Should the Town Centre be made pedestrian only?

The last question in each group is the same. Discuss what you think is the purpose of the first two questions in each group.

5 Write a questionnaire on a topic of your own choice, using different types of question. Sometimes the wording can be misunderstood: try the question out on a few people and adjust the wording where necessary.

PRACTICAL
WORK

This chapter started with some assertions. One was that 'Not eating green vegetables makes people unhealthy.'
Find out how much truth you think there is in this statement.

You should include in your answer

- how you measured the quantity of green vegetables people eat
- how you decided what constitutes a person's state of health
- how you collected any evidence and who you collected it from
- any diagrams illustrating your evidence
- your conclusions
- any assumptions that you made
- how reliable you think your conclusions are.

CIRCUMFERENCE AND AREA OF A CIRCLE

Ken has entered for a 50 km sponsored cycle ride. Out of idle curiosity he wonders how many pedal strokes this will involve. This is a complicated problem and Ken would probably give up, but it can be solved by finding out how many times the wheels turn for each turn of the pedals and how far the bike moves forward for each turn of the wheels.

To find out how far the bike moves forward for each turn of the wheels we need to know the distance round the outside of each tyre. This can be found by measuring, but it is difficult to measure round a curved wheel accurately. It is much easier if we can calculate the distance round the wheel by first measuring the distance across the wheel. To solve this and many simple everyday problems we need to know the basic facts about a circle.

DIAMETER, RADIUS AND CIRCUMFERENCE

When you use a pair of compasses to draw a circle, the place where you put the point is the *centre* of the circle. The line that the pencil draws is the *circumference* of the circle. All the points on the circumference are the same distance from the centre.

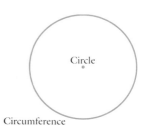

Circumference

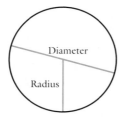

Any straight line joining the centre to a point on the circumference is a *radius*.

A straight line across the full width of a circle (going through the centre) is a *diameter*.

The diameter is twice as long as the radius. If d cm is the length of a diameter and r cm is the length of a radius, we can write this as a formula, i.e. $d = 2r$

168

In questions **1** to **6** write down the length of the diameter of the circle whose radius is given.

1

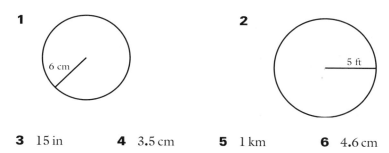

2

6 cm

5 ft

3 15 in   **4** 3.5 cm   **5** 1 km   **6** 4.6 cm

7 For this question you will need some thread and a cylinder (a soft drink can, the cardboard tube from a roll of kitchen paper).

Measure across the top of the cylinder to get a value for the diameter. Wind the thread 10 times round the can. Measure the length of thread needed to do this and then divide your answer by 10 to get the value of the circumference.

If C stands for the length of the circumference and d for the length of the diameter find, approximately, the value of $C \div d$.

(Note that you can also use the label from a cylindrical tin. If you are careful you can reshape it and measure the diameter and then unroll it to measure the circumference.)

8 Compare the results from the whole class for the value of $C \div d$.

INTRODUCING π From the last exercise you will see that, for any circle,

$$\text{circumference} \approx 3 \times \text{diameter}$$

The number that you have to multiply the diameter by to get the circumference is slightly larger than 3.

This number is unlike any number that you have met so far. It cannot be written down exactly, either as a fraction or as a decimal:

as a fraction it is approximately, but *not exactly*, $\frac{22}{7}$;

as a decimal it is approximately 3.142, which is correct to 3 decimal places.

Over the centuries mathematicians have spent a lot of time trying to find the true value of this number. The ancient Chinese used 3. Three is also the value given in the Old Testament (I Kings 7:23). The Egyptians (c. 1600 BC) used $4 \times (8/9)^2$. Archimedes (c. 225 BC) was the first person to use a sound method for finding its value and a mathematician called Van Ceulen (1540–1610) spent most of his life finding it to 35 decimal places !

Now, with a computer to do the arithmetic we can find its value to as many decimal places as we choose: it is a never ending, never repeating, decimal fraction. To as many figures as we can get across the page, the value of this number is

3.141 592 653 589 793 238 462 643 383 279 502 884 197 169 399 375

Because we cannot write it down exactly we use the Greek letter π (pi) to stand for this number. We can then write a formula connecting the circumference and diameter of a circle in the form $C = \pi d$.
But $d = 2r$ so we can rewrite this formula as

$$C = 2\pi r$$

where C is the circumference and r is the radius.

Because π is close to 3, we can always use 3 as a value for π to work out estimated answers.

CALCULATING THE CIRCUMFERENCE

EXERCISE 9B

The radius of a circle is **3.8** m.

a Use $\pi \approx 3$ to estimate the circumference of this circle.

b Use the π button on your calculator to work out a more accurate value. Give your answer correct to **3** significant figures.

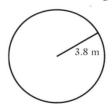

3.8 m

a Using $C = 2\pi r$ and $\pi \approx 3$ gives $C \approx 2 \times 3 \times 3.8 = 22.8$
i.e. $C \approx 23$

b Using $C = 2\pi r$
gives $C = 2 \times \pi \times 3.8$

Press **2** **×** **π** **×** **3** **.** **8** **=**

$= 23.87\ldots$
$= 23.9$ (correct to 3 s.f.)

Circumference $= 23.9$ m (correct to 3 s.f.)

The approximate answer agrees closely with this,
so the more accurate answer is likely to be correct.

In questions **1** to **12**

a Use $\pi \approx 3$ to estimate the circumference of a circle with the given radius.

b Use the π button on your calculator, giving your answers correct to 3 significant figures, to find the circumference.

1 2.3 m	**4** 53 mm	**7** 4.8 m	**10** 28 mm
2 4.6 cm	**5** 250 mm	**8** 1.8 m	**11** 1.4 m
3 2.9 cm	**6** 36 ft	**9** 7 yd	**12** 35 in

Find the circumference of a circle of diameter 12.6 mm. Use the π button on your calculator. Check that your answer is sensible by finding an approximate answer using π as 3.

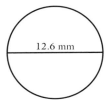

12.6 mm

Using $C = 2\pi r$,

$$r = \tfrac{1}{2} \text{ of } 12.6 = 6.3$$
$$C = 2 \times \pi \times 6.3$$
$$= 39.58 \ldots$$
$$= 39.6 \ (\text{correct to 3 s.f.})$$

Circumference $= 39.6$ mm (correct to 3 s.f.)

> Instead we could have used $C = \pi d$
> Then $C = \pi \times 12.6$
> $= 39.6$ (correct to 3 s.f.)

Check: If $\pi \approx 3$ and the diameter ≈ 13 mm the circumference is approximately 3×13 mm $= 39$ mm.

Using the π button on your calculator, and giving your answers correct to 2 significant figures, find the circumference of a circle of

13 radius 154 mm	**16** radius 34.6 cm
14 diameter 28 in	**17** diameter 511 mm
15 radius 7.7 m	**18** diameter 630 ft

In the problems that follow use the π button on your calculator, and give your answers correct to 3 significant figures. Check that your answers are sensible by estimating them using π as 3.

Find the perimeter of the given semicircle.
(The prefix 'semi' means half.)

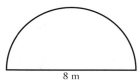

8 m

The complete circumference of the circle is $2\pi r$

The curved part of the semicircle is $\frac{1}{2} \times 2\pi r$

$$= \frac{1}{2} \times 2 \times \pi \times 4\,\text{m}$$
$$= 12.56\ldots\text{m}$$

The perimeter $=$ curved part $+$ straight edge
$$= (12.56 + 8)\,\text{m}$$
$$= 20.56\ldots\text{m}$$
$$= 20.6\,\text{m}\ (\text{correct to 3 s.f.})$$

Find the perimeter of each of the following shapes.

19

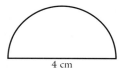

4 cm
A semicircle

20

5 cm   120°

(This is one third of a circle
because 120° is $\frac{1}{3}$ of 360°.)

21

3 in

(This is called a quadrant:
it is one quarter of a circle.)

22

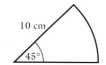

10 cm   45°

A 'slice' of a circle is called
a sector. $\left(\frac{45}{360} = \frac{1}{8},\right.$
so this sector is $\frac{1}{8}$ of a circle. $\left.\right)$

23

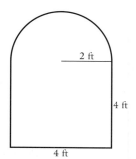

2 ft

4 ft

4 ft

A window frame with
a semi-circular top

24

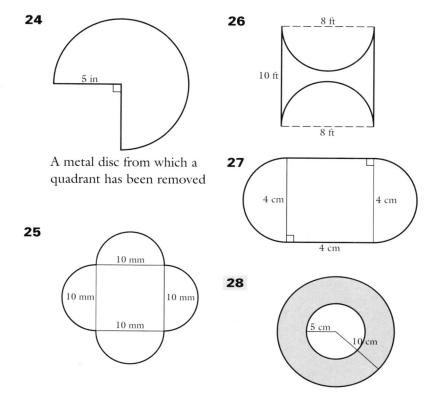

A metal disc from which a quadrant has been removed

25

10 mm

10 mm   10 mm

10 mm

26

8 ft

10 ft

8 ft

27

4 cm   4 cm

4 cm

28

5 cm

10 cm

A circular flower-bed has a diameter of 1.5 m. A metal edging is to be placed round it. Find the length of edging needed and the cost of the edging if it is sold by the metre (i.e. you can only buy a whole number of metres) and costs 60 p a metre.

Using $C = \pi d,$
$$C = \pi \times 1.5$$
$$= 4.712\ldots$$

1.5 m

Length of edging needed $= 4.71$ m (correct to 3 s.f.)

Note that if you use $C = 2\pi r,$ you must remember to halve the diameter.

As the length is 4.71 m we have to buy 5 m of edging.
$$\text{Cost} = 5 \times 60\,\text{p}$$
$$= 300\,\text{p}\quad\text{or}\quad £3$$

29 Measure the diameter, in millimetres, of a 2 p coin. Use your measurement to find the circumference of a 2 p coin.

30 Repeat question **29** with a 10 p coin and a 1 p coin.

31 A circular table-cloth has a diameter
of 1.4 m.
How long is the hem of the cloth?

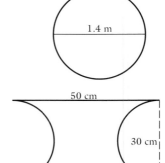

32 A rectangular sheet of metal
measuring 50 cm by 30 cm has a
semicircle of radius 15 cm cut from
each short side as shown.
Find the perimeter of the shape
that is left.

33 A bicycle wheel has a radius of 28 cm. What is the circumference of
the wheel?

34 Hank has made a circular table that has a diameter of 1.2 m.
He wants to buy plastic edging to go round it.
Using 3 as an approximate value for π he calculates that he needs
$3 \times 1.2\,m = 3.6\,m$ of edging. Will he have enough? Give reasons
for your answer.

35 How far does a bicycle wheel of radius 28 cm travel in one complete
revolution? How many times will the wheel turn when the bicycle
travels a distance of 352 m?

36 A cylindrical tin has a radius of 2 cm.
What length of paper is needed to put
a label on the tin if the edges just meet?

37 A square sheet of metal has sides of length 30 cm. A quadrant (one
quarter of a circle) of radius 15 cm is cut from each of the four
corners. Sketch the shape that is left and find its perimeter.

For the following problems, give your answers as accurately as you think
is sensible.

38 A boy flies a model aeroplane on the end of a wire 10 m long. If he
keeps the wire horizontal, how far does his aeroplane fly in one
revolution?

39 If the aeroplane described in question **38** takes 1 second to fly
10 m, how long does it take to make one complete revolution? If
the aeroplane has enough power to fly for 1 minute, how many
turns can it make?

40 A cotton reel has a diameter of 2 cm. There are 500 turns of thread on the reel. How long is the thread?
Is your answer likely to be larger or smaller than the actual length on the reel? Give a reason.

41 A bucket is lowered into a well by unwinding rope from a cylindrical drum. The drum has a radius of 20 cm and with the bucket out of the well there are 10 complete turns of the rope on the drum. When the rope is fully unwound the bucket is at the bottom of the well. How deep is the well?

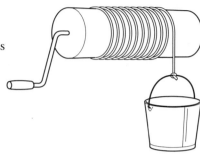

42 A garden hose is 100 m long. For storage it is wound onto a cylindrical hose reel of diameter 45 cm. How many turns of the reel are needed to wind up the hose?

43 The cage which takes miners up and down the shaft of a coal mine is raised and lowered by a steel cable wound round a circular drum of diameter 3 m. It takes 10 revolutions of the drum to lower the cage from ground level to the bottom of the shaft. How deep is the shaft?

FINDING THE RADIUS OF A CIRCLE GIVEN THE CIRCUMFERENCE

If a circle has a circumference of 24 cm, we can find its radius by using the formula $C = 2\pi r$. This formula gives

$$24 = 2 \times \pi \times r$$

which is an equation that can be solved for r.

By taking π as 3, the formula $C = 2\pi r$ can be used to estimate r,

i.e. $C \approx 2 \times 3 \times r$ giving $C \approx 6r$

$\therefore$ when $C = 24$ we have $24 \approx 6r$ so $r \approx 4$.

EXERCISE 9C

1 By taking π as 3 estimate the radius of a circle whose circumference is

    a 60 cm   **b** 72 in   **c** 40 m   **d** 582 cm

The circumference of a circle is 36 cm.
Find the radius of this circle.

Using $C = 2\pi r$ gives

$36 = 2 \times \pi \times r$

$18 = \pi \times r$. | Dividing both sides by 2. |

$\dfrac{18}{\pi} = r$ | Dividing both sides by π. |

$r = 5.729\ldots$

$r = 5.73$ (correct to 3 s.f.)

Therefore the radius is 5.73 cm correct to 3 s.f.

Check: Using $\pi \approx 3$, diameter $\approx 36 \div 3$ cm, so radius ≈ 6 cm

For the remaining questions in this exercise, use the π button on your calculator, and give your answers correct to 3 significant figures.

2 Find the radius of each circle given in question **1**.

3 Find the radius of a circle whose circumference is

    a 275 cm   **b** 462 mm   **c** 831 in   **d** 87.4 m

4 Find the diameter of a circle whose circumference is

    a 550 mm   **b** 52 cm   **c** 391 yd   **d** 76 ft

5 A roundabout at a major road junction is to be built. It has to have a minimum circumference of 188 m. What is the corresponding minimum diameter?

6 A bicycle wheel has a circumference of 200 cm. What is the radius of the wheel?

7 A car has a turning circle whose circumference is 63 m. What is the narrowest road that the car can turn round in without going on the pavement?

8 When the label is taken off a tin of soup it is found to be 32 cm long.
If there was an overlap of 1 cm when the label was on the tin, what is the radius of the tin?

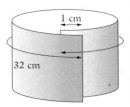

9 The diagram shows a quadrant of a circle.
If the curved edge is 15 cm long,
what is the length of a straight edge?

10 A tea cup has a circumference of 24 cm. What is the radius of the
cup? Six of these cups are stored edge to edge in a straight line on a
shelf. What length of shelf do they occupy?

11 The shape in the diagram is made up of a
semicircle and a square.
Find the length of a side of this square.

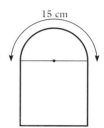

12 The curved edge of a sector of angle 60° is 10 cm.
Find the radius and perimeter of the sector.

13 Make a cone from a sector of a circle as follows:
On a sheet of paper draw a circle of radius
8 cm. Draw two radii at an angle of 90°.
Make a tab on one radius as shown. Cut
out the larger sector and stick the straight
edges together. What is the circumference
of the circle at the bottom of the cone?

14 A cone is made by sticking together the straight edges of the sector
of a circle, as shown in the diagram.

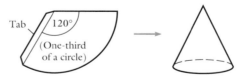

The circumference of the circle at the bottom of the finished cone is
10 cm. What is the radius of the circle from which the sector was cut?

15 A gardener wants to create a circular flower-bed with a circumference
of 14 m. He marks out the edge by putting a stake in the centre and
using a string with a peg tied at the other end. He works out the
length of the string from the approximation 14 ÷ 6 m ≈ 2 m.
Will this give a circumference of more or less than 14 m?
Justify your answer and comment on the method.

THE AREA OF A CIRCLE

A printer uses sheets of paper measuring 590 mm by 420 mm to print circular labels of radius 3.3 cm. He would like to know the largest number of these labels that can be printed on one sheet and the area of paper wasted in this case.

- To answer these questions the printer needs to know how to find the area of a circle.

The formula for finding the area of a circle is

$$A = \pi r^2$$

You can see this if you cut a circle up into sectors and place the pieces together as shown to get a shape which is roughly rectangular.

Consider a circle of radius r whose circumference is $2\pi r$.

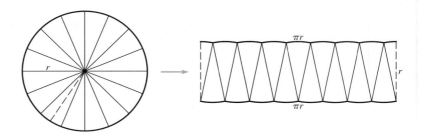

Area of circle = area of 'rectangle'
= length × width
= $\pi r \times r = \pi r^2$

Note: πr^2 involves length × length so represents an area (π is a number) whereas $2\pi r$ is concerned with a single length so gives another length.

EXERCISE 9D

Give your answers correct to 3 significant figures.

Find the area of a circle of radius 2.5 cm.

Using $A = \pi r^2$ with $r = 2.5$
gives $A = \pi \times (2.5)^2$
$= 19.63\ldots$
$= 19.6$ (correct to 3 s.f.)

Area is 19.6 cm² correct to 3 s.f.

2.5 cm

In questions **1** to **3** find first an estimate for the area, then find the area.
How do these values compare?

1 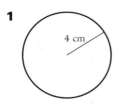 4 cm

2 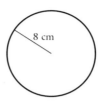 8 cm

3 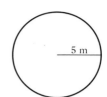 5 m

Find the areas of the following circles.

4 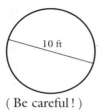 10 ft

(Be careful !)

6 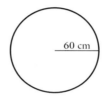 60 cm

8 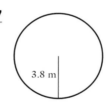 3.5 km

5 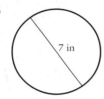 7 in

7  3.8 m

9   80 m

This is a *sector* of a circle. Find its area.

  45° 3 m

$$\frac{45^{9}}{{}_{72}360} = \frac{9^{7}}{{}_{8}72} = \frac{1}{8}$$

$\therefore$ area of sector $= \dfrac{1}{8}$ of area of circle of radius 3 m

$$\text{Area of sector} = \frac{1}{8} \text{ of } \pi r^2$$

$$= \frac{1}{8} \times \pi \times 9 \, \text{m}^2$$

$$= 3.534\ldots = 3.53 \, \text{m}^2 \; (\text{correct to 3 s.f.})$$

Find the areas of the following shapes.

10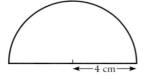

←—4 cm—→

11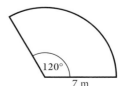

120°

7 m

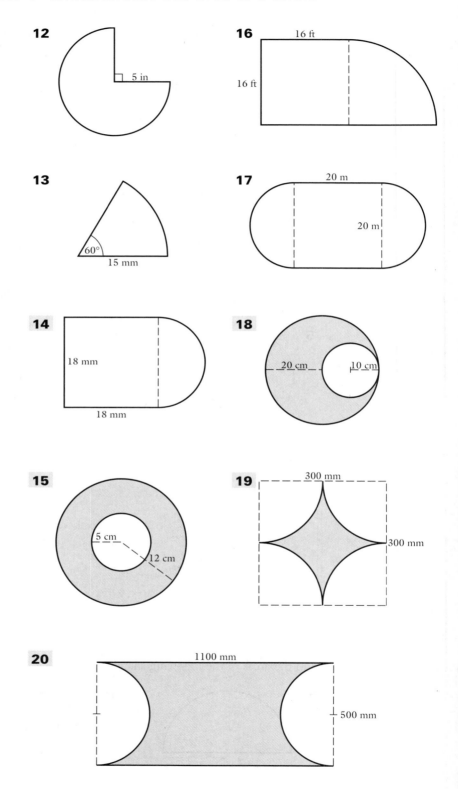

12

13

   60°

   15 mm

14

18 mm

18 mm

15

5 cm

12 cm

16

16 ft

16 ft

17

20 m

20 m

18

20 cm   10 cm

19

300 mm

300 mm

20

1100 mm

500 mm

5 in

A circular table has a radius of 75 cm. Find the area of the table top. The top of the table is to be varnished. One tin of varnish covers $4\,m^2$. Will one tin be enough to give the table top three coats of varnish?

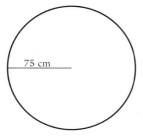

75 cm

Area of table top is πr^2
$= \pi \times 75 \times 75\,cm^2$
$= 17\,671.4\ldots cm^2$
$= 17\,670\,cm^2$ (correct to 4 s.f.)
$= 17\,670 \div 100^2\,m^2$
$= 1.767\,m^2$ (correct to 4 s.f.)

For three coats, enough varnish is needed to cover
$$3 \times 1.767\,m^2 = 5.301\ldots m^2$$
$$= 5.30\,m^2 \text{ (correct to 3 s.f.)}$$

So one tin of varnish is not enough.

21 The minute hand on a clock is 15 cm long. What area does it pass over in 1 hour?

22 What area does the minute hand described in question **21** cover in 20 minutes?

23 The diameter of a 2 p coin is 26 mm. Find the area of one of its 'flat' faces.

24 The hour hand of a clock is 10 in long. What area does it pass over in 1 hour?

25 A circular lawn has a radius of 5 m. The contents of a bottle of lawn weedkiller are sufficient to cover $50\,m^2$. Is one bottle enough to treat the whole lawn?

26 The largest possible circle is cut from a square of paper measuring 10 in by 10 in. What area of paper is left?

27 Circular place mats of diameter 8 cm are made by stamping as many circles as possible from a rectangular strip of card measuring 8 cm by 64 cm. How many mats can be made from the strip of card and what area of card is wasted?

28 A wooden counter top is a rectangle measuring 2800 mm by 450 mm. There are three circular holes in the counter, each of radius 100 mm. Find the area of the wooden top.

29 The surface of the counter top described in question **28** is to be given four coats of varnish. If one tin of varnish covers $3.5 \, \text{m}^2$, how many tins will be needed?

30 Take a cylindrical tin of food with a paper label:

Measure the diameter of the tin and use it to find the length of the label. Find the area of the label. Now find the total surface area of the tin (two circular ends and the curved surface).

MIXED EXERCISE

Give your answer to 3 significant figures.
(Remember to give an estimate for each answer first.)

EXERCISE 9E

1 Find the circumference of a circle of radius 2.8 mm.

2 Find the radius of a circle of circumference 60 m.

3 Find the circumference of a circle of diameter 12 cm.

4 Find the area of a circle of radius 2.9 m.

5 Find the diameter of a circle of circumference 280 mm.

6 Find the area of a circle of diameter 25 cm.

7 Find the perimeter of the quadrant in the diagram.

8 mm

8 Find the area of the sector in the diagram.

45°
4.5 cm

9 Find the perimeter of the sector in the diagram.

120°
10 cm

10 Find the area of the shaded part
of the diagram.

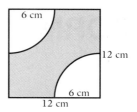

Count Buffon's experiment

Count Buffon was an eighteenth-century scientist who carried out
many probability experiments. The most famous of these is his
'Needle Problem'. He dropped needles onto a surface ruled with
parallel lines and considered the drop successful when the needle fell
across a line and unsuccessful when the needle fell between two lines.
His amazing discovery was that the number of successful drops
divided by the number of unsuccessful drops was an expression
involving π.

You can repeat his experiment and get a good approximation for the
value of π from it.

Take a matchstick or a similar small stick and measure its length.
Now take a sheet of paper measuring about 50 cm each way and fill
the sheet with a set of parallel lines whose distance apart is equal to
the length of the stick. With the sheet on the floor drop the stick on
to it from a height of about 1 m. Repeat this about a hundred times
and keep a tally of the number of times the stick touches or crosses a
line and of the number of times it is dropped. Do not count a drop if
the stick misses the paper. Then find the value of

$$\frac{2 \times \text{number of times it is dropped}}{\text{number of times it crosses or touches a line}}$$

a Can you provide Ken with the answer to the problem he posed at
the beginning of the chapter? Justify your answer.
(Assume that one pedal stroke gives one complete turn of the wheels.)

b What happens if Ken uses a gear that gives two turns of the wheels
for each pedal stroke?

c Find out how the gears on a racing bike affect the ratio of the
number of pedal strokes to the number of turns of the wheels.
Discuss the assumptions made in order to answer parts **a** and **b**.
Write a short report on how these assumptions affect the
reasonableness of your answers.

FORMULAS

A formula is a general instruction for finding one quantity in terms of other quantities. You might, for example, see this instruction for cooking a joint of beef in a microwave oven: allow 6 minutes per 400 g. If you know the weight of the joint of beef, this formula can be used to calculate the cooking time. The formula does, however, assume that you can work out how many 400 g there are in the weight of your joint.

- We could try to overcome this problem by expressing the formula more fully in words, for example,

    Find the weight of your joint in kilograms.
    Multiply this number by 1000 and divide the result by 400.
    Then multiply the answer by 6 to give the cooking time in minutes.

    This is lengthy and probably not much clearer.

- We can get over this problem by using letters for the unknown numbers, for example, t for the number of minutes and W for the number of kilograms. We can then use symbols to give the formula as

$$t = W \times 1000 \div 400 \times 6.$$

We can shorten this further by simplifying $W \times 1000 \div 400 \times 6$,

i.e. $\quad \dfrac{W}{1} \times \dfrac{1000}{1} \times \dfrac{1}{400} \times \dfrac{6}{1} = 15 \times W$.

This gives $t = 15 \times W$ or even more briefly, $t = 15W$

Provided you understand the conventions used, algebra is a powerful tool for expressing general instructions very concisely.

EXERCISE 10A

Discuss what is wrong about each of the following statements.

1 The length of the room is 2.

2 The length of the room is l metres and $l = 2$ metres.

3 There are n cans on the shelf and $n = 5$ cans.

4 The length of the room is l and $l = 2$.

5 Now write each of the statements in questions **1** to **4** so that they make sense.

**CONSTRUCTING
A FORMULA**

Discussion of the examples in **Exercise 10A** shows that when we use letters, it is important that we are clear about what the letters stand for.

It is sensible to keep to the convention of using letters for unknown *numbers*, for example,

> if there is an unknown number of cans on a shelf, we could use a for the *number*. We can then say that there are a cans on the shelf, that is, a is the number of cans.
> If we use a for the cans as well, we would have to talk about a on the shelf. This is not a clear way to describe the situation.

Similarly, if a length is unknown, that is, it is an unknown number of units, we use a letter for the unknown number only; the letter does not include the units. We can then say that the length is b cm, rather than talk about the length being b.

If we stick to this convention, any expression involving letters is a relationship between numbers only, and the ordinary rules of arithmetic apply.

The other convention we use when working with letters is that when two letters are multiplied together, or a number is multiplied by a letter, we omit the multiplication symbol,

> e.g.   we write $a \times b$ as ab,
> and $3 \times a$ as $3a$.

EXERCISE 10B

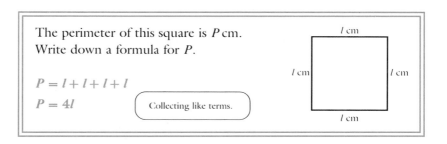

The perimeter of this square is P cm.
Write down a formula for P.

$P = l + l + l + l$

$P = 4l$ (Collecting like terms.)

In each of the following diagrams the perimeter is P cm.
Write down a formula for P starting with $P = $.

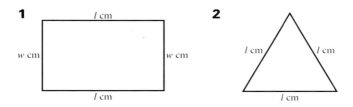

1 l cm w cm w cm l cm

2 l cm l cm l cm

3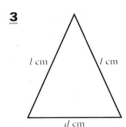

4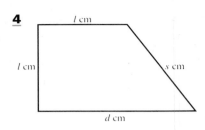

If G is the number of girls in a class and B is the number of boys, write down a formula for the total number, T, of children in the class.

$$T = G + B$$

---

5 I buy x lb of apples and y lb of pears. Write down a formula for W if W lb is the weight of fruit that I have bought.

6 If l m is the length of a rectangle and b m is the breadth, write down a formula for P if the perimeter of the rectangle is P m.

7 I start a game with N marbles and lose L marbles. Write down a formula for the number, T, of marbles that I finish with.

8 Peaches cost n pence each. Write down a formula for N if the cost of 10 peaches is N pence.

9 Oranges cost x p each and I buy n of these oranges. Write down a formula for C where C p is the total cost of the oranges.

10 I have a piece of string which is l cm long. I cut off a piece which is d cm long. Write down a formula for L if the length of string which is left is L cm.

11 A rectangle is $2b$ m long and b m wide. Write down a formula for P where P m is the perimeter of the rectangle.

12 Write down a formula for A where A m^2 is the area of the rectangle described in question **11**.

13 A lorry weighs T tonnes when empty. Steel girders weighing a total of S tonnes are loaded on to the lorry. Write down a formula for W where W tonnes is the weight of the loaded lorry.

14 A train travels p km in one direction and then it comes back q km in the opposite direction. If it is then r km from its starting point, write down a formula for r. (Assume that $p > q$.)

15 Two points have the same y-coordinate. The x-coordinate of one point is a and the x-coordinate of the other point is b. If d is the distance between the two points, write down a formula for d given that a is less than b. Draw a sketch to illustrate this problem.

16 These are the coordinates of some points on a straight line.

x-coordinate	1	2	3	4	5
y-coordinate	2	3	4	5	6

Find the formula for y in terms of x.

17 In the sequence,

$$3, \quad 5, \quad 7, \quad 9, \quad 11, \quad 13, \ldots$$

the first term is 3, the second term is 5, and so on.
If the nth term is u_n, (e.g. the third term is u_3) we can place these terms in a table, i.e.

n	1	2	3	4	5	
u_n	3	5	7	9	11	

Find a formula for u_n.

18 A letter costs x pence to post. The cost of posting 20 such letters is £q. Write down a formula for q. (Be careful – look at the units given.)

19 One grapefruit costs y pence. The cost of n such grapefruit is £L. Write down a formula for L.

20 A rectangle is l m long and b cm wide. The area is A cm^2. Write down a formula for A.

21 On my way to work this morning the train I was travelling on broke down. I spent t hours on the train and s minutes walking. Write down a formula for T if the total time that my journey took was T hours.

USING
BRACKETS

In this rectangle, the expression
for length has two terms and we
use brackets to keep them together.

$(3x - 2)$ cm

x cm

The perimeter of the rectangle is twice the length added to twice the
width,

i.e.   if P cm is the perimeter then

$$P = 2 \times (3x - 2) + 2 \times x.$$

Now $2 \times x$ can be written as $2x$, and in the same way, we can leave out
the multiplication sign in $2 \times (3x - 2)$ and write $2(3x - 2)$.

Therefore $P = 2(3x - 2) + 2x$

Now $2(3x - 2)$ means 'twice everything in the bracket',

i.e. $2(3x - 2) = 2 \times 3x - 2 \times 2$
$$= 6x - 4$$

(This process is called 'multiplying out the bracket'.)

Therefore the formula becomes $P = 6x - 4 + 2x$

i.e. $P = 8x - 4$

EXERCISE 10C

Multiply out   **a** $3(4x + 2)$ **b** $2(x - 1)$

a $3(4x + 2) = 12x + 6$ $3 \times 4x = 3 \times 4 \times x = 12x$

b $2(x - 1) = 2x - 2$

Multiply out the brackets.

1 $2(x + 1)$ **5** $2(4 + 5x)$ **9** $3(6 - 4x)$

2 $3(3x - 2)$ **6** $2(6 + 5a)$ **10** $5(x - 1)$

3 $5(x + 6)$ **7** $8(3 - 2x)$ **11** $7(2 - x)$

4 $4(3x - 3)$ **8** $4(4x - 3)$ **12** $5(a + b)$

To simplify an expression containing brackets we first multiply out the brackets and then collect like terms.

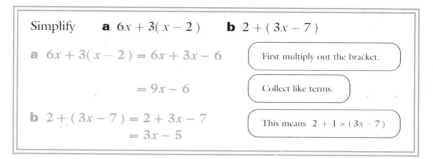

Simplify   **a** $6x + 3(x - 2)$ **b** $2 + (3x - 7)$

a $6x + 3(x - 2) = 6x + 3x - 6$ First multiply out the bracket.

$\qquad\qquad\qquad = 9x - 6$ Collect like terms.

b $2 + (3x - 7) = 2 + 3x - 7$ This means $2 + 1 \times (3x - 7)$

$\qquad\qquad\qquad = 3x - 5$

Simplify the following expressions.

13 $2x + 4(x + 1)$ **18** $3x + 3(x - 5)$

14 $3 + 5(2x + 3)$ **19** $3(x + 1) + 4$

15 $2(x + 4) + 3(x + 5)$ **20** $6(2x - 3) + 5(x - 1)$

16 $6(2x - 3) + 2x$ **21** $3x + (2x + 5)$

17 $4 + (3x - 1)$ **22** $7 + 2(2x + 5)$

Simplify   **a** $4x - 2(x + 3)$ **b** $5 - (x + 4)$

a $4x - 2(x + 3)$ means $4x$ take away 2 xs *and* 2 threes

$\qquad 4x - 2(x + 3) = 4x - 2x - 6$

$\qquad\qquad\qquad\quad = 2x - 6$

b $5 - (x + 4) = 5 - x - 4$

$\qquad\qquad\quad = 1 - x$

Simplify the following expressions.

23 $3x - 2(3x + 4)$ **26** $5x - 4(2 + x)$ **29** $40 - 2(1 + 5w)$

24 $5 - 4(5 + x)$ **27** $7a - (a + 6)$ **30** $6y - 3(3y + 4)$

25 $7c - (c + 2)$ **28** $10 - 4(3x + 2)$ **31** $8 - 3(2 + 5x)$

32 Stuart started the day with three unopened tubes of sweets. He ate all the sweets in one tube and 5 sweets from another tube. If x is the number of sweets in each unopened tube, write down an expression for the number of sweets he has   **a** eaten   **b** left.

33 The perimeter of this rectangle is $8x$ cm.
Find the formula for y in terms of x.

$(2x + 1)$ cm

y cm

MULTIPLICATION OF DIRECTED NUMBERS

From Book 7A we know that $\left.\begin{array}{c} +(+a) \\ -(-a) \end{array}\right\} = +a$ and $\left.\begin{array}{c} +(-a) \\ -(+a) \end{array}\right\} = -a$

Consider the expression $6x - (x - 3)$.

Now $6x - (x - 3)$ means '$6x$ take away x *and* take away -3',

i.e. $\qquad 6x - (x - 3) = 6x - x - (-3)$

We know that $-(-3) = +3$, so

$\qquad 6x - (x - 3) = 6x - x + 3$

Similarly $8x - 3(x - 2)$ means $8x - [3(x - 2)]$

Therefore $\qquad 8x - 3(x - 2) = 8x - [3x - 6]$

i.e. $\qquad 8x - 3(x - 2) = 8x - 3x - (-6)$

$\qquad\qquad\qquad\qquad = 8x - 3x + 6$

We can also interpret $8x - 3(x - 2)$ to mean $8x - 3 \times x - 3 \times -2$

Comparing $\qquad 8x - 3 \times x - 3 \times -2$

with $\qquad\qquad 8x \quad - \quad 3x \quad + \quad 6$

we see that $\quad -3 \times x = -3x$, i.e. $(-3) \times (+x) = -3x$

and that $\quad -3 \times -2 = +6$.

In general, $\left.\begin{array}{c} (+a) \times (+b) \\ (-a) \times (-b) \end{array}\right\} = +ab$ and $\left.\begin{array}{c} (-a) \times (+b) \\ (+a) \times (-b) \end{array}\right\} = -ab$

i.e.

> when two numbers are multiplied together, both signs the same give a positive answer, different signs give a negative answer.

EXERCISE 10D

Calculate
   a $(+3) \times (-4)$
   c $(-5) \times (-2)$
   b -3×4
   d $-5(-4)$

a $(+3) \times (-4) = -12$ Different signs give a negative answer.

b $-3 \times 4 = -12$ Remember, if a number does not have a positive or negative sign, it is positive.

c $(-5) \times (-2) = 10$ The same signs gives a positive answer.

d $-5(-4) = 20$ $-5(-4)$ means $-5 \times (-4)$.

Calculate

1 $(-3) \times (+5)$ **2** $(+4) \times (-2)$ **3** $(-7) \times (-2)$

4 $(+4) \times (+1)$ **7** $(-6) \times (+3)$ **10** $(-6) \times (-3)$

5 $(+6) \times (-7)$ **8** $(-8) \times (-2)$ **11** $(-3) \times (-9)$

6 $(-4) \times (-3)$ **9** $(+5) \times (-1)$ **12** $(-2) \times (+8)$

13 $7 \times (-5)$ **17** $-6(4)$ **21** $3(-2)$

14 $-6(-4)$ **18** $-2(-4)$ **22** 5×3

15 -3×5 **19** $-(-3)$ **23** $6 \times (-3)$

16 $5 \times (-9)$ **20** $4 \times (-2)$ **24** $-5(-4)$

25 $6 \times \left(-\frac{1}{2}\right)$ **27** $\left(+\frac{2}{3}\right) \times (+9)$ **29** $\frac{3}{4}(-4)$

26 $-3\left(+\frac{3}{4}\right)$ **28** $-\frac{3}{8} \times \frac{2}{3}$ **30** $\left(-\frac{1}{4}\right) \times \left(-\frac{2}{3}\right)$

Division is the reverse of multiplication. Therefore the rules for dividing with directed numbers are the same as those for multiplying them,

e.g. $(-4) \div (-2) = +2$ and $6 \div (-3) = -2$

Calculate

31 $-4 \div (-2)$ **33** $12 \div (-3)$ **35** $(-3.6) \div (-1.8)$

32 $-12 \div 3$ **34** $\frac{1}{2} \div \left(-\frac{1}{4}\right)$ **36** $(-1.2) \div 0.3$

EXERCISE 10E

> Multiply out $-4(3x - 4)$
>
> $-4(3x - 4) = -12x + 16$

Multiply out the following brackets.

1 $-6(x - 5)$ **6** $-7(x + 4)$

2 $-5(3c + 3)$ **7** $-3(2d - 2)$

3 $-2(5e - 3)$ **8** $-2(4 + 2x)$

4 $-(3x - 4)$ **9** $-7(2 - 3x)$

5 $-8(2 - 5x)$ **10** $-(4 - 5x)$

Simplify $4(x - 3) - 3(2 - 3x)$

$4(x - 3) - 3(2 - 3x) = 4x - 12 - 6 + 9x$ Multiply out the brackets first.

$= 13x - 18$ Collect like terms

Simplify

11 $5x + 4(5x + 3)$

16 $9 - 2(4g - 2)$

12 $42 - 3(2c + 5)$

17 $4 - (6 - x)$

13 $2m + 4(3m - 5)$

18 $10f + 3(4 - 2f)$

14 $7 - 2(3x + 2)$

19 $7 - 2(5 - 2s)$

15 $x + (5x - 4)$

20 $7x + 3(4x - 1)$

21 $7(3x + 1) - 2(2x + 4)$

26 $6x + 2(3x - 7)$

22 $5(2x - 3) - (x + 3)$

27 $20x - 4(3 + 4x)$

23 $2(4x + 3) + (x - 5)$

28 $4(x + 1) + 5(x + 3)$

24 $7(3 - x) - (6 - 2x)$

29 $3(2x - 3) - 5(x + 6)$

25 $5 + 3(4x + 1)$

30 $5(6x - 3) + (x + 4)$

31 $3x + 2(4x + 2) + 3$

36 $4(x - 1) + 5(2x + 3)$

32 $4(x - 1) - 5(2x + 3)$

37 $5 - 4(2x + 3) - 7x$

33 $7x + 8x - 2(5x + 1)$

38 $3(x + 6) - (x - 3)$

34 $4(x - 1) - 5(2x - 3)$

39 $3(x + 6) - (x + 3)$

35 $8(2x - 1) - (x + 1)$

40 $4(x - 1) + 5(2x - 3)$

The width of a rectangle is x cm and the length is 4 cm more than the width. The perimeter is P cm. Find a formula for P.

The width is x cm,
so the length is $(x + 4)$ cm.
$\therefore$ $P = 2x + 2(x + 4)$
$= 2x + 2x + 8$
i.e. $P = 4x + 8$

$(x + 4)$ cm

x cm

41

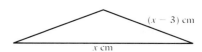

$(x - 3)$ cm

x cm

The length of the base of an isosceles triangle is x cm and the two equal sides are 3 cm shorter. The perimeter of the triangle is P cm. Find a formula for P.

42 I think of a number, x, add 5 to it and treble the result. If the result is y, find a formula for y.

43 N sweets are divided among Amy, Madge and Amir.
Amy has x sweets, Madge has 3 less than Amy and Amir has twice as many as Madge.

  a Write down an expression for the number of sweets Madge has.

  b Write down an expression for the number of sweets Amir has.

  c Write down a formula for N.

44 A group of n children are taken to the theatre. For each child, the cost of travel is x pence and the theatre ticket costs three times as much as the travel. Each child is given an ice cream which costs 20 p less than the travel cost. The total cost is £ P.
Find a formula for P.

45 The terms in a sequence are formed from the rule 'double the position number of a term and subtract 2'.
Hence the 1st term is $2(1) - 2 = 0$,
the 2nd term is $2(2) - 2 = 2$, and so on.

  a Write down the first six terms of the sequence.

  b If n is the position number of a term, find a formula for u_n, where u_n is the nth term of the sequence.

INDICES

Using index notation, we know that 2×2 can be written as 2^2. We can also use index notation with letters that stand for unknown numbers,

e.g. we can write $a \times a$ as a^2,
and $b \times b \times b$ can be written as b^3.

In the same way, d^5 means $d \times d \times d \times d \times d$
and $3b^4$ means $3 \times b \times b \times b \times b$

Notice that the index applies only to the number it is attached to:
$2x^2$ means that x is squared, 2 is not.

If we want to square 2, we write $(2x)^2$ or $2^2 x^2$.

Simplify $3a \times 2a$

$$3a \times 2a = 3 \times a \times 2 \times a$$
$$= 6 \times a \times a$$
$$= 6a^2$$

> Remember that numbers can be multiplied in any order,
> i.e. $3 \times a \times 2 \times a = 3 \times 2 \times a \times a$
> $= 6 \times a^2$

Write in index notation

1 $x \times x$

2 $p \times p \times p$

3 $s \times s \times s$

4 $t \times t \times t \times t$

5 $2a \times 4a$

6 $5d \times 3d$

7 $4x \times 3x$

8 $3y \times y$

9 $a \times 2a \times 3a$

10 $4n \times n \times 2n$

11 $2p \times 2p \times 3p$

12 $5s \times 3s \times 4$

Simplify $2a \times 3b \times a$

$$2a \times 3b \times a = 2a \times a \times 3b$$
$$= 2a^2 \times 3b$$
$$= 6a^2 b$$

> $2a^2 \times 3b = 2 \times a^2 \times 3 \times b$
> $= 2 \times 3 \times a^2 \times b$

Simplify

13 $2b \times 3a \times 4b$

14 $4x \times 2x \times 3y$

15 $p \times 2q \times p$

16 $3s \times r \times s$

17 $2x \times y \times y$

18 $4t \times 2t \times s$

19 $2p \times 2r \times p$

20 $4a \times 3b \times 2b$

21 $4x \times y \times 2y$

22 $s \times t \times 2s$

23 $5r \times r \times 2s$

24 $3w \times 2u \times 5w$

A rectangle is $3a$ cm long and a cm wide. The area is A cm^2.
Find a formula for A.

$3a$ cm
a cm

$$A = 3a \times a$$
$$A = 3a^2$$

> The area of a rectangle
> = length × width

The area of each rectangle is A cm^2. Find a formula for A.

25
$5a$ cm
$2a$ cm

26
$7t$ cm
t cm

27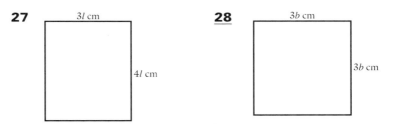

3*l* cm

4*l* cm

28

3*b* cm

3*b* cm

The volume of each cuboid is $V\,\text{cm}^3$. Find a formula for V.

29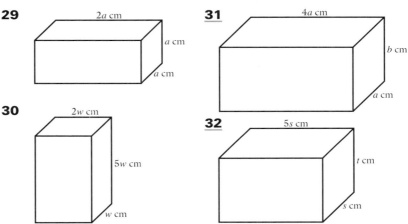

2*a* cm

a cm

a cm

31

4*a* cm

b cm

a cm

30

2*w* cm

5*w* cm

w cm

32

5*s* cm

t cm

s cm

SUBSTITUTING NUMBERS INTO FORMULAS

The diagram shows a pattern made with floor tiles, each of which is 3 times as long as it is wide, laid round a central square.

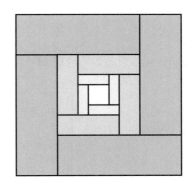

If l cm is the width of any one of these tiles, then its length is $3l$ cm and its area, $A\,\text{cm}^2$, is given by the formula

$$A = 3l^2$$

If the width of the smallest of these rectangles is 5 cm, we can find its area directly from the formula by substituting 5 for l,

i.e.   when $l = 5$, $A = 3 \times 5^2$
$$= 3 \times 25 = 75$$

Therefore the area of the smallest rectangle is $75\,\text{cm}^2$.

We can use this formula to find the area of any rectangle in the pattern if we know its width.

> If $v = u + at$, find v when $u = 2$, $a = \frac{1}{2}$ and $t = 4$.
>
> $v = u + at$
>
> When $u = 2$, $a = \frac{1}{2}$ and $t = 4$,
>
> $v = 2 + \frac{1}{2} \times 4$
>
> $\quad = 2 + 2 = 4$
>
> Remember, multiplication and division before addition and subtraction.

1 If $N = T + G$, find N when $T = 4$ and $G = 6$.

2 If $T = np$, find T when $n = 20$ and $p = 5$.

3 If $P = 2(l + b)$, find P when $l = 6$ and $b = 9$.

4 If $L = x - y$, find L when $x = 8$ and $y = 6$.

5 If $N = 4(I - s)$, find N when $I = 7$ and $s = 2$.

6 If $S = n(a + b)$, find S when $n = 20$, $a = 2$ and $b = 8$.

7 If $V = lbw$, find V when $l = 4$, $b = 3$ and $w = 2$.

8 If $A = \dfrac{PRT}{100}$, find A when $P = 100$, $R = 3$ and $T = 5$.

9 If $w = u(v - t)$, find w when $u = 5$, $v = 7$ and $t = 2$.

10 If $s = \frac{1}{2}(a + b + c)$, find s when $a = 5$, $b = 7$ and $c = 3$.

> If $v = u - at$, find v when $u = 5$, $a = -2$, $t = -3$.
>
> When $u = 5$, $a = -2$ $t = -3$,
>
> Notice that where negative numbers are substituted for letters they have been put in brackets. This makes sure that only one operation at a time is carried out.
>
> $v = u - at$
>
> $v = 5 - (-2) \times (-3)$
>
> $\quad = 5 - (+6)$
>
> $\quad = 5 - 6$
>
> $\quad = -1$

11 If $N = p + q$, find N when $p = 4$ and $q = -5$.

12 If $C = RT$, find C when $R = 4$ and $T = -3$.

13 If $z = w + x - y$, find z when $w = 4$, $x = -3$ and $y = -4$.

14 If $r = u(v - w)$, find r when $u = -3$, $v = -6$ and $w = 5$.

15 Given that $X = 5(T - R)$, find X when $T = 4$ and $R = -6$.

16 Given that $P = d - rt$, find P when $d = 3$, $r = -8$ and $t = 2$.

17 Given that $v = l(a + n)$, find v when $l = -8$, $a = 4$ and $n = -6$.

18 If $D = \dfrac{a - b}{c}$, find D when $a = -4$, $b = -8$ and $c = 2$.

19 If $Q = abc$, find Q when $a = 3$, $b = -7$ and $c = -5$.

20 If $I = \frac{2}{3}(x + y - z)$, find I when $x = 4$, $y = -5$ and $z = -6$.

21 Oranges cost n pence each and a box of 50 of these oranges costs C pence.

  a Write down a formula for C.

  b Use your formula to find the cost of a box of oranges if each orange costs $12\,$p.

22 Lemons cost n pence each and a box of 50 lemons costs £L.

  a Write down a formula for L (be careful with the units).

  b Use your formula to find the cost of a box of these lemons when they cost $10\,$p each.

23 Given $N = 2(n - m)$, find **a** N when $n = 6$ and $m = 4$

                                  b N when $n = 7$ and $m = -3$

                                    c n when $N = 12$ and $m = -4$.

24 If $z = x - 3y$, find **a** z when $x = 3\frac{1}{2}$ and $y = \frac{3}{4}$

                             b z when $x = \frac{3}{8}$ and $y = -1\frac{1}{2}$

                             c x when $z = \frac{1}{4}$ and $y = \frac{7}{8}$

25 If $P = 10r - t$, find **a** P when $r = 0.25$ and $t = 10$

                             b P when $r = 0.145$ and $t = 15.6$

                             c t when $P = 18.5$ and $r = 0.026$

26 A rectangular box is $3l\,$cm long, $2l\,$cm wide and $l\,$cm deep. The volume of the box is $V\,$cm^3.

  a Write down a formula for V.

  b Use your formula to find the volume of a box measuring $5\,$cm deep.

27 A rectangle is a cm long and $(2a - 10)$ cm wide.

    a Write down a formula for P if P cm is the perimeter of the rectangle.

    b Use your formula to find the perimeter of a rectangle 6.9 cm wide.

    c What is the smallest integer that a can represent for the value of P to make sense?

28 The length of a rectangle is twice its width. If the rectangle is x cm wide, write down a formula for P if its perimeter is P cm. Use your formula to find the width of a rectangle that has a perimeter of 36 cm.

29 A roll of paper is L m long. N pieces each of length r m are cut off the roll. If the length of paper left is P m, write down a formula for P. A roll of paper 20 m long had 10 pieces, each of length 1.5 m cut from it. Use your formula to find the length of paper left.

FINDING A FORMULA FOR THE nTH TERM OF A SEQUENCE

In the sequence 1, 4, 9, 16, 25,...
the 1st term is 1, the 2nd term is 4, the 3rd term is 9, and so on.
If we use n for the position number of a term and u_n for that term, we can arrange the terms in a table,

i.e.

n	1	2	3	4	5	
u_n	1	4	9	16	25	

Now we can see that the nth term, u_n, is equal to n^2.
Therefore the formula for the nth term of this sequence is $u_n = n^2$

EXERCISE 10H

Find a formula for u_n.

1

n	1	2	3	4	5	
u_n	2	3	4	5	6	

3

n	1	2	3	4	5	
u_n	2	4	6	8	10	

2

n	1	2	3	4	5	
u_n	4	5	6	7	8	

4

n	1	2	3	4	5	
u_n	0	1	2	3	4	

5 For each sequence given in question **1** to **4**, write down, in words, the rule for generating the sequence.

6 The terms of a sequence are generated by starting with 2 and adding 3 each time.

    a Write the first five terms in a table like those given in questions **1** to **4**.

    b Find a formula for u_n in terms of n.

Questions **7** to **10** describe how the terms of a sequence are generated. For each question repeat parts **a** and **b** of question **6**.

    7 Double the position number and add one.

    8 Subtract 5 from the position number.

    9 Multiply the position number by itself and subtract one.

  10 Add the position number to the next position number.

  11 Look at this pattern of squares.

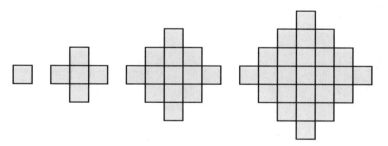

    a Copy and continue this table.

nth pattern	1	2	3	4	5	
Number of squares	1	5				

    b If N is the number of squares in the 10th pattern, find N.

  12 Repeat question **11** for the number of sides in these patterns.

The formula for the nth term of a sequence is given by $u_n = \dfrac{n+1}{2n}$

a Copy and continue this table, giving values of u_n correct to
5 decimal places where necessary.

n	1	2	3	4	5	
u_n	1	0.75	0.666 67			

b What seems to be happening to the values of the terms as the value
of n gets larger?

c To be more certain about your answer to part **b**, you need to find
many more terms. This can be done easily using a spreadsheet
program on a computer. An empty spreadsheet looks like this:

	A	B	C	D	E	F
1						
2						
3						
4						
5						
6						
7						
8						
9						
10						
11						
12						

We can enter numbers and words into each cell (rectangle) and we
can also enter formulas, which the spreadsheet then calculates with
numbers. The letters and the numbers at the edge are used to
identify the position of a cell.

In cell A1, we enter n.

In cell A2, we enter 1, in A3 we enter 2, and so on.

In cell B1, we enter nth term $= (n+1)/2n$.

In B2, we enter $(A2 + 1)/(2*A2)$. (The symbol * means
multiply.) This is the formula for the nth term, telling the
spreadsheet to use the number in cell A2 to calculate the value of
$(n+1)/2n$.

Using the appropriate command (usually called FILL), you can
get the spreadsheet to repeat this formula down the B column,
automatically replacing A2 with A3, A3 with A4, and so on.
The screen will then look something like the diagram opposite.

	A	B	C
1	n	n th term $= (n+1)/2n$	
2	1		1
3	2		0.75
4	3		0.66666666666667
5	4		0.625
6	5		0.6
7	6		0.58333333333333
8	7		0.57142857142857
9	8		0.5625
10	9		0.55555555555556
11	10		0.55
12	11		0.54545454545455
13	12		0.54166666666667
14	13		0.53846153846154
15	14		0.53571428571429
16	15		0.53333333333333

Now we get a clearer picture of what is happening to the terms; to get an even clearer picture we can reproduce the columns further down the spreadsheet.

Use a spreadsheet to investigate what happens to the terms in these sequences as n gets larger   **i** $u_n = \dfrac{2n - 1}{10n}$ **ii** $u_n = \dfrac{2n^2 + 3}{n^2}$

INVESTIGATION

This is the pattern from page 195. It is formed round a central square from rectangles each of which is 3 times as long as it is wide.

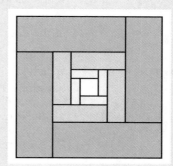

a The width of the smallest rectangle is a cm. Write down a formula for A, where A cm^2 is its area.

b Find a formula for B, where B cm^2 is the area of each rectangle in the next ring of 4 rectangles.

c Find a formula for C, where C cm^2 is the area of each rectangle is the third, outer ring of the pattern.

d The pattern is continued by adding further rings of rectangles.

Find a formula for N, where N cm^2 is the area of a rectangle in the n th ring.

REFLECTIONS, TRANSLATIONS AND ROTATIONS

Keith has bought a new panelled hardwood door for the lounge. He rests it against the wall ready to hang it. A large mirror is fixed to a perpendicular wall.

The image of the door in the mirror is a reflection.

When he moves the door along the wall and fits it, the door is being translated.

When the door is fixed it rotates about a vertical line through the hinges. In all three cases: reflection, translation and rotation, the size and shape of the door remain unchanged. All that changes is its position.

EXERCISE 11A   Which kind of movement is suggested by each of the descriptions in questions **1** to **7** – a reflection, a translation or a rotation?

1 A car travelling along a straight road.

2 A car travelling round a roundabout.

3 Pushing a coin across the counter to pay for a magazine.

4 Turning a page of this book.

5 Looking at yourself in the bathroom mirror.

6 Can you think of everyday situations/happenings that can be described as   **a** a reflection   **b** a translation   **c** a rotation? Try to think of at least three of each type.

REFLECTIONS

Consider a piece of paper, with a drawing on it, lying on a table. Stand a mirror upright on the paper and the reflection can be seen as in the picture.

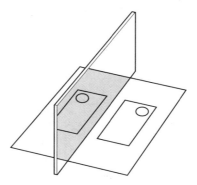

If we did not know about such things as mirrors, we might imagine that there were two pieces of paper lying on the table like this:

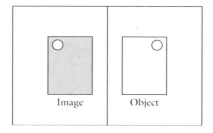

Image   Object

The *object* and the *image* together form a symmetrical shape and the *mirror line* is the axis of symmetry.

EXERCISE 11B

In this exercise it may be helpful to use a small rectangular mirror, or you can use tracing paper to trace the object and turn the tracing paper over, to find the shape of the image.

Copy the objects and mirror lines (indicated by dotted lines) onto squared paper and draw the image of each object.

1
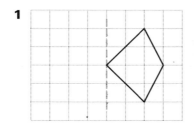

2
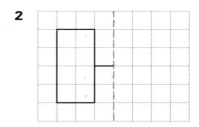

3

5

4

6

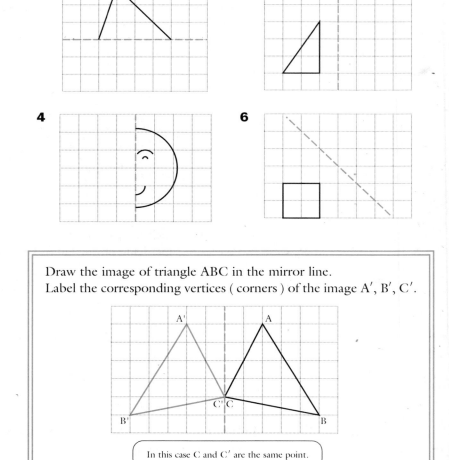

Draw the image of triangle ABC in the mirror line.
Label the corresponding vertices (corners) of the image A', B', C'.

In this case C and C' are the same point.

In questions **7** to **14** copy the object and the mirror line onto squared paper. Draw the image.

Label the vertices of the object A, B, C, etc. and label the corresponding vertices of the image A', B', C', etc.

7

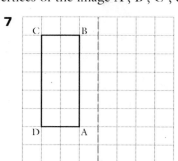

8

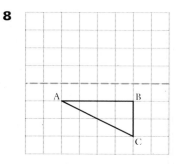

In mathematical reflection, though not in real life, the object can cross the mirror line, e.g.

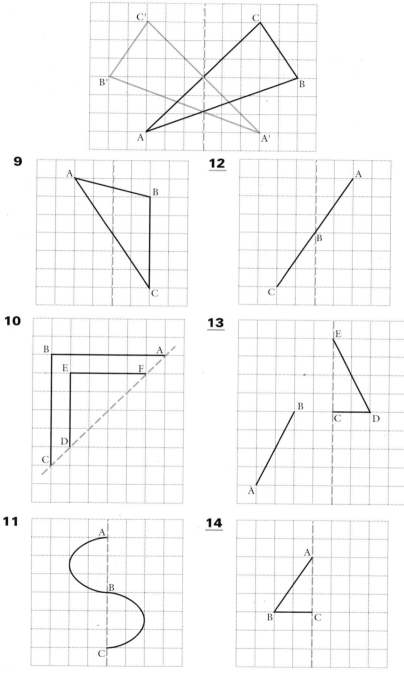

9

10

11

12

13

14

15 Which points in questions **7** to **14** are labelled with two letters?
What is special about their positions?

16 In the diagram for question **9**, join A and A′.

    a Measure the distances of A and A′ from the mirror line. What do you notice?

    b At what angle does the line AA′ cut the mirror line?

17 Repeat question **16** on other suitable diagrams, in each case joining each object point to its image point. What conclusions do your draw?

In questions **18** and **19** use 1 cm to 1 unit.

18 Draw axes, for x from -5 to 5 and for y from 0 to 5. Draw triangle ABC by plotting A(1, 2), B(3, 2) and C(3, 5). Draw the image triangle A′B′C′ when triangle ABC is reflected in the y-axis.

19 Draw axes, for x from 0 to 5 and for y from -2 to 2. Draw triangle PQR where P is (1, -1), Q is (5, -1) and R is (4, 0). Draw the image triangle P′Q′R′ when triangle PQR is reflected in the x-axis.

20 Draw $\triangle ABC$ where $\widehat{B} = 90°$.
Stand a mirror along each edge in turn and draw the shape formed by the original triangle and its image.

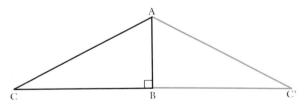

Describe each shape as accurately as you can.

For example, if the mirror is placed along the edge AB the resulting shape is an isosceles triangle where $AC = AC'$ and A is vertically above B, the midpoint of CC'.

21 a Repeat question **20** for
    i an isosceles triangle where $\widehat{B} = 90°$
    ii an isosceles triangle where $\widehat{B}$ is less than $90°$.
    iii an equilateral triangle.

    b Draw a rectangle ABCD in which $AB = 2AD$. What shape is formed by the original shape and its image if a mirror is placed along AB?

In questions **22** and **23** draw axes for x and y from -5 to 5.

22 Draw square PQRS: P(1, 1), Q(4, 1), R(4, 4), S(1, 4). Draw square P′Q′R′S′: P′(−2, 1), Q′(−5, 1), R′(−5, 4), S′(−2, 4). Draw the mirror line so that P′Q′R′S′ is the reflection of PQRS and describe the mirror line.

23 Draw lines AB and PQ: A(2, −1), B(4, 4), P(−2, −1), Q(−5, 4). Is PQ a reflection AB ? If it is, draw the mirror line. If not, give a reason.

TRANSLATIONS

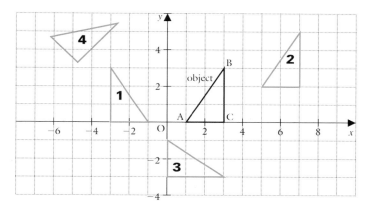

Imagine a triangle ABC cut out of card and lying in the position shown. We can reflect △ABC in the y-axis by picking up the card, turning it over and putting it down again in position 1.

Starting again from its original position, we can change its position by sliding the card over the surface of the paper to position 2, 3 or 4. Some of these movements can be described in a simple way, others are more complicated.

Consider the movements in this diagram.

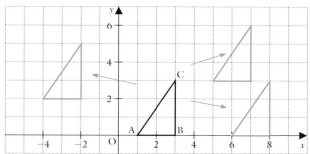

All these movements are of the same type. The side AB remains parallel to the x-axis in each case and the triangle continues to face in the same direction. This type of movement is called a *translation*.

Although not a reflection we still use the words *object* and *image*.

EXERCISE 11C

1 In the following diagram, which image of △ABC is given by a translation?

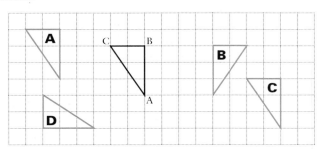

2 In the following diagram, which images of △ABC are given by a translation, which by a reflection and which by neither?

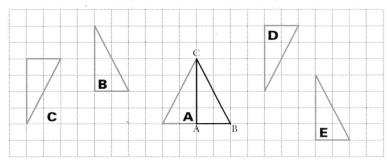

3 Repeat question **2** with the first diagram on page 207.

Draw sketches to illustrate the following translations.

4 An object is translated 6 cm to the left.

5 An object is translated 4 units parallel to the *x*-axis, to the right.

6 An object is translated 3 m due north.

7 An object is translated 5 km south-east.

8 An object is translated 3 units parallel to the *x*-axis to the right and then 4 units parallel to the *y*-axis upwards.

9

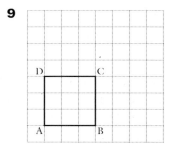

a Square ABCD is translated parallel to AB a distance equal to AB. Sketch the diagram and draw the image of ABCD. Mark it P.

b Square ABCD is translated parallel to AC a distance equal to AC. Sketch the diagram and draw the image of ABCD. Mark it Q.

10 Copy the diagram of rectangle ABCD onto squared paper.
A translation maps ABCD to the image A′B′C′D′. In each of the
following cases sketch A′B′C′D′ and give the coordinates of A′.

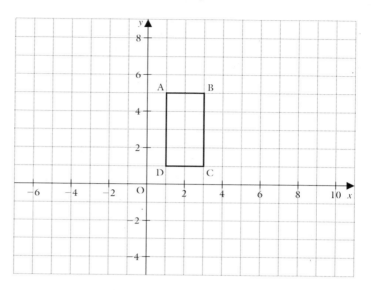

a The translation is 6 units to the right followed by 3 units up.
Mark the image P.

b The translation is 6 units to the left followed by 3 units down.
Mark the image Q.

c The translation is 5 units down followed by 3 units to the left.
Mark the image R.

d The translation is 2 units up followed by 6 units to the left.
Mark the image S.

e What translation will map   **i** Q to R   **ii** P to Q?

11 Draw axes for x and y from 0 to 9.
Draw △ABC with A(3, 0), B(3, 3), C(0, 3)
and △A′B′C′ with A′(8, 2), B′(8, 5), C′(5, 5).

Is △A′B′C′ the image of △ABC under a translation? If so,
describe the translation using the method given in the previous
question.

Join AA′, BB′ and CC′.
What type of quadrilateral is AA′B′B?
Give reasons for your answer.

Name other quadrilaterals of the same type in the figure.

**ORDER OF
ROTATIONAL
SYMMETRY**

A shape has rotational symmetry when it can be rotated, or turned, about a centre point, called the *centre of rotation*, and still look the same.

If a shape needs to be turned through a third of a complete turn to look the same, then it will need two more such turns to return to its original position. Therefore, starting from its original position, it takes three turns, each one-third of a revolution, to return to its starting point.

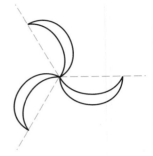

It has *rotational symmetry of order 3.*

EXERCISE 11D

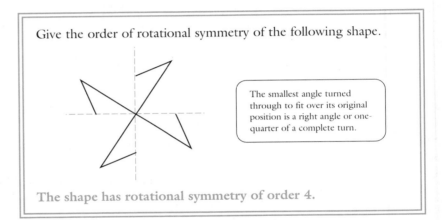

Give the order of rotational symmetry of the following shape.

The smallest angle turned through to fit over its original position is a right angle or one-quarter of a complete turn.

The shape has rotational symmetry of order 4.

1   Give the order of rotational symmetry for each shape.

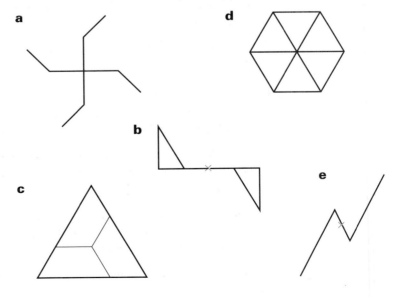

Copy and complete the diagram, given that there is rotational symmetry of order 4.

You may find it helpful to trace the shape, then turn the tracing through 90° about the point marked with a cross (place the point of your compasses or a pencil in the centre) and then repeat twice more.

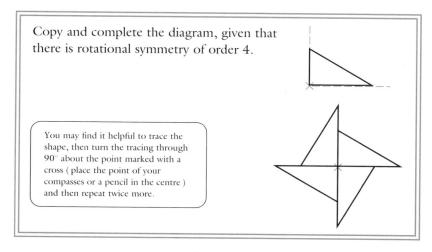

Each of the diagrams in questions **2** to **7** has rotational symmetry of the order given and ✕ marks the centre of rotation.
Copy and complete the diagrams. (Tracing paper may be helpful.)

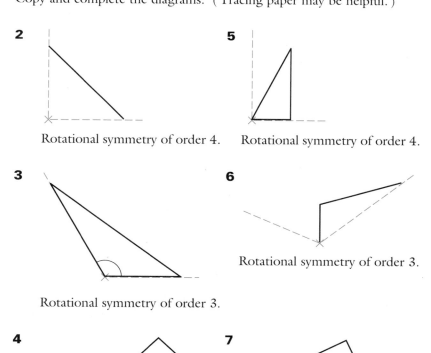

2

Rotational symmetry of order 4.

5

Rotational symmetry of order 4.

3

Rotational symmetry of order 3.

6

Rotational symmetry of order 3.

4

Rotational symmetry of order 2.

7

Rotational symmetry of order 2.

8 In questions **2** to **7**, give the size of the angle, in degrees, through which each shape is turned.

EXERCISE 11E

Some shapes have both line symmetry and rotational symmetry:

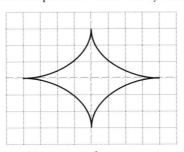

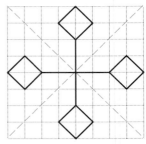

Two axes of symmetry
Rotational symmetry order 2

Four axes of symmetry
Rotational symmetry order 4

Which of the following shapes have

a rotational symmetry only   **b** line symmetry only   **c** both?

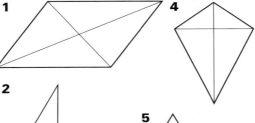

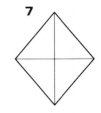

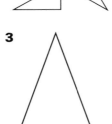

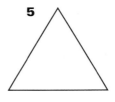

10 The capital letter **X** has line symmetry (two axes) and rotational symmetry (of order 2). Investigate the other letters of the alphabet.

11 Make up three shapes which have rotational symmetry only. Give the order of symmetry and the angle of turn, in degrees.

12 Make up three shapes with line symmetry only. Give the number of axes of symmetry.

13 Make up three shapes which have both line symmetry and rotational symmetry.

ROTATIONS

A 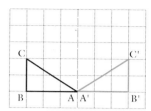   **B** 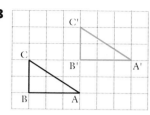   **C**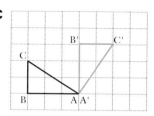

So far, in transforming an object we have used reflection, as in **A**, and translation, as in **B**, but for **C** we need a rotation.

In this case we are rotating △ABC about A through 90° clockwise (↻). We could also say △ABC was rotated through 270° anticlockwise (↺).

The point A is called the *centre of rotation*.

For a rotation of 180° we do not need to say whether it is clockwise or anticlockwise.

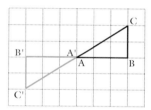

EXERCISE 11F

Give the angle of rotation when △ABC is mapped to △A′B′C′.

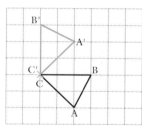

The angle of rotation is 90° anticlockwise.

In questions **1** to **4** give the angle of rotation when △ABC is mapped to △A′B′C′. If the angle is not obvious you may find using tracing paper helps.

1 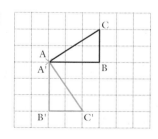   **2**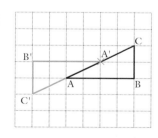

3 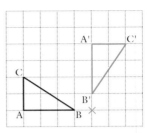   **4**

In questions **5** to **10** state the centre of rotation and the angle of rotation. △ABC is the object in each case.

5 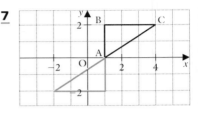   **7**

6 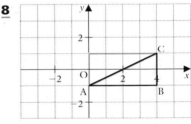   **8**

9 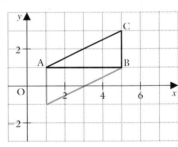   **10**

Copy the diagrams in questions **11** to **18** using 1 cm to 1 unit. Draw the images of the given objects under the rotations described. Using tracing paper may help.

11

Centre of rotation (0, 0)
Angle of rotation 90° anticlockwise.

12

Centre of rotation (−1, 0)
Angle of rotation 180°.

13

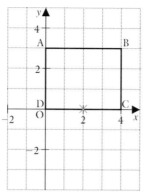

Centre of rotation (2, 0)
Angle of rotation 90°
anticlockwise

16

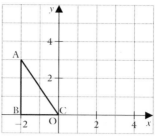

Centre of rotation (2, 0)
Angle of rotation
90° clockwise.

14

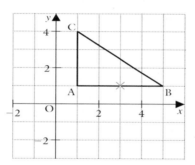

Centre of rotation (3, 1)
Angle of rotation 180°.

17

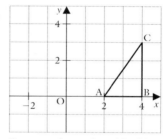

Centre of rotation (0, 0)
Angle of rotation 90°
anticlockwise.

15

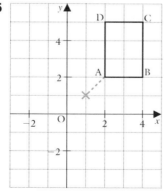

Centre of rotation (1, 1)
Angle of rotation 180°.
(As the centre of rotation
is not a point on the
object, join it to A first.)

18

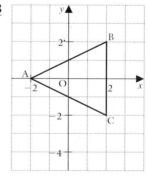

Centre of rotation (0, 0)
Angle of rotation 180°.

19 △ABC is rotated about O through 180° to give the image, △A′B′C′. Copy and complete the diagram, using 1 cm to 1 unit.

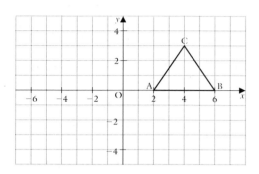

a What is the shape of the path traced out by C as it moves to C′?

b Measure OC and OC′. How do they compare?
Repeat with OB and OB′.

20

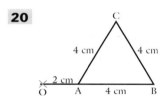

Draw the diagram accurately.
Then draw accurately, using a protractor, the image of △ABC under a rotation of 60° anticlockwise about O.

MIXED EXERCISE

EXERCISE 11G

1 Copy the objects and mirror lines on to squared paper, and draw the reflection of each object.

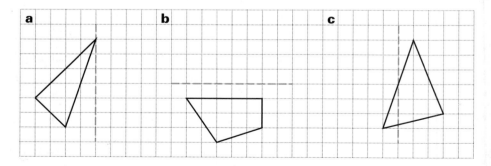

2 Draw axes, for *x* from −8 to 8 and for *y* from 0 to 8. Draw △ABC by plotting A(2, 1), B(5, 7) and C(7, 3). Draw the image A′B′C′ when ABC is reflected in the *y*-axis.

3 In the diagram, which images of △LMN are given by

　a a translation　　**b** a reflection　　**c** a rotation?

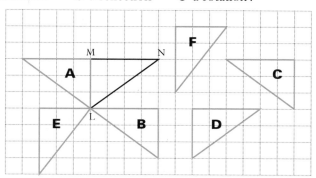

4 Which of these shapes have

　a rotational symmetry　　**b** line symmetry　　**c** neither?

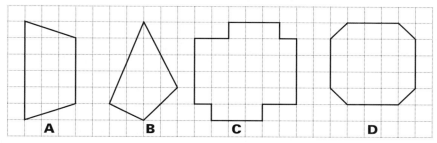

5 Give the angle of rotation
when △ABC is mapped to △A′B′C′.

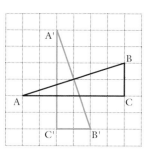

6 Copy the diagram and draw the image of rectangle ABCD if it is
rotated 90° anticlockwise about the point (2, 1).

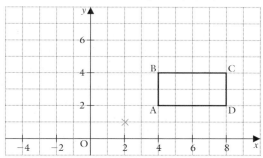

In questions **7** and **8** describe fully the transformation that maps the black shape into the coloured image.

7

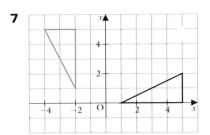

8

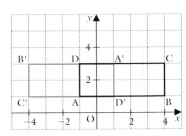

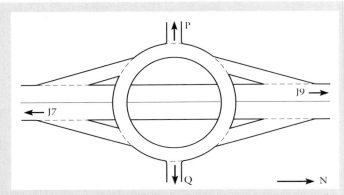

The sketch, which is not drawn to scale, shows Junction 8 on a British motorway where traffic drives on the left. A roundabout above the motorway enables traffic to enter and leave it.

a Make an accurate copy of the sketch, using tracing paper if necessary. Does this layout have
  i line symmetry
  ii rotational symmetry?
  Justify your answers by drawing any line of symmetry and giving any centre of rotation. If there is rotational symmetry give the order.

b Bianca approaches the junction from Q and wishes to use the motorway to get to Junction 9. Show her route using a solid line. Mark it B. (Think carefully about which way we drive round roundabouts in the UK.)

c Harlan approaches Junction 8 from Junction 7 and wishes to drive to P. Show his route with a dashed line. Mark it H.

d When the junction was built the designer claimed that it could easily be converted so that traffic could drive on the right hand side of the road as they do on the continent. Make another copy of the junction and repeat parts **b** and **c** to show how traffic would flow if it changed from driving on one side of the road to driving on the other side.

e

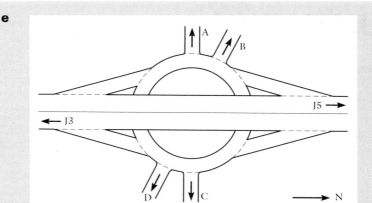

This diagram shows Junction 4 on the same motorway. Make an accurate copy and use it to show

  i the route of a motorist approaching the junction from B who wants to drive south on the motorway

  ii the route of a motorist approaching the junction from C who wants to drive to A.

f Repeat part **e** if the road system is changed so that traffic drives on the right instead of on the left.

SUMMARY 3

SHAPES

Polygons

A polygon is a plane figure bounded by straight lines, e.g.

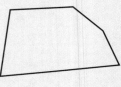

A *regular polygon* has all angles equal and all sides the same length.
This is a regular hexagon.

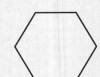

The *sum of the exterior angles* of any polygon is 360°.

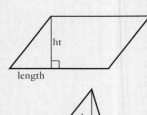

The *sum of the interior angles* of a polygon depends on the number of sides.
For a polygon with *n* sides, this sum is $(180n - 360)°$ or $(2n - 4)$ right angles.

PARALLELOGRAMS AND TRIANGLES

The *area* of a parallelogram is given by
length × height.

The area of a triangle is given by
$\frac{1}{2}$ base × height.

When we talk about the height of a triangle or of a parallelogram, we mean the perpendicular height.

CIRCLES

The names of the parts of a circle are shown in the diagram.
The diameter of a circle is twice the radius.

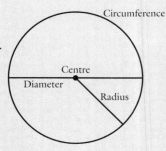

The length of the *circumference* is given by $C = 2\pi r$,
where *r* units is the radius of the circle and $\pi = 3.1415\ldots$

The area of a circle is given by $A = \pi r^2$

SQUARE ROOTS

$\sqrt{20}$ means the square root of 20, and is that number which, when multiplied by itself, gives 20.

For example, $\sqrt{4} = 2$ because $2 \times 2 = 4$.

DIRECTED NUMBERS

When two numbers of the same sign are multiplied together, the result is positive.

When two numbers of different signs are multiplied together, the result is negative.

For example, $(+2) \times (+3) = +6$ and $(-2) \times (-3) = +6$

$(+2) \times (-3) = -6$ and $(-2) \times (+3) = -6$

The same rules apply to division, for example,

$(+8) \div (+2) = +4$ and $(-8) \div (-2) = +4$

$(+8) \div (-2) = -4$ and $(-8) \div (+2) = -4$

STATISTICS

A *hypothesis* is a statement that has not been shown to be true or untrue.

Scatter graphs
We get a scatter graph when we plot points showing values of one quantity against corresponding values of another quantity.

When the points are scattered about a straight line, we can draw that line by eye; the line is called the *line of best fit*.

We use the word *correlation* to describe the amount of scatter about this line.

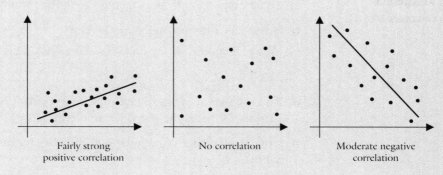

Fairly strong positive correlation

No correlation

Moderate negative correlation

TRANSFORMATIONS

Reflection in a mirror line

When an object is reflected in a mirror line, the object and its image form a symmetrical shape with the mirror line as the axis of symmetry.

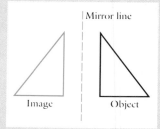

Translation

When an object is translated it moves without being turned or reflected to form an image.

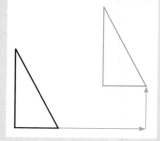

Rotation

When an object is rotated about a point to form an image, the point about which it is rotated is called the *centre of rotation* and the angle it is turned through is called the *angle of rotation*.

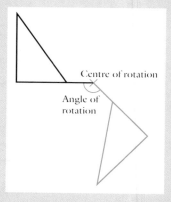

REVISION EXERCISE 3.1
(Chapters 6 to 8)

1 a How many sides has

    i a hexagon   **ii** an octagon   **iii** a rhombus?

  b How many interior angles are there in

    i a pentagon   **ii** a parallelogram   **iii** a regular hexagon?

2 a Find the size of each exterior angle of a regular polygon with 15 sides. What is the size of each interior angle?

  b How many sides has a regular polygon if each exterior angle is $12°$?

3 Find the size of each angle marked with a letter.

a 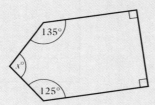 **b**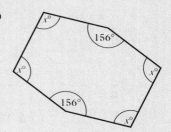

4 Find the area of each shape.

a 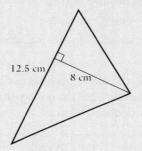 **b**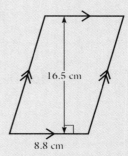

5 a The area of a rectangular washer is $8.5\,\text{cm}^2$. If it is $3.4\,\text{cm}$ long, how wide is it?

   b The area of a triangular-shaped gasket is $14.88\,\text{cm}^2$. If the height of the triangle is $4.8\,\text{cm}$, find the length of its base.

6 Use squared paper to draw axes for x and y from 0 to 8 using 1 square for 1 unit.
Find the area of triangle ABC as a number of square units if the coordinates of the three corners are A(2, 4), B(6, 4) and C(2, 1).

7 Find the area of each shape in square centimetres. All measurements are given in centimetres.

a 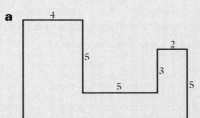 **b**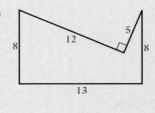

8 On squared paper draw x- and y-axes and scale each axis from -5 to 5. Use 1 square as 1 unit on both axes. Plot the points A(-3, 1), B(3, 1), C(1, -4), D(-5, -4) and draw the parallelogram ABCD. Find the area of ABCD in square units.

9

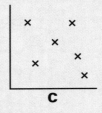

A   B   C   D

a Which of the above scatter graphs shows
 i strong negative correlation
 ii moderate positive correlation
 iii no correlation?

b Describe the correlation illustrated in the remaining graph.

10 The table shows the average number of hours per week spent in training by six boys selected as potentially good sprinters, and the best times they recorded for 100 m during the last weeks of the season.

Sprinter	Ed	Brendan	Hank	Pete	Roger	Colin
Average number of hours per week	9	3	8	15	4	10
Best time (seconds)	11.2	11.8	11.5	10.9	12.1	11.4

a Show this information on a scatter graph. Use a horizontal scale of 8 cm for 1 second, and mark this axis from 10.5 sec to 12.5 sec.
Use 1 cm for 1 hour on the vertical axis and mark it from 0 to 16.

b Does this scatter graph suggest that increasing the number of hours spent in training could well result in better times? How would you describe the correlation between these two quantities?

c George wants to join the group and is potentially as good as anyone already included. If he trains for 12 hours a week about what 'best time' can he expect?

REVISION
EXERCISE 3.2
(Chapters 9 to 11)

1 **a** Find the circumference of a circle of radius 3.4 cm.

b Find the area of a circle whose diameter is 1.2 m.

c **i** Find the perimeter of the quadrant given in the diagram.

 ii What is the area of this quadrant?

5 cm

2 The largest possible circle is cut from a square of paper of side 12 cm.

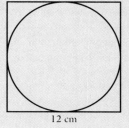

12 cm

   a Write down the diameter of the circle and find

     i its circumference   **ii** its area.

   b What is the area of the original sheet?

   c What area of paper is wasted?

   d Express the amount of paper wasted as a percentage of the original sheet.

3 Calculate   **a** $(-4) \times (+6)$ **c** $3(-3)$

               b $(-3) \times (-2)$ **d** $-4(-5)$

4 Multiply out the brackets and simplify

   a $5(x-3)$ **d** $7x - 3(2x+1)$

   b $4x + 2(3x-1)$ **e** $-4(a-2)$

   c $9 - 2(x-7)$ **f** $3(x-2) - 4(x+3)$

5 **a** A rectangle is l cm long and b cm wide. Write down a formula for P where P cm is the perimeter of the rectangle.

   b

n	1	2	3	4	5
u_n	4	5	6	7	8

   Find a formula for u_n in terms of n.

6 **a** If $s = \frac{1}{2}(a+b+c)$ find s when $a = 6$, $b = 7$ and $c = 9$.

   b If $N = abc$ find N when $a = 5$, $b = -2$ and $c = -6$.

7

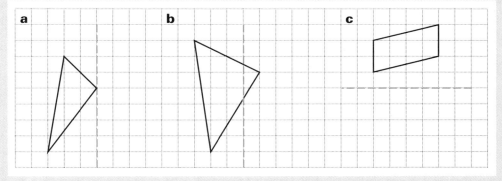

Copy the objects and mirror lines onto squared paper, and draw the image of each object, when it is reflected in the mirror line.

8 In the diagram, which images of triangle LMN are given by

a a translation

b a reflection

c a rotation

d none of these ?

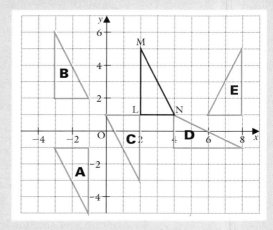

1 Is it possible for each exterior angle of a regular polygon to be

a 30°   **b** 35° ?

If your answer is 'yes' state how many sides the polygon has. If it is 'no', justify your answer.

2 Find the value of x.

a

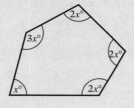

b

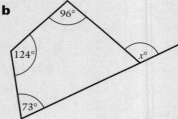

3 a Find, without using a calculator, the square root of

i 81   **ii** 144   **iii** 0.01

b Find, without using a calculator

i 1.1^2 **ii** $(0.9)^2$ **iii** $(0.06)^2$

c Use a calculator to find, correct to 3 significant figures, the square root of

i 40   **ii** 12.4

d Find the length of the side of a square that has an area of 500 mm².

4 For each of the following shapes the area is given, together with one dimension. Find the measurement marked with a letter.

a

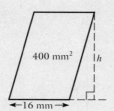

4.9 cm

15.68 cm²

b

c

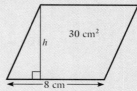

24 cm

h

156 cm²

b

400 mm²

|*h*

←16 mm→

d

h

30 cm²

←——— 8 cm ———→

5 In each case describe the correlation you think there may be between the two quantities. Explain your answer.

a The heights and weights of the girls in Year 8 of a school.

b The cost of a household electricity bill and the distance of the household north from the most southerly point in the country.

c Ability in foreign languages and ability in football.

d The monthly sale of umbrellas in London and its monthly rainfall.

e Ability in maths and ability in music.

f The sale of hot-water bottles during the winter months in Scotland and the outside temperature there.

6 a Find the diameter of a circle which has a circumference of 76 cm. What is the area of this circle?

b For the given shape, find
   i the perimeter
   ii the area.

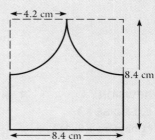

←4.2 cm→

8.4 cm

←——— 8.4 cm ———→

7 The minute hand of the City Hall clock is 1.6 m long while the hour hand is 1.1 m long. How far does the tip of each hand move in 1 hour?

8 a Write in index notation   **i** $5a \times 3a$ **ii** $3a \times 4b \times 2b$

  b Simplify   **i** $3 + (5x - 4)$ **ii** $7x + 5x - 3(4x + 3)$

9 a The area of this rectangle is A cm^2.
Find a formula for A.

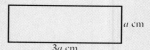

$3a$ cm a cm

  b If $X = 2a + bc$ find X when $a = 5$, $b = 3$ and $c = 8$.

  c Find a formula for u_n in terms of n.

n	1	2	3	4	5
u_n	3	6	9	12	15

10

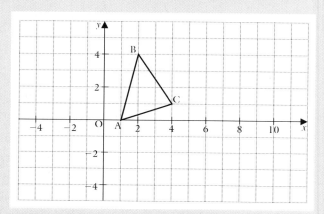

Draw diagrams to illustrate the following translations of $\triangle ABC$.

a 2 units to the right followed by 4 units down.
Label the new triangle P.

b 3 units down followed by 5 units to the left.
Label the new triangle Q.

c 5 units to the right followed by 2 units up.
Label the new triangle R.

In each case give the coordinates of the point to which A is translated.

REVISION EXERCISE 3.4
(Chapters 1 to 11)

1 a Write as a single expression in index form

    i $3^4 \times 3^3$ **ii** $5^9 \div 5^6$ **iii** $a^3 \times a^3 \times a^5$ **iv** $b^4 \div b^4$

  b Find the value of   **i** 8^0 **ii** 10^{-2} **iii** 4^{-3}

2 In the Christmas raffle for the local motor club 800 tickets are sold. If Gemma buys 10 tickets what is the probability that she will not win first prize?

3 Calculate

a $\frac{9}{16} \times \frac{4}{7} \times \frac{14}{27}$ **c** $4\frac{1}{4} \times \frac{4}{9}$ **e** $2\frac{5}{8} \div 12\frac{1}{4}$

b $2\frac{2}{5} \times 20$ **d** $3\frac{5}{8} \div \frac{3}{16}$ **f** $1\frac{3}{4} \div 2\frac{1}{3}$

4 **a** Write 80 cm as a fraction of 3 m.

   b Find 37.5% of 240 cm².

   c Increase £280 by 45%.

   d Decrease £480 by 25%.

5 **a** Simplify the ratio 1.2 m : 40 cm.

   b If $3 : 2 = 4.5 : x$ find x.

6 **a** Find the size of each exterior angle of a regular polygon with 20 sides.

   b How many sides has a regular polygon if the interior angle is $157\frac{1}{2}°$?

7 **a** Use your calculator to find
     i the square of 0.77   **ii** the square root of 0.048.
     Give each answer correct to 3 significant figures.

   b Find
     i the area of triangle ABC
     ii the length of CD.

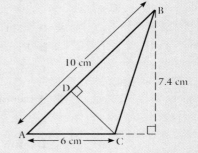

8 **a** The circumference of a coin is 7 cm. Eight similar coins are placed together in a line. How far will they stretch?

   b How many of these coins, arranged in a line, are needed so that they extend a distance that is at least 20 cm?

9 **a** Simplify   **i** $3p \times 3p$ **ii** $2x \times y \times 3x$

   b Simplify

     i $7(x - 3)$ **ii** $10 - 4(x + 4)$ **iii** $9x + 4 - 2(3x - 7)$

   c If $A = \dfrac{PRT}{100}$ find the value of A when $P = 300$, $R = 3$ and $T = 6$.

10 Copy the diagram onto squared paper and draw the image when triangle ABC is rotated through 90° anticlockwise with centre of rotation at the point (4, 1).

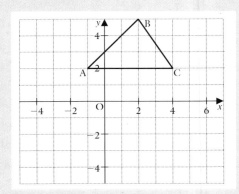

1 a Write these numbers as ordinary numbers.

    i 5.42×10^{4} **ii** 1.84×10^{-2}

b Correct each number to 1 significant figure and hence give a rough value to $\dfrac{789 \times 0.123}{37.42}$

2 An ordinary dice is rolled 600 times. How many times would you expect to get   **a** 1   **b** 6   **c** a score that is even ?

3 Find

  a $35 \times 1\frac{5}{7}$ **b** $7\frac{3}{5} \times 7\frac{1}{2}$ **c** $4\frac{1}{7} \div 2\frac{5}{12}$ **d** $\frac{5}{12}$ of 66 kg

4 a Express 350 g as a fraction of 1.5 kg.

  b Express as a percentage   **i** $\frac{23}{50}$ **ii** 0.155

  c Find 8.2% of 9.2 km.

5 A bag of 60 coins weighs 300 g.

  a What is the weight of a bag of 100 similar coins ?

  b How many coins are there in a bag weighing 255 g ?

    (Neglect the weight of the bag.)

6 a Is it possible for each interior angle of a regular polygon to be

    i 115°   **ii** 165° ?

    Where it is possible, give the number of sides.

  b Find the size of the angle marked $p°$.

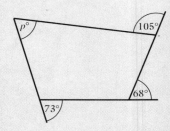

7 a Draw x- and y-axes on squared paper. Scale the x-axis from 0 to 12 and the y-axis from 0 to 6, using 1 square as 1 unit on both axes.

b Draw the parallelogram ABCD with these vertices: A(1, 1), B(2, 5), C(5, 5) and D(4, 1).

c Draw the parallelogram PQRS with these vertices: P(6, 5), Q(9, 5), R(12, 1) and S(9, 1).

d Without using a calculator or finding areas by counting squares, compare the areas of the two parallelograms. Justify your answer.

8 The graph illustrates information relating to the unit cost of the product a company manufactures and the number of units produced.

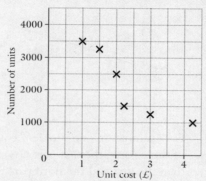

a Would you describe this correlation as
   i positive or negative   **ii** strong or moderate ?

b Does the evidence support the statement, 'The more we produce the lower the unit cost' ?

9 A length of wire 120 cm long is cut into two pieces, one of which is three times as long as the other.

a Find the length of each piece of wire.

Each piece is now bent to form a circle.

b For each circle find   **i** its diameter   **ii** its area.
   (Give your answers correct to 3 significant figures.)

c Find, correct to the nearest whole number
   i the diameter of the large circle divided by the diameter of the small circle
   ii the area of the large circle divided by the area of the small circle.
   What simple relation is there between your answers to part **c** ?

10 a Oranges cost n pence each. Write down a formula for N if N pence is the cost of 5 oranges.

b Calculate   **i** $(-4) + (-6)$ **ii** $(-4) \times (-6)$ **iii** $(-5) \times (+3)$

LINEAR EQUATIONS

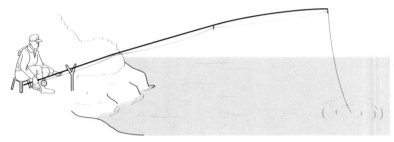

In Book 7A, Chapter 21, we found that we could solve many simple problems by forming an algebraic equation and solving it.

For example, in one problem you were told that a fishing rod consisted of three parts; the first part was x ft long, the second part was 1 ft longer than the first and the third part was 1 ft longer than the second part.

The lengths of the three parts were therefore x ft, $(x+1)$ ft and $(x+2)$ ft. Since the sum of the three lengths was 27 ft we could form the equation $x + x + 1 + x + 2 = 27$.
This simplified to $3x = 24$ from which we found that $x = 8$.

It followed that the lengths of the three sections were 8 ft, 9 ft and 10 ft.

You can remind yourself about the basic principles to use when solving equations by referring to the Summary and Revision Exercise 1.7 at the beginning of this book.

In this chapter we learn how to solve harder equations. This in turn enables us to solve more difficult problems.

EQUATIONS CONTAINING BRACKETS

If we know that the perimeter of this rectangle is 22 cm we can find the length by forming the equation

$$4 + 2(x+1) = 22$$

If we wish to solve an equation like this one, containing brackets, we first multiply out the brackets and then collect like terms.

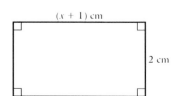

232

EXERCISE 12A

Solve the equation $4 + 2(x + 1) = 22$

$$4 + 2(x + 1) = 22$$

Remember that $2(x + 1)$ means $2 \times x + 2 \times 1$

$$4 + 2x + 2 = 22$$
$$2x + 6 = 22$$

Take 6 from each side $2x = 16$

Divide both sides by 2 $x = 8$

Check: If $x = 8$,

left-hand side $= 4 + 2(8 + 1) = 4 + 2 \times 9 = 22$

Right-hand side $= 22$, so $x = 8$ is the solution.

Solve the following equations.

1 $6 + 3(x + 4) = 24$ **6** $28 = 4(3x + 1)$

2 $3x + 2 = 2(2x + 1)$ **7** $4 + 2(x - 1) = 12$

3 $5x + 3(x + 1) = 14$ **8** $7x + (x - 2) = 22$

4 $5(x + 1) = 20$ **9** $1 - 4(x + 4) = x$

5 $2(x + 5) = 6(x + 1)$ **10** $8x - 3(2x + 1) = 7$

11 $16 - 4(x + 3) = 2x$ **16** $16 - 2(2x - 3) = 7x$

12 $5x - 2(3x + 1) = -6$ **17** $3x - 2 = 5 - (x - 1)$

13 $4x - 2 = 1 - (2x + 3)$ **18** $7x + x = 4x - (x - 1)$

14 $4 = 5x - 2(x + 4)$ **19** $3 - 6(2x - 3) = 33$

15 $9x - 7(x - 1) = 0$ **20** $6x = 2x - (x - 4)$

21 $3(x + 2) + 4(2x + 1) = 6x + 20$

22 $9(2x - 1) + 2(3x + 4) = 20x + 3$

23 $3(x + 2) + 4(2x - 1) = 5(x - 2)$

24 $2(2x + 1) + 4 = 6(3x - 6)$

25 $6x + 4 + 5(x + 6) = 12$

26 $3 - 4(2x + 3) = -25$

27 $15 + 5(x - 7) = x$

28 $6x - 2 - 3(x - 4) = 13$

29 $6(x - 2) - (2x - 1) = 2$

30 $4(2x - 5) + 6 = 2$

The width of a rectangle is x cm. Its length is 4 cm more than its width. The perimeter is 48 cm. What is the width?

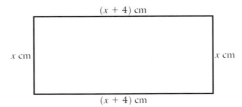

The width is x cm so the length is $(x + 4)$ cm.

$$\therefore \qquad x + (x + 4) + x + (x + 4) = 48$$
$$4x + 8 = 48$$

Take 8 from each side $\qquad 4x = 40$

Divide both sides by 4 $\qquad x = 10$

Therefore the width is 10 cm.

A choc-ice costs x pence and a cone costs 3 pence less.
One choc-ice and two cones together cost £1.44 or 144 pence.
How much is a choc-ice?

A choc-ice costs x pence and a cone costs $(x - 3)$ pence.

$$\therefore \qquad x + 2(x - 3) = 144$$
$$x + 2x - 6 = 144$$
$$3x - 6 = 144$$

Add 6 to each side $\qquad 3x = 150$

Divide each side by 3 $\qquad x = 50$

Therefore a choc-ice costs 50 pence.

> Check by working through the question to make sure that the given information is satisfied.

Solve the following problems by forming an equation in each case. Explain, either in words or on a diagram, what each letter stands for and always end by answering the question asked.
(Not all of these problems result in an equation containing brackets.)

31 I think of a number, double it and then add 14. The result is 36. If x is the number, form an equation and solve it to find x.

32 I think of a number and add 6. The result is equal to twice the first number.

   a If x is the first number, express the result in terms of x in two different ways.

   b Form an equation in x and solve it.

33 In triangle ABC, AB $=$ AC. The perimeter is 24 cm.

   a Form an equation.

   b Solve your equation and hence find the length of AB.

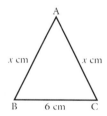

34 A bun costs x pence and a cake costs 5 pence more than a bun. Four cakes and three buns together cost £3.21.

   a Express the cost of a cake in terms of x.

   b Write down an expression for the cost of **i** 4 cakes **ii** 3 buns

   c Form an equation in x and solve it. How much does one bun cost?

35 A bus started from the terminus with x passengers. At the first stop another x passengers got on and 3 got off. At the next stop, 8 passengers got on. There were then 37 passengers. How many passengers were there on the bus to start with?

36 Packets of small screws cost x pence each and packets of large screws cost twice as much. Greg buys three packets of the small screws and five packets of the large screws for which he pays £11.05.

   a Find, in terms of x, the price of a packet of large screws.

   b Find, in terms of x, the cost of

     i three packets of small screws

     ii five packets of large screws.

   c Form an equation in x and solve it. How much does a packet of large screws cost?

37 Jane has x pence and Michael has 6 pence less than Jane. Together they have 30 pence. How much has Jane?

38 The first angle measures $x°$, the second angle (going clockwise) is twice the first, the third is $30°$ and the fourth is $90°$. Find the first angle.

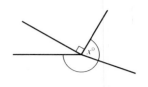

39 In triangle ABC, $\widehat{A} = x°$, $\widehat{B} = 70°$ and $\widehat{C}$ is $20°$ more than $\widehat{A}$. Draw a diagram and use it to form an equation. Solve the equation and hence find $\widehat{A}$.

40 Seventy-four sweets are divided among Anne, Mary and John. Mary has six more than Anne and John has three times as many as Mary.

a If Anne has x sweets express, in terms of x, the number that Mary has and the number that John has.

b Form an equation in x and solve it. How many sweets does each person have?

41 I think of a number, take away 7 and multiply the result by 3, giving 15. What is the number?

42 I think of a number x, take 4 away and double the result. Then I start again with x, treble it and subtract 12. The two results are the same. What is the number?

MULTIPLICATION AND DIVISION OF FRACTIONS

Remember that to multiply fractions, the numerators are multiplied together and the denominators are multiplied together:

i.e.
$$\frac{3}{4} \times \frac{5}{7} = \frac{3 \times 5}{4 \times 7} = \frac{15}{28}$$

Also
$$\frac{1}{6} \text{ of } x \text{ means } \frac{1}{6} \times x = \frac{1}{6} \times \frac{x}{1} = \frac{x}{6} \qquad [1]$$

Remember that to divide by a fraction, that fraction is turned upside down and multiplied:

i.e.
$$\frac{2}{3} \div \frac{5}{7} = \frac{2}{3} \times \frac{7}{5} = \frac{14}{15}$$

and
$$x \div 6 = \frac{x}{1} \div \frac{6}{1} = \frac{x}{1} \times \frac{1}{6} = \frac{x}{6} \qquad [2]$$

Comparing [1] and [2] we see that

$$\frac{1}{6} \text{ of } x, \quad \frac{1}{6}x, \quad x \div 6 \quad \text{and} \quad \frac{x}{6} \quad \text{are all equivalent.}$$

EXERCISE 12B

Simplify **a** $12 \times \dfrac{x}{3}$ **b** $\dfrac{2x}{3} \div 8$

a $12 \times \dfrac{x}{3} = \dfrac{\cancel{12}^{4}}{1} \times \dfrac{x}{\cancel{3}_{1}}$

$= 4x$

b $\dfrac{2x}{3} \div 8 = \dfrac{2x}{3} \div \dfrac{8}{1}$

$= \dfrac{\cancel{2x}^{1}}{3} \times \dfrac{1}{\cancel{8}_{4}} = \dfrac{x}{12}$

> Remember that $2x = 2 \times x$

Simplify

1 $4 \times \dfrac{x}{8}$

2 $\dfrac{1}{2} \times \dfrac{x}{3}$

3 $9 \times \dfrac{x}{6}$

4 $\dfrac{1}{3}$ of $2x$

5 $\dfrac{2x}{3} \times \dfrac{6}{5}$

6 $\dfrac{1}{5}$ of $10x$

7 $\dfrac{2}{5} \times \dfrac{3x}{4}$

8 $\dfrac{3}{4} \times 2x$

9 $\dfrac{2}{3}$ of $9x$

10 $\dfrac{x}{2} \times \dfrac{2}{3}$

11 $\dfrac{5x}{2} \div 4$

12 $\dfrac{4x}{9} \div 8$

13 $\dfrac{x}{3} \div \dfrac{1}{6}$

14 $\dfrac{x}{4} \div \dfrac{1}{2}$

15 $\dfrac{2x}{3} \div \dfrac{5}{6}$

16 $\dfrac{3}{4} \times \dfrac{2x}{5}$

17 $\dfrac{4x}{9} \div \dfrac{2}{3}$

18 $\dfrac{3}{5}$ of $15x$

19 $\dfrac{3x}{2} \div \dfrac{1}{6}$

20 $\dfrac{5x}{3} \times \dfrac{6}{25}$

Simplify **a** $\dfrac{3x}{5} \times \dfrac{10x}{13}$ **b** $\dfrac{5x}{9} \div \dfrac{20x}{3}$

a $\dfrac{3x}{\cancel{5}_{1}} \times \dfrac{\cancel{10x}^{2}}{13} = \dfrac{3 \times 2 \times x \times x}{13}$

$= \dfrac{6x^{2}}{13}$

b $\dfrac{5x}{9} \div \dfrac{20x}{3} = \dfrac{\cancel{5x}^{1}}{\cancel{9}_{3}} \times \dfrac{\cancel{3}^{1}}{\cancel{20x}_{4}}$

$= \dfrac{1}{12}$

> Remember that to divide by a fraction, we turn it upside down and multiply.

Simplify

21 $\dfrac{x}{2} \times \dfrac{x}{3}$

22 $\dfrac{3x}{2} \times \dfrac{x}{4}$

23 $\dfrac{4a}{7} \times \dfrac{a}{8}$

24 $\dfrac{4b}{9} \times 18b$

25 $\dfrac{25x}{7} \times \dfrac{21x}{5}$

26 $\dfrac{x}{2} \div \dfrac{x}{3}$

27 $\dfrac{9x}{4} \div \dfrac{9x}{8}$

28 $12b \div \dfrac{4b}{5}$

29 $\dfrac{16x}{3} \div \dfrac{4x}{9}$

EQUATIONS CONTAINING FRACTIONS

Peter doesn't know the capacity of his fuel tank. When the gauge shows that it is quarter full, he empties the contents into a measuring container. This shows 7.25 litres. Peter can find the capacity of his tank by multiplying 7.25 by 4; he can also form an equation and find out how to solve it.

This second method will help with more complicated situations that cannot be sorted out by using simple arithmetic.

If the capacity of the tank is x litres, then

$$\tfrac{1}{4} \text{ of } x = 7.25$$

Now $\tfrac{1}{4}$ of x is $\dfrac{x}{4}$

so the equation is $\dfrac{x}{4} = 7.25$

To solve this equation, we need to find x.

We know what $\tfrac{1}{4}$ of x is, so to find x we need to make $\dfrac{x}{4}$ four times larger,

i.e.   we need to multiply both sides by 4.

$$\frac{x}{\cancel{4}} \times \frac{\cancel{4}^{1}}{1} = 7.25 \times 4$$
$$x = 29$$

The capacity of Peter's tank is **29** litres.

EXERCISE 12C

Solve the equation $\dfrac{x}{3} = 2$

> As $\dfrac{x}{3}$ means $\tfrac{1}{3}$ of x, to find x we need to make $\dfrac{x}{3}$ three times larger.

$$\frac{x}{3} = 2$$

Multiply each side by 3 $\qquad \dfrac{x}{\cancel{3}} \times \dfrac{\cancel{3}^{1}}{1} = 2 \times 3$

$$x = 6$$

Solve the following equations.

1 $\dfrac{x}{5} = 3$ **4** $\dfrac{x}{8} = 9$ **7** $\dfrac{x}{3} = 6$

2 $\dfrac{x}{2} = 4$ **5** $7 = \dfrac{x}{9}$ **8** $\dfrac{x}{4} = 4$

3 $\dfrac{x}{6} = 8$ **6** $\dfrac{x}{7} = 8$ **9** $8 = \dfrac{x}{12}$

A bag of sweets was divided into five equal shares.
David had one share and he got 8 sweets.
How many sweets were there in the bag?

Let x stand for the number of sweets in the bag.

One share is $\frac{1}{5}$ of x $\qquad \therefore \frac{1}{5}$ of $x = 8$

$$\frac{x}{5} = 8$$

Multiply each side by 5. $\qquad x = 40$

Therefore there were 40 sweets in the bag.

In questions **10** to **14** let x stand for the appropriate unknown number. Then form an equation and solve it. Do not just give the value of x, make sure you answer the question asked.

10 Tracey Brown came first in the Newtown Golf Tournament and won £100. This was $\frac{1}{3}$ of the total prize money paid out. Find the total prize money.

11 Peter lost 8 marbles in a game. This number was one-fifth of the number that he started with. Find how many he started with.

12 The width of a rectangle is 12 cm. This is one-quarter of its length. Find the length of the rectangle.

13 I think of a number, halve it and the result is 6. Find the number that I first thought of.

14 The length of a rectangle is 8 cm and this is $\frac{1}{3}$ of its perimeter. Find its perimeter.

Solve the equation $\dfrac{3x}{4} = 12$

$$\frac{3x}{4} = 12$$

Multiply both sides by 4 $\qquad \dfrac{3x}{4} \times \dfrac{4}{1} = 12 \times 4$

$$3x = 48$$

Divide both sides by 3 $\qquad x = 48 \div 3$

$$x = 16$$

Check: $\text{LHS} = \dfrac{3x}{4} = \dfrac{3}{4} \times \dfrac{16}{1} = 12 = \text{RHS}$

Note that LHS means left-hand side and RHS means right-hand side.

Solve the equations

15 $\dfrac{2x}{3} = 8$ **17** $\dfrac{5x}{7} = 1$ **19** $\dfrac{4x}{7} = 12$

16 $\dfrac{3x}{5} = 9$ **18** $\dfrac{5x}{8} = 20$ **20** $\dfrac{2x}{9} = 5$

Solve the equation $\dfrac{2x}{5} = \dfrac{1}{3}$

$$\dfrac{2x}{5} = \dfrac{1}{3}$$

Multiply each side by 5 $\dfrac{2x}{\cancel{5}} \times \dfrac{\cancel{5}}{1} = \dfrac{1}{3} \times \dfrac{5}{1}$

$$2x = \dfrac{5}{3}$$

Divide each side by 2 $x = \dfrac{5}{3} \div 2$

$$x = \dfrac{5}{3} \times \dfrac{1}{2}$$

$$x = \dfrac{5}{6}$$

Check: LHS $= \dfrac{2x}{5} = \dfrac{\cancel{2}}{\cancel{5}} \times \dfrac{\cancel{5}}{\cancel{6}} = \dfrac{1}{3} =$ RHS

Solve the following equations.

21 $\dfrac{3x}{2} = \dfrac{1}{4}$ **24** $\dfrac{2x}{3} = \dfrac{4}{5}$ <u>**27**</u> $\dfrac{5x}{7} = \dfrac{3}{4}$

22 $\dfrac{4x}{3} = \dfrac{1}{5}$ **25** $\dfrac{6x}{5} = \dfrac{2}{3}$ <u>**28**</u> $\dfrac{3x}{7} = \dfrac{2}{5}$

23 $\dfrac{2x}{9} = \dfrac{1}{3}$ **26** $\dfrac{3x}{8} = \dfrac{1}{2}$ <u>**29**</u> $\dfrac{4x}{7} = \dfrac{3}{14}$

Solve questions **30** to **35** by forming equations and solving them.

30 Tim Silcocks came first in the local Tennis Tournament and won £200. What was the total prize money if Tim won two-thirds of it?

31 The width of a rectangular flower-bed is 18 m. This is $\frac{5}{6}$ of its lengths. Find the length of the bed.

32 The width of a rectangle is $\frac{2}{3}$ its length. If the rectangle is $\frac{5}{6}$ ft wide, how long is it?

33 The capacity of a can of oil is x gallons. It is $\frac{3}{4}$ full. When this oil is poured into a separate container it is found to be $\frac{1}{2}$ gallon. Find the capacity of the can.

34 When a screw, $\frac{5}{8}$ in long, is screwed into a piece of wood until the top of its head is flush with the surface of the wood, it has penetrated through $\frac{7}{8}$ of the thickness of the wood. If the wood is w inches thick, form an equation in w and solve it. How thick is the wood?

35 The length of a rectangle is $1\frac{1}{3}$ times its width. Find the perimeter of the rectangle if the length is 160 mm.

**EQUATIONS
INVOLVING
DECIMALS**

If a problem, or equation, is given in decimals, the answer should also be given in decimal form. Decide yourself whether or not you need to use a calculator.

EXERCISE 12D

Solve the equation $0.55x = 21$

$$0.55x = 21$$

Divide both sides by 0.55

$$0.55 \times x \div 0.55 = 21 \div 0.55$$

$$\frac{0.55^{\prime}}{1} \times \frac{x}{1} \times \frac{1}{{}_{\prime}0.55} = 38.18\ldots$$

$$x = 38.18\ldots$$

$$x = 38.2 \ (\text{correct to 3 s.f.})$$

Solve the following equations. Give answers that are not exact correct to 3 significant figures.

1 $0.75x = 6$

2 $0.3x = 9$

3 $3.5x = 1.4$

4 $2.8x = 5$

5 $3.7x = 7.2$

6 $0.5x = 12$

7 $0.4x = 19$

8 $5.4x = 4.32$

9 $7.8x = 4$

10 Melanie buys $3.8\,\text{m}^2$ of floor covering for £22.23. If the cost of $1\,\text{m}^2$ is £x form an equation in x and solve it. How much does the floor covering cost per square metre?

11 The weight of $8.5\,\text{cm}^3$ of gold is $164.05\,\text{g}$. If the weight of $1\,\text{cm}^3$ is x grams form an equation in x and solve it. What is the weight of $1\,\text{cm}^3$?

12 In a French journal the area of a farm that is for sale is given as 88.4 hectares. An estate agent in the south of England advertises the same farm as having an area of 218.3 acres. If there are x acres in 1 hectare form an equation in x and solve it. How many acres are equivalent to 1 hectare?

Solve the equation $3.8x + 5 = 9.2$, giving your answer correct to 3 significant figures.

$$3.8x + 5 = 9.2$$

$$3.8x = 4.2 \qquad \text{Subtracting 5 from each side.}$$

$$x = \frac{4.2}{3.8} \qquad \text{Dividing both sides by 3.8.}$$

$$= 1.105\ldots$$

$$x = 1.11 \ (\text{correct to 3 s.f.})$$

Solve the following equations, giving answers that are not exact correct to 3 significant figures.

13 $4.2x - 3 = 9.6$

14 $6.3x + 4.5 = 36$

15 $2.2x - 3 = 7.5$

16 $1.9x + 2.4 = 8.9$

17 $5.5x - 4 = 9.2$

18 $3.7x + 1.2 = 16$

19 $4.9x - 5.2 = 4.7$

20 $5.2x + 3.7 = 9.3$

HARDER
EQUATIONS
INVOLVING
FRACTIONS

Some problems result in an equation involving the sum or difference of two or more fractions.

Angela took part in the Vosper Tennis Tournament. She won the Singles title and, with Tony, also won the Mixed Doubles. Her prize money for the Singles was $\frac{1}{6}$ of the total prize money and her prize money for the Mixed Doubles was $\frac{1}{24}$ of the total prize money. Altogether she won £1500.

What was the total prize money awarded to all the winners?

Let the total prize money be £x.

Then the amount, in £s, Angela won in the Singles is

$$\frac{1}{6} \times \frac{x}{1} = \frac{x}{6}$$

the amount, in £s, Angela won in the Mixed Doubles is

$$\frac{1}{24} \times \frac{x}{1} = \frac{x}{24}$$

But the sum of her prize money is £1500

$$\therefore \qquad \frac{x}{6} + \frac{x}{24} = 1500$$

Both 6 and 24 divide into 24, so multiplying both sides by 24 eliminates all fractions from the equation. Notice that there are three terms, *each* of which is multiplied by 24,

i.e. $$\frac{24^{4}}{1} \times \frac{x}{6} + \frac{24^{1}}{1} \times \frac{x}{24} = 24 \times 1500$$

$$4x + x = 24 \times 1500$$

$$5x = 24 \times 1500$$

Dividing both sides by 5 $$x = \frac{24 \times 1500^{300}}{5}$$

$$= 7200$$

The total prize money for the tournament is therefore £7200.

EXERCISE 12E

Solve the equation $\dfrac{x}{5} + \dfrac{1}{2} = 1$

> Both 5 and 2 divide into 10, so if we multiply each side by 10 we can eliminate all fractions from this equation before we start to solve for x.

$$\frac{x}{5} + \frac{1}{2} = 1$$

Multiply both sides by 10 $\qquad \dfrac{\overset{2}{\cancel{10}}}{1} \times \dfrac{x}{\cancel{5}} + \dfrac{\overset{5}{\cancel{10}}}{1} \times \dfrac{1}{\cancel{2}} = 10$

$$2x + 5 = 10$$

Take 5 from each side $\qquad\qquad\qquad 2x = 5$

Divide each side by 2 $\qquad\qquad\qquad x = 2\dfrac{1}{2}$

Solve the following equations.

1 $\dfrac{x}{4} - 1 = 4$ $\qquad\qquad$ **5** $\dfrac{x}{3} - \dfrac{1}{3} = 4$

2 $\dfrac{x}{5} + \dfrac{3}{4} = 2$ $\qquad\qquad$ **6** $\dfrac{x}{6} + 2 = 3$

3 $\dfrac{x}{3} + \dfrac{1}{4} = 1$ $\qquad\qquad$ **7** $\dfrac{x}{3} + \dfrac{5}{6} = 2$

4 $\dfrac{x}{5} - \dfrac{3}{4} = 2$ $\qquad\qquad$ **8** $\dfrac{x}{3} - \dfrac{2}{9} = 4$

9 I think of a number, take $\dfrac{1}{3}$ of it and then add 4. The result is 7. Find the number I first thought of.

10 I think of a number and divide it by 3. The result is 2 less than the number I first thought of. Find the number I first thought of.

11 I think of a number and add $\dfrac{1}{3}$ of it to $\dfrac{1}{2}$ of it. The result is 10. Find the number I first thought of.

12 When $\dfrac{1}{2}$ is subtracted from $\dfrac{2}{3}$ of a number the result is 4. Find the number.

13 I think of a number. When $\dfrac{3}{7}$ is subtracted from $\dfrac{3}{4}$ of the number the result is 1. Find the number.

14 Benjamin Findley won the singles competition of a local bowls tournament, for which he got $\dfrac{1}{5}$ of the total prize money. He also won the pairs competition for which he got $\dfrac{1}{20}$ of the prize money. He won £250 altogether. How much was the total prize money?

Solve the equation $\dfrac{x}{2} = \dfrac{1}{6} + \dfrac{x}{3}$

$$\dfrac{x}{2} = \dfrac{1}{6} + \dfrac{x}{3}$$

2, 3 and 6 all divide into 6, so multiplying each side by 6 will eliminate all fractions from this equation.

Multiply each side by 6

$$\dfrac{x}{2} \times \dfrac{6^3}{1} = \dfrac{1}{6} \times \dfrac{6^1}{1} + \dfrac{x}{3} \times \dfrac{6^2}{1}$$

$$3x = 1 + 2x$$

Take $2x$ from each side

$$x = 1$$

Check: $\text{LHS} = \dfrac{1}{2}$ $\qquad \text{RHS} = \dfrac{1}{6} + \dfrac{1}{3}$

$$= \dfrac{1}{6} + \dfrac{2}{6} = \dfrac{3}{6} = \dfrac{1}{2}$$

Solve the following equations.

15 $\dfrac{x}{3} + \dfrac{1}{4} = \dfrac{1}{2}$

16 $\dfrac{x}{5} - \dfrac{x}{6} = \dfrac{1}{15}$

17 $\dfrac{x}{4} + \dfrac{2}{3} = \dfrac{x}{3}$

18 $\dfrac{5x}{6} + \dfrac{x}{8} = \dfrac{3}{4}$

19 $\dfrac{3x}{5} + \dfrac{2}{9} = \dfrac{11}{15}$

20 $\dfrac{x}{3} - \dfrac{x}{12} = \dfrac{1}{4}$

21 $\dfrac{2x}{5} - \dfrac{3}{10} = \dfrac{x}{4}$

22 $\dfrac{5x}{12} - \dfrac{1}{3} = \dfrac{x}{8}$

23 A square was reduced in size, more in one direction than the other, to give a rectangle whose length was $\dfrac{5}{8}$ the length of the square and whose width was $\dfrac{3}{5}$ the width of the square. The perimeter of the rectangle is $2\dfrac{3}{4}$ inches. Find the length of the sides of the square.

24 I think of a number. When $\dfrac{1}{15}$ of this number is subtracted from $\dfrac{2}{5}$ of the number the answer is $\dfrac{5}{9}$. What number did I first think of?

Solve the following equations.

25 $\dfrac{3x}{4} + \dfrac{1}{3} = \dfrac{x}{2} + \dfrac{5}{8}$

26 $\dfrac{2x}{7} - \dfrac{3}{4} = \dfrac{x}{14} + \dfrac{1}{4}$

27 $\dfrac{3}{11} - \dfrac{x}{2} = \dfrac{2x}{11} + \dfrac{1}{4}$

28 $\dfrac{3}{5} - \dfrac{x}{9} = \dfrac{2}{15} - \dfrac{2x}{45}$

29 $\dfrac{4}{7} + \dfrac{2x}{9} = \dfrac{15}{9} - \dfrac{4x}{21}$

30 $\dfrac{5x}{7} - \dfrac{2}{3} = \dfrac{3}{7} - \dfrac{x}{3}$

INEQUALITIES

In Book 7A, Chapter 17, we used a number line to decide which of two positive or negative whole numbers is the larger.

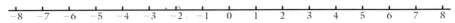

We used the inequality sign $>$ instead of writing 'is greater than' and the sign $<$ instead of writing 'is smaller than'.

Using the number line together with these symbols, we can see that

$$-1 > -4, \qquad 3 > -1, \qquad -5 < -2 \qquad \text{and} \qquad -3 < 1$$

Consider the statement: there are p metres of carpet left on a roll and this must be less than 30. In mathematical form we can write this statement as

$$p < 30$$

This is an *inequality* as opposed to $p = 30$ which is an equality or an equation.

This inequality is true when p stands for any number less than 30. Thus there is a range of numbers that p can stand for and we can illustrate this range on a number line.

```
 ┌────────────────────○
 27   28   29   30   31   32   33
```

The circle at the right hand end of the range is 'open', because 30 is not included in the range.

EXERCISE 12F

> Use inequalities to give the following statements in mathematical form.
>
> **a** The Cornflower Grill guarantee that the weight, w grams, of one of their steaks will be more than 150 g.
>
> **b** When Anna is given letters to address, the number of letters, n, must be less than 250.
>
> **a** $w > 150$ **b** $n < 250$

In questions **1** to **5** use inequalities to give each statement in mathematical form.

1 The government of Mostova passed a law that the number, n, of children per family should be less than 3.

2 In a freezer the temperature, $x\,°C$, should be lower than $-6\,°C$.

3 Before Charleston School will run a course in motor mechanics the number, x, of pupils agreeing to follow the course must be more than 10.

4 Before Townlink will accept a parcel they check that its weight, w kg, is more than 5 kg.

5 A notice on the 'Big Wheel' at a theme park says that the height, x metres, of any passenger taking the ride must be more than 1 metre.

In questions **1** and **3**, the letter can only stand for a positive whole number – we cannot have $2\frac{1}{2}$ people ! In questions **4** and **5**, the letter can stand for any positive number – a parcel can weigh 4.5 kg, whereas in question **2**, x could be any number, positive or negative.

Use a number line to illustrate the range of values of x for which $x < -1$.

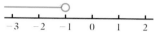

Use a number line to illustrate the range of values of x for which each of the following inequalities is true. Assume that x can stand for any number, whole or otherwise.

6 $x > 7$ **9** $x > 0$ <u>**12**</u> $x < 5$

7 $x < 4$ **10** $x < -2$ <u>**13**</u> $x < 0$

8 $x > -2$ **11** $x > \frac{1}{2}$ <u>**14**</u> $x < 1.5$

15 State which of the inequalities given in questions **6** to **14** are satisfied when x is

     a 2   **b** -3 **c** 0   **d** 1.5   **e** 0.0005

16 For each question from **6** to **14** give a value of x that satisfies the inequality and is

     a a whole number   **b** not a whole number.

In questions **17** to **20** express each statement as an inequality. Give two values that satisfy each inequality.

17 Before a wholesaler will deliver to a retailer, the value, £x, of the order must exceed £500.

18 Sid's lorry is licensed to carry any weight, w tons, provided it is less than 7.5 tons.

19 The Barland Bank does not make charges on a current account provided that the amount in the account, £x, is always greater than £500.

20 Before Spender & Co will open a store in an area the population, x, within 10 miles must be more than 250 000.

21 Consider the true inequality $3 > 1$

    a Add 2 to each side.   **c** Take 5 from each side.

    b Add -2 to each side.   **d** Take -4 from each side.

In each case state whether or not the inequality remains true.

22 Repeat question **21** with the inequality $-2 > -3$

23 Repeat question **21** with the inequality $-1 < 4$

24 Write down a true inequality of your own choice. Add or subtract the same number from each side and see if the inequality remains true.

SOLVING INEQUALITIES

From the last exercise we can see that

> an inequality remains true when the *same* number is added to, or subtracted from, *both* sides.

The temperature today is $2\,°C$ lower than it was yesterday. It is also less than $3\,°C$.
If $x\,°C$ was the temperature yesterday, we can form the inequality

$$x - 2 < 3$$

Solving this inequality means finding the range of values of x for which it is true.
Adding 2 to each side gives $x < 5$

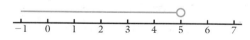

We have now solved the inequality and have found that the temperature yesterday was less than $5\,°C$.

EXERCISE 12G

Solve the following inequalities and illustrate your solutions on a number line.

1 $x - 4 < 8$ **4** $x - 3 > -1$ **7** $x - 3 < -6$

2 $x + 2 < 4$ **5** $x + 4 < 2$ **8** $x + 7 < 0$

3 $x - 2 > 3$ **6** $x - 5 < -2$ **9** $x + 2 < -3$

Solve the inequality $4 - x < 3$

$$4 - x < 3$$

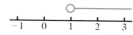

We put the x term on the side with the greater number of xs to start with, i.e. on the right-hand side because $0 > -1$.

Add x to each side $4 < 3 + x$
Take 3 from each side $1 < x$ or $x > 1$

Solve the following inequalities and illustrate your solutions on a number line.

10 $4 - x > 6$ **13** $5 < x + 5$ **16** $3 - x > 2$

11 $2 < 3 + x$ **14** $5 - x < 8$ **17** $6 < x + 8$

12 $7 - x > 4$ **15** $2 > 5 + x$ **18** $2 + x < -3$

19 $2 > x - 3$ **22** $3 - x < 3$ **25** $3 > -x$

20 $4 < 5 - x$ **23** $5 < x - 2$ **26** $4 - x > -9$

21 $1 < -x$ **24** $7 > 2 - x$ **27** $5 - x < -7$

28 The temperature yesterday was $3\,°C$ lower than the day before. It was also less than $-4\,°C$. If $x\,°C$ was the temperature the day before, form an inequality in x and solve it.

29 Consider the true inequality $12 < 36$
    a Multiply each side by 2   **d** Divide each side by 6
    b Divide each side by 4   **e** Multiply each side by -2
    c Multiply each side by 0.5   **f** Divide each side by -3
  In each case state whether or not the inequality remains true.

30 Repeat question **29** with the true inequality $36 > -12$

31 Repeat question **29** with the true inequality $-18 < -6$

32 Repeat question **29** with a true inequality of your own choice.

33 Can you multiply both sides of an inequality by any one number and be confident that the inequality remains true?

> An inequality remains true when both sides are multiplied or divided by the same *positive* number.

Multiplication or division of an inequality by a negative number must be avoided, because it destroys the truth of the inequality.

Solve the inequality $2x - 4 > 5$ and illustrate the solution on a number line.

$$2x - 4 > 5$$

Add 4 to both sides $\qquad 2x > 9$

Divide both sides by 2 $\qquad x > 4\tfrac{1}{2}$

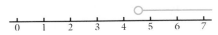

Solve the inequalities and illustrate the solutions on a number line.

34 $3x - 2 < 7$ **36** $4x - 1 > 7$ **38** $5 + 2x < 6$

35 $1 + 2x > 3$ **37** $3 + 5x < 8$ **39** $3x + 1 > 5$

Solve the inequality $3 - 2x \leqslant 5$ and illustrate the solution on a number line. ($\leqslant$ means 'less than or equal to')

> As with equations, we collect the letter term on the side with the greater number to start with. In this case we collect on the right, because $-2 < 0$.

$$3 - 2x \leqslant 5$$

Add $2x$ to each side $\qquad 3 \leqslant 5 + 2x$

Take 5 from each side $\qquad -2 \leqslant 2x$

Divide each side by 2 $\qquad -1 \leqslant x$ i.e. $x \geqslant -1$

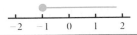

> A solid circle is used for the end of the range because -1 *is* included.

Solve the inequalities and illustrate the solution on a number line.

40 $3 \leqslant 5 - 2x$ **44** $4 \geqslant 9 - 5x$ **48** $2x + 1 \geqslant 5 - x$

41 $5 \geqslant 2x - 3$ **45** $10 < 3 - 7x$ **49** $3x + 2 \leqslant 5x + 2$

42 $4 - 3x \leqslant 10$ **46** $8 - 3x \geqslant 2$ **50** $1 - x > 2x - 2$

43 $x - 1 > 2 - 2x$ **47** $2x + 1 \leqslant 7 - 4x$ **51** $2x - 5 > 3x - 2$

PROBLEMS
INVOLVING TWO
INEQUALITIES
At any one time the number, q, of players on the field of play in a rugby
match must be less than 31.

Expressed as an inequality, this becomes $q < 31$.

But q cannot be a negative number so $q \geqslant 0$, that is $0 \leqslant q$.

We can combine these two inequalities to give the double inequality

$$0 \leqslant q < 31$$

q can take the values $0, 1, 2, 3, \ldots, 30$

EXERCISE 12H

Find, where possible, the range of values of x which satisfy both of
the inequalities   **a** $x \geqslant 2$ and $x > -1$

   b $x \leqslant 2$ and $x > -1$

   c $x \geqslant 2$ and $x < -1$

a

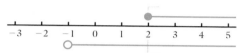

> Illustrating the ranges on a number line, we can see that *both* inequalities
> are satisfied for values on the number line where the ranges overlap.

$\therefore$ $x \geqslant 2$ and $x > -1$ are both satisfied for $x \geqslant 2$

b

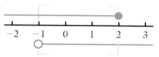

$x \leqslant 2$ and $x > -1$ are both satisfied for $-1 < x \leqslant 2$

c

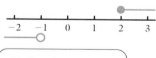

> The lines do not overlap.

There are no values of x for which $x \geqslant 2$ and $x < -1$

Find, where possible, the range of values of x for which the two
inequalities are both true.

1 a $x > 2$ and $x > 3$ **3 a** $x \leqslant 4$ and $x > -2$

   b $x \geqslant 2$ and $x \leqslant 3$ **b** $x \geqslant 4$ and $x < -2$

   c $x < 2$ and $x > 3$ **c** $x \leqslant 4$ and $x < -2$

2 a $x \geqslant 0$ and $x \leqslant 1$ **4 a** $x < -1$ and $x > -3$

   b $x \leqslant 0$ and $x \leqslant 1$ **b** $x < -1$ and $x < -3$

   c $x < 0$ and $x > 1$ **c** $x > -1$ and $x < -3$

Solve each of the following pairs of inequalities and then find the range of values of x which satisfy both of them.

5 $x - 4 < 8$ and $x + 3 > 2$ **9** $5x - 6 > 4$ and $3x - 2 < 7$

6 $3 + x \leqslant 2$ and $4 - x \geqslant 1$ **10** $3 - x > 1$ and $2 + x > 1$

7 $x - 3 \leqslant 4$ and $x + 5 \geqslant 3$ **11** $1 - 2x \leqslant 3$ and $3 + 4x < 11$

8 $2x + 1 > 3$ and $3x - 4 < 2$ **12** $0 > 1 - 2x$ and $2x - 5 \leqslant 1$

13 For each question from **5** to **8**, give three whole number values of x (if possible) which satisfy the inequalities.

The maximum load a lorry is registered to carry is 8 tonnes. The minimum load that will make any journey worthwhile is 4 tonnes. Use inequalities to show the load, w tonnes, on the lorry when a worthwhile journey is to be made. Illustrate the inequalities on a number line.

If the load on the lorry is w tonnes then $4 \leqslant w \leqslant 8$

In questions **14** to **20** express the range of values using inequalities. Illustrate each inequality on a number line.

14 The number, n, of newspapers that George can carry when he leaves the newsagent must not be more than 50.

15 Robin's basic weekly wage is £269 but he can earn up to £90 a week extra in overtime payments. His total weekly wage is £p.

16 To run a coach trip to Scarborough, using a 53-seater coach, the organiser must have a minimum of 28 fare-paying passengers.

     a Eventually p people make the trip.

     b The trip had to be cancelled because only q people agreed to go.

17 The minimum speed on a stretch of road is 45 mph and the maximum speed is 65 mph. The speed at which a car travels is s mph.

18 Tim is told by a friend that his mark, n, in a test is at least 37 and at most 96.

19 The maximum number of pupils that can go on a school trip abroad is 47. If the trip goes ahead, which inequality best describes the statement that n pupils go?

a $n \geqslant 47$ **b** $n > 47$ **c** $0 < n \leqslant 47$ **d** $0 \leqslant n < 47$

20 The minimum number of passengers that need to be booked to fly on a 184-seater aeroplane to cover the overhead costs is 103. Which inequality best describes the number, x, of passengers flying when a profit is made?

a $103 < x < 184$ **c** $103 \leqslant x \leqslant 184$

b $103 \leqslant x < 184$ **d** $103 < x \leqslant 184$

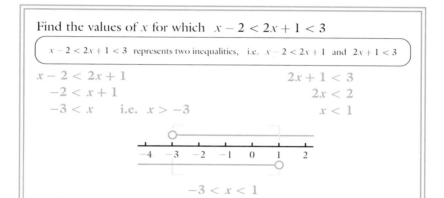

Find the range of values of x for which the following inequalities are true.

21 $x + 4 > 2x - 1 > 3$ <u>**26**</u> $x - 3 < 2x + 1 < 5$

22 $x - 3 \leqslant 2x \leqslant 4$ <u>**27**</u> $2x < x - 3 < 4$

23 $3x + 1 < x + 4 < 2$ <u>**28**</u> $4x - 1 < x - 4 < 2$

24 $2 - x < 3x + 2 < 8$ <u>**29**</u> $4 - 3x < 2x - 5 < 1$

25 $2 - 3x \leqslant 4 - x \leqslant 3$ <u>**30**</u> $x < 3x - 1 < x + 1$

31 The number, x, of apples in a 1 kg bag is described by the inequalities $5 \leqslant x \leqslant 10$. Say, with reasons, whether or not each of the following numbers is a possible value for x.

a 2   **b** 7.5   **c** 12   **d** -4 **e** 8

32 The acceptable temperature, $x\,°C$, in a cold store is described by the inequalities $-3 < x \leqslant 2$. Say, with reasons, whether or not each of the following temperatures would be acceptable as the temperature of the cold store.

a $2.4\,°C$ **b** $-3\,°C$ **c** $0\,°C$ **d** $-2.7\,°C$ **e** $1\,°C$

1 Solve the following equations.

   a $6x - 7 = 11 - 3x$ **c** $11x - 7(x - 4) = 0$

   b $5x + (3x - 4) = 20$ **d** $4(x - 3) = 3(2x + 5)$

2 Solve the equations.

   a $\dfrac{3x}{5} = 15$ **c** $\dfrac{4x}{3} = \dfrac{7}{15}$

   b $0.25x = 1.6$ **d** $\dfrac{3x}{4} - \dfrac{x}{2} = 6$

3 Solve the equations

   a $\dfrac{x}{7} - 2 = 7$ **c** $\dfrac{x}{4} = \dfrac{2}{3} + \dfrac{x}{6}$

   b $\dfrac{x}{3} - \dfrac{3}{4} = \dfrac{7}{12}$ **d** $\dfrac{3x}{5} + \dfrac{x}{4} = \dfrac{3}{15}$

4 Simplify

   a $\dfrac{3}{4}$ of $24x$ **d** $\dfrac{15a}{7} \times \dfrac{14a}{25}$

   b $\dfrac{5x}{3} \times \dfrac{7}{20}$ **e** $\dfrac{3x}{8} \div \dfrac{9x}{4}$

   c $\dfrac{4x}{9} \div \dfrac{5}{18}$ **f** $\dfrac{12x}{5} \div \dfrac{3}{10x}$

5 Solve these equations giving your answers correct to 3 significant figures.   **a** $5.4x = 7$ **b** $3.2x + 4 = 8.7$

6 The lengths of the three sides of a triangle are x cm, $(x + 3)$ cm and $(x + 7)$ cm. If the perimeter of this triangle is 47.5 cm form an equation in x and solve it. Hence find the length of

   a its shortest side   **b** its longest side.

7 Solve the following inequalities and illustrate your solutions on a number line.

   a $x - 5 < 9$ **c** $5x - 3 \leqslant 12$

   b $5 - x \geqslant 4$ **d** $5 > 3 - 4x$

8 **a** Find, if possible, the range of values of x for which the two inequalities $x \geqslant 4$ and $x \leqslant 6$ are both true.

   b Find the range of values of x for which the inequalities $5 - 4x < 2x - 5 < 1$ are true.

9 When $\frac{5}{8}$ of a number is subtracted from $\frac{3}{4}$ of that number the result is 2. Find the number.

10 Frank takes between 3 and 4 minutes to walk to the nearest bus stop. The last time he did this it took t minutes. Which inequality best describes his last walk to the nearest bus top?

    a $0 < t < 4$ **b** $0 < t < 3$ **c** $3 < t < 4$

PUZZLES

1 Elson thought of a simple way of finding out his grandfather's age without asking him. Elson asked his grandfather to do the following calculations. 'Think of your age, add 5, and double your answer. Now take away your age and tell me what answer you have.' Elson's grandfather replied that his answer was 78. 'That means you must be 68 years old', said Elson.
How did Elson work out his grandfather's age from the number he was given?
Form an algebraic equation which proves that the method will always work with anyone's age.

2 Beverly was late arriving for her maths lesson which was due to start at 9.30 a.m. 'Do you know what time it is?' asked her teacher as she walked into the classroom. 'Yes, Mrs Petit', replied Beverly, who was very good at mathematics and a bit cheeky into the bargain. 'Add one-quarter of the number of hours from midnight until now to half the number of hours from now until midnight, and you have the number of hours it is after midnight now.' What time did Beverly arrive for her maths lesson? How late was she?

STRAIGHT LINE GRAPHS

13

The diagram shows a straight line drawn on x- and y-axes.

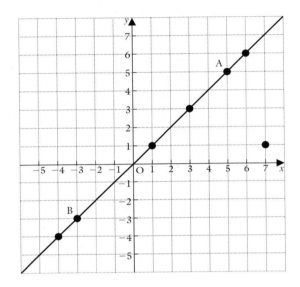

The points A and B are on the line and,

for A, the x-coordinate is 5 and the y-coordinate is 5.

For B, we see that the

x-coordinate is -3 and the y-coordinate is also -3.

Looking at other points on the line, we can see that the y-coordinate is always equal to the x-coordinate. From this we can deduce that *any* point on the line is such that its y-coordinate is equal to its x-coordinate.

However, if we take any point that is *not* on this line, we find that the y-coordinate is *not* equal to the x-coordinate.

- So for points on this line we can give a formula for y,

  i.e. $$y = x$$

- We can also see that, for points not on the line, $y \neq x$.

  (The symbol $\neq$ means 'is not equal to'.)

EXERCISE 13A

1 The diagram shows some points on a line drawn through the origin.

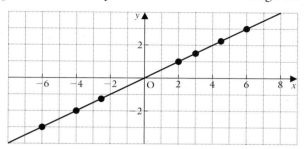

a Copy and complete the table for the x- and y-coordinates of the points marked.

x	-6	-4	-2.5		3	4.5	
y			-1.25				

b Give an instruction in words for finding the y-coordinate of a point on this line from its x-coordinate.

c Find a formula for y.

d Does this formula apply to the coordinates of any point on the line? Discuss the reasons for your answer.

e Does this formula apply to the coordinates of any point *not* on the line? Discuss the reasons for your answer.

2 a Copy this graph onto squared paper.

b On your copy, mark about 6 points on the line.

c Make a table, like the one in question **1**, showing the x- and y-coordinates of each of your points.

d Use your completed table to find a formula for y.

e Discuss which points your formula applies to, and to which points it does not.

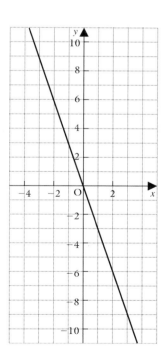

**THE EQUATION
OF A LINE**

The examples on the previous page show that

- it is possible to give an instruction for finding the y-coordinate from the x-coordinate of any point on a line
- this instruction can be given as a formula in the form $y = \ldots$

> The formula that gives the y-coordinate of a point on the line in terms of the x-coordinate is called the *equation of the line*.

In the example above **Exercise 13A**, the *equation of the line* is $y = x$. We often briefly refer to *the line* $y = x$.

EXERCISE 13B

Find the y-coordinates of points on the line $y = -5x$ whose x-coordinates are $\quad$ **a** 4 $\qquad$ **b** -3

> The equation $y = -5x$ is a formula for finding the y-coordinate from the x-coordinate.

a When $x = 4$, $\quad y = -5x$ gives
$$y = -5 \times 4 = -20$$
b When $x = -3$, $y = -5x$ gives
$$y = (-5) \times (-3) = 15$$

> This means that the points $(4, -20)$ and $(-3, 15)$ are points on the line $y = -5x$.

---

1 Find the y-coordinates of points on the line $y = x$ which have x-coordinates of

a 2	**c** 7	**e** -1	**g** -8
b 3	**d** 12	**f** -6	**h** -20

2 Find the y-coordinates of points on the line $y = -x$ which have x-coordinates of

$\quad$ **a** 4 $\qquad$ **b** -2 $\qquad$ **c** $3\frac{1}{2}$ $\qquad$ **d** $-4\frac{1}{2}$ $\qquad$ **e** 6.1 $\qquad$ **f** -8.3

Find the x-coordinate of the point on the line $y = -5x$ whose y-coordinate is 15.

When $y = 15$, $\quad y = -5x$ gives
$$15 = -5x$$

> This is an equation which we can solve. Remember: $(+15) \div (-5)$ gives a negative answer.

$$-3 = x,$$
i.e. $\qquad x = -3$

---

3 Find the x-coordinates of points on the line $y = -x$ which have y-coordinates of

   a 7   **b** -2 **c** $5\frac{1}{2}$ **d** -4.2

4 Find the y-coordinates of points on the line $y = 2x$ which have x-coordinates of

   a 5   **b** -4 **c** $3\frac{1}{2}$ **d** -2.6

5 Find the x-coordinates of points on the line $y = -3x$ which have y-coordinates of

   a 3   **b** -9 **c** 6   **d** -4

6 Find the y-coordinates of points on the line $y = \frac{1}{2}x$ which have x-coordinates of

   a 6   **b** -12 **c** $\frac{1}{2}$ **d** -8.2

7 If the points $(-1, a)$, $(b, 15)$ and $(c, -20)$ lie on the straight line with equation $y = 5x$, find the values of a, b and c.

8 If the points $(3, a)$, $(-12, b)$ and $(c, -12)$ lie on the straight line with equation $y = -\dfrac{2x}{3}$, find the values of a, b and c.

9 Using squared paper and 1 square to 1 unit on each axis, plot the points $(-2, -6)$, $(1, 3)$, $(3, 9)$ and $(4, 12)$. What is the equation of the straight line which passes through these points?

10 Using squared paper and 1 square to 1 unit on each axis, plot the points $(-3, 6)$, $(-2, 4)$, $(1, -2)$ and $(3, -6)$. What is the equation of the straight line which passes through these points?

11 Using the same scale on each axis, plot the points $(-6, -4)$, $(-3, -2)$, $(6, 4)$ and $(12, 8)$. What is the equation of the straight line which passes through these points?

12 Which of the points $(-2, -4)$, $(2.5, 4)$, $(6, 12)$ and $(7.5, 10)$ lie on the line $y = 2x$?

13 Which of the points $(-5, -15)$, $(-2, 6)$, $(1, -3)$ and $(8, -24)$ lie on the line $y = -3x$?

14 Which of the points $(2, 2)$, $(-2, 1)$, $(3, 0)$, $(-4.2, -2)$, $(-6.4, -3.2)$ lie

   a above the line $y = -\frac{1}{2}x$ **b** below the line $y = -\frac{1}{2}x$?

PLOTTING A LINE
FROM ITS
EQUATION

If we want to draw the graph of $y = 3x$ for values of x from -3 to $+3$, then we need to find the coordinates of some points on the line.

As we know that it is a straight line, two points are enough. However, it is sensible to find three points and use the third point as a check on the working. It does not matter which three points we find, so we will choose easy values for x, well spread over the range -3 to 3.

If $x = -3$, $y = 3 \times (-3) = -9$

If $x = 0$, $y = 3 \times 0 = 0$

If $x = 3$, $y = 3 \times 3 = 9$

These are easier to use
if we write them in a table, i.e.

x	-3	0	3
y	-9	0	9

Now we can plot these
points and draw a straight
line through them.

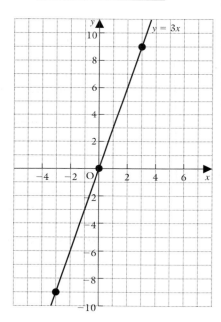

EXERCISE 13C

Keep your graphs: you will need them for the next exercise.

For questions **1** to **6** draw one pair of axes. Use the same scale on both axes, taking values of x between -4 and 4, and values of y between -6 and 6.

Make a table for each equation using at least three x-values and record the corresponding y-values.

Draw the graphs of the given equations on the axes. Write the equation of each line somewhere on the line.

1 $y = x$ **3** $y = \frac{1}{2}x$ **5** $y = \frac{1}{3}x$

2 $y = 2x$ **4** $y = \frac{1}{4}x$ **6** $y = \frac{2}{3}x$

In questions **7** to **12**, draw the graphs of the given equations on the same set of axes.

7 $y = -x$

8 $y = -2x$

9 $y = -\frac{1}{2}x$

10 $y = -\frac{1}{4}x$

11 $y = -\frac{1}{3}x$

12 $y = -\frac{2}{3}x$

13 In questions **1** to **12** the equation of the line is of the form $y = mx$. Discuss the effect that the value of m has on the position of the line.

Discussion from question **13** shows that

> the graph of an equation of the form $y = mx$, is a straight line that
>
> - passes through the origin
> - gets steeper as m increases
> - makes an acute angle with the positive x-axis if m is positive, i.e. slopes up when moving from left to right
> - makes an obtuse angle with the positive x-axis if m is negative, i.e. slopes down when moving from left to right.

GRADIENT OF A STRAIGHT LINE

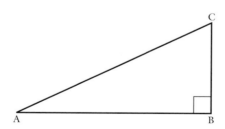

The *gradient* or slope of a line is defined as the amount the line rises vertically divided by the distance moved horizontally,

i.e.   the gradient, or slope, of AC $= \dfrac{BC}{AB}$

If a line is drawn on a set of axes, we can find the gradient by taking any two points on the line. Then the gradient of the line is given by

$$\frac{\text{the increase in } y\text{-value}}{\text{the increase in } x\text{-value}}$$

The points O, B and C are three points on the line $y = x$.

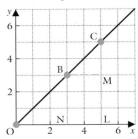

The gradient of OC $= \dfrac{CL}{OL} = \dfrac{5}{5} = 1$

The gradient of OB $= \dfrac{BN}{ON} = \dfrac{3}{3} = 1$

The gradient of BC $= \dfrac{CM}{BM} = \dfrac{5-3}{5-3} = \dfrac{2}{2} = 1$

These show that, whichever two points are taken, the gradient of the line $y = x$ is 1.

This means that, when working out the gradient of a line, we can choose *any* two points, P and Q say, on a line.
However it is important that we use the points in the same order,
i.e. if we choose to find the increase in the y-value from P to Q, we must also find the increase in the x-value from P to Q.

For example,

using the points P and Q to find
the gradient of this line, we have

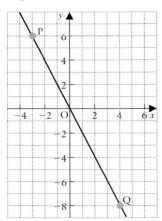

$$\dfrac{\text{the increase in } y\text{-value from P to Q}}{\text{the increase in } x\text{-value from P to Q}}$$

$$= \dfrac{y\text{-coordinate of } Q - y\text{-coordinate of P}}{x\text{-coordinate of } Q - x\text{-coordinate of P}}$$

$$= \dfrac{(-8)-(6)}{(4)-(-3)}$$

$$= \dfrac{-14}{7} = -2$$

1 For each graph drawn for questions **1** to **12** of **Exercise 13C**, choose two points on the line and calculate the gradient of the line.

2 In question **1**, what relationship is there between the gradient of each line and its equation?

3 Copy and complete the following table and use it to draw the graph of $y = 2.5x$.

x	-3	0	4
y			

Choose your own pair of points to find the gradient of this line. How does the value of the gradient compare with the equation of the line?

4 Copy and complete the following table and use it to draw the graph of $y = 0.5x$.

x	-6	-2	4
y			

Choose your own pair of points to find the gradient of this line. How does the value of the gradient compare with the equation of the line?

5 Find the gradient of these lines.

   a $y = 5x$ **c** $y = 12x$

   b $y = -7x$ **d** $y = -\frac{1}{4}x$

6 Rearrange these equations so that each is in the form $y = \ldots$ Hence determine whether the straight lines with the following equations have positive or negative gradients.

   a $3y = -x$ **b** $5y = 12x$

These exercises confirm the conclusion on page 261, namely that

- the larger the value of m the steeper is the slope
- lines with positive values for m make an acute angle with the positive x-axis
- lines with negative values for m make an obtuse angle with the positive x-axis.

In addition we can now see that

the value of m gives the gradient of the line.

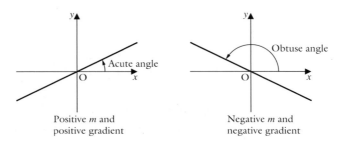

Positive m and
positive gradient

Negative m and
negative gradient

EXERCISE 13E

For each of the following pairs of lines, state which line is the steeper. Sketch both lines on the same diagram.

1 $y = 5x, \; y = \frac{1}{5}x$

2 $y = 2x, \; y = 5x$

3 $y = \frac{1}{2}x, \; y = \frac{1}{3}x$

4 $y = -2x, \; y = -3x$

5 $y = 10x, \; y = 7x$

6 $y = -\frac{1}{2}x, \; y = -\frac{1}{4}x$

7 $y = -6x, \; y = -3x$

8 $y = 0.5x, \; y = 0.75x$

Determine whether each of the following straight lines makes an acute angle or an obtuse angle with the positive x-axis.

9 $y = 4x$

10 $y = -3x$

11 $y = 3.6x$

12 $y = \frac{1}{3}x$

13 $y = 10x$

14 $y = 0.5x$

15 $y = -6x$

16 $y = -\frac{2}{3}x$

17 $y = -\frac{1}{10}x$

18 Estimate the gradient of each of the lines shown in the sketch. (The same scale is used on each axis.)

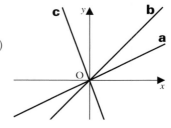

LINES THAT DO NOT PASS THROUGH THE ORIGIN

When we are given a line drawn on a graph, we can find its gradient and where it crosses the y-axis. We can also make a table showing the coordinates of some points on the line. From this table we may be able to see the formula which gives y when we know x, so we may be able to find the equation of the line. This process is shown in the worked example in the next exercise. We can then see if there is a connection between the equation of the line and its gradient and where it crosses the y-axis.

EXERCISE 13F

a Find the gradient of this line.

b Where does the line cross the y-axis?

c Copy and complete the table for some points on the line.

x	-4	-2	2	5
y	-10			8

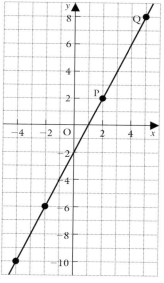

d Find the equation of the line.

e Is there a relationship between the equation of the line and the answers to parts **a** and **b**?

a Using the points P and Q, the gradient $= \frac{6}{3} = 2$

b The line crosses the y-axis two units below the origin, i.e. at $(0, -2)$

c Using the points marked on the graph,

x	-4	-2	2	5
y	-10	-6	2	8

d From the table we see that $y = 2x - 2$. If you cannot see the formula, try adding some more points to your table.

The equation of the line is $y = 2x - 2$.

e The gradient is the same as the number of xs in the equation and -2 describes where the line cuts the y-axis.

For questions **1** to **4**

a Find the gradient of the line.

b Write down the value of y where the line crosses the y-axis.

c Make a table showing the x- and y-coordinates of 4 points on the line.

d Find the equation of the line.

e Compare your answers to parts **a** and **b** with the numbers on the right-hand side of the equation.

1
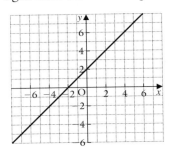

3
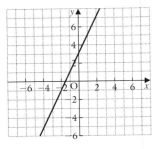

2
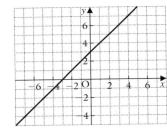

4
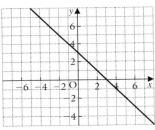

For questions **5** to **8** draw the graph of the line whose equation is given, using the stated values of x to find the points on the line. Use squared paper and 1 square for 1 unit on each axis, using values of x ranging from -8 to 8 and values of y ranging from -10 to 10.

Hence find the gradient of the line and the value of y where it crosses the y-axis. Compare these values with the numbers on the right-hand side of the equation of the line.

5 $y = 3x + 1$; x-values $-3, 1, 3$

Use your graph to find y when x is   **a** -2 **b** 2

6 $y = -3x + 4$; x-values $-2, 2, 4$

Use your graph to find y when x is   **a** -1 **b** 3

7 $y = \frac{1}{2}x + 4$; x-values $-8, 0, 6$

Use your graph to find   **a** y when x is -2 **b** x when y is 6

8 $y = -2x - 7$; x-values $-6, -2, 1$

Use your graph to find   **a** y when x is -1 **b** x when y is 6

THE EQUATION
$y = mx + c$

The results of **Exercise 13F** show that we can 'read' the gradient of a straight line and the value of y where it crosses the y-axis from its equation.

For example, the line with equation $y = 3x - 4$ has a gradient of 3 and crosses the y-axis 4 units below the origin.

The value of y where a line crosses the y-axis is called the y-*intercept*.

> In general we see that the equation $y = mx + c$ gives a straight line where m is the gradient of the line and c is the y-intercept.

EXERCISE 13G

> Write down the gradient, m, and the y-intercept, c, for the straight line with equation $y = 5x - 2$
>
> For the line $\quad y = 5x - 2 \quad\quad m = 5 \quad\quad$ and $\quad\quad c = -2$

Write down the gradient, m, and y-intercept, c, for each straight line.

1 $y = 4x + 7$ **4** $y = -4x + 5$ **7** $y = \frac{3}{4}x + 7$

2 $y = \frac{1}{2}x - 4$ **5** $y = 7x + 6$ **8** $y = 4 - 3x$

3 $y = 3x - 2$ **6** $y = \frac{2}{5}x - 3$ **9** $y = -3 - 7x$

> Sketch the straight line with equation $y = 5x - 7$
>
>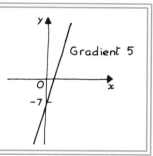
>
> A sketch does not need scaled axes, but the important features of the graph must be marked, i.e. where it crosses the y-axis and its gradient.

Sketch the straight lines with the given equations.

10 $y = 2x + 5$ **15** $y = 4x + 2$

11 $y = 7x - 2$ **16** $y = -5x - 3$

12 $y = \frac{1}{2}x + 6$ **17** $y = 3x + 7$

13 $y = -2x - 3$ **18** $y = \frac{3}{4}x - 2$

14 $y = -\frac{2}{3}x + 8$ **19** $y = \frac{1}{3}x - 5$

Sketch the straight line with equation $y = 2 - 3x$

First rearrange the equation in the form $y = mx + c$, i.e. with the x term first followed by the number on its own.

$y = -3x + 2$

Now we can read the gradient and the y-intercept.

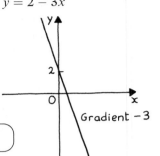

Gradient -3

Sketch the straight lines with the given equations.

20 $y = 4 - x$

21 $y = 3 - 2x$

22 $y = 8 - 4x$

23 $y = -3 - x$

24 $y = 2(x + 1)$

25 $y = 3(x - 2)$

26 $y = -5(x - 1)$

27 $y = 3(4 - x)$

28 $2y = (2x + 3)$

29 $3y = (x - 4)$

INTERSECTING LINES

The point where two lines cross is called the *point of intersection* of the lines.

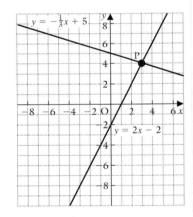

The point of intersection of the two lines in the diagram is P, and at P, $x = 3$ and $y = 4$.

We say that the lines $\left. \begin{array}{l} y = -\frac{1}{3}x + 5 \\ y = 2x - 2 \end{array} \right\}$ intersect at the point $(3, 4)$.

Now $(3, 4)$ is a point on *both* lines,

i.e. $x = 3$ and $y = 4$ satisfies *both* $y = -\frac{1}{3}x + 5$ and $y = 2x - 2$.

We say that $\left. \begin{array}{l} x = 3 \\ y = 4 \end{array} \right\}$ satisfy $\left. \begin{array}{l} y = -\frac{1}{3}x + 5 \\ y = 2x - 2 \end{array} \right\}$ *simultaneously*.

EXERCISE 13H

For questions **1** to **3** use squared paper and prepare axes for values of x ranging from -8 to 8 and values of y ranging from -10 to 10. Use 1 square for 1 unit on each axis.

Draw the graphs of the lines whose equations are given. Hence find the coordinates of the point of intersection of the two lines.

1 $y = 2x - 1$ **2** $y = \frac{1}{2}x + 1$ **3** $y = 4x + 5$

     $y = x + 2$ $y = -x - 2$ $y = x - 4$

For questions **4** to **6** use *graph* paper and 1 cm for 1 unit on each axis. Draw the graphs of the lines whose equations are given for values of x ranging from -8 to 8 and values of y ranging from -10 to 10.

Hence find the value of x and the value of y which satisfy the two equations simultaneously, giving your answers correct to 1 decimal place where necessary.

4 $y = -x + 6$ **5** $y = \frac{2}{3}x + 4$ **6** $y = -2x + 3$

     $y = x + 3$ $y = -\frac{1}{2}x - 1$ $y = -x + 2.5$

USING A GRAPHICS CALCULATOR

Plotting graphs to find their points of intersection is time consuming; using a graphics calculator takes out the hard work.

Calculators vary, but the general principles are similar. First the range has to be set; this specifies the scales on the axes. Then the graph function button is used: this displays 'Y = ' and then the rest of the equation is entered.

The diagram shows the display on a graphics calculator when the graphs $y = -\frac{1}{3}x + 5$ *and* $y = 2x - 2$ are plotted.

The range was set to
x min: -2.5
x max: 5
scale: 1
y min: -1
y max: 7
scale: 1

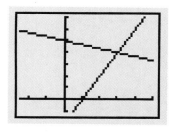

The 'TRACE' button can then be used to give the coordinates of the point of intersection. A cursor appears on the display which can then be moved until it is over the point where the graphs cross.

The display will look something like this:

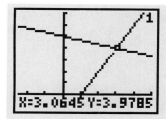

Use the lines given in **Exercise 13H** to practice using a graphics calculator.

**LINES PARALLEL
TO THE AXES**

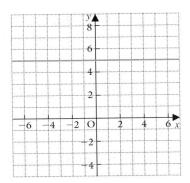

The line in the diagram is parallel to the x-axis and at every point on it, the y-coordinate is 5.

Therefore an instruction for finding the y-coordinate of any point on this line in terms of its x-coordinate could be 'whatever the x-coordinate is, the y-coordinate is 5',

i.e. $y = 5$ for all values of x or, even more briefly, $y = 5$.

Any line parallel to the x-axis can be described in a similar way,

> therefore an equation of the form $y = c$ gives a line that is parallel to the x-axis and distant c units from it.

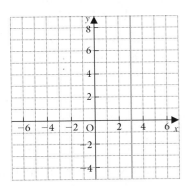

The line in this diagram is parallel to the y-axis, and the x-coordinate of every point on it is 3. Hence $x = 3$ is all that is needed to give the relationship between the x- and y-coordinates of all the points on this line.

> Therefore $x = b$ is the equation of a straight line parallel to the y-axis at a distance b units from it.

Draw, on the same diagram, the graphs of
$x = -3$, $x = 5$, $y = -2$ and $y = 4$.

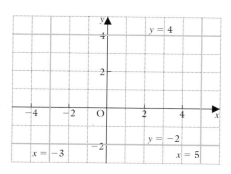

In the following questions take both x and y in the range -8 to $+10$.
Use squared paper and 2 squares for 1 unit on each axis.

1 Draw the straight line graphs of the following equations in a single
diagram: $x = 2$, $x = -5$, $y = \frac{1}{2}$, $y = -3\frac{1}{2}$

2 Draw the straight line graphs of the following equations on a single
diagram: $y = -5$, $x = -3$, $x = 6$, $y = 5.5$
Name the shape of the area enclosed by these lines.

3 On one diagram, draw graphs to show the following equations:
$x = 5$, $y = -5$, $y = 2x$
Write down the coordinates of the three points where these lines
intersect. What kind of triangle do they form?

4 On one diagram, draw the graphs of the straight lines with equations
$x = 4$, $y = -\frac{1}{2}x$, $y = 3$
Write down the coordinates of the three points where these lines
intersect. What kind of triangle is formed by these lines?

5 On one diagram, draw the graphs of the straight lines with equations
$y = 2x + 4$, $y = -5$, $y = 4 - 2x$
Write down the coordinates of the three points where these lines
intersect. What kind of triangle is enclosed by the lines?

PARALLEL AND PERPENDICULAR LINES

The line $y = 5x - 1$ has gradient 5.

Any line parallel to $y = 5x - 1$ also has gradient 5, i.e.

> two lines are parallel when they have the same gradient

Now consider the triangle ABC.

The gradient of the line through A and B is $\frac{4}{3}$.

Rotating triangle ABC about A by 90° anticlockwise gives triangle ADE where D is the point $(3, -3)$.

Now AD is perpendicular to AB and the gradient of the line through A and D is $-\frac{3}{4}$.

Gradient AB × gradient AD $= -1$.

This is true for any two perpendicular lines,

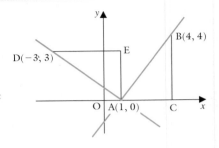

i.e.   > when two lines are perpendicular, the product of their gradients is -1

EXERCISE 13J

> Find the equation of the line through $(0, -5)$ that is parallel to the line $y = 3 - 2x$.
>
> The equation is $y = -2x - 5$.   The line has gradient -2 as it is parallel to $y = 3 - 2x$ and it cuts the y-axis at the point where $y = -5$. So $m = -2$ and $c = -5$.

Find the equation of the line parallel to the given line that passes through the given point.

1 $y = x - 3$, $(0, 2)$ **3** $y = -\frac{1}{2}x + 4$ $(0, 1)$ **5** $y = 1 - 4x$, $(0, -1)$

2 $y = 3x + 1$, $(0, 3)$ **4** $y = 3 - x$, $(0, 2)$ **6** $y = \frac{2}{5}x - 2$, $(0, -\frac{1}{5})$

> Find the equation of the line through $(0, -5)$ that is perpendicular to the line $y = 3 - 2x$.
>
> If the perpendicular line has gradient m,
> then $(-2)(m) = -1$ so $m = \frac{1}{2}$.   The line has gradient $\frac{1}{2}$ and it cuts the y-axis at the point where $y = -5$. So $m = \frac{1}{2}$ and $c = -5$.
> The equation is $y = \frac{1}{2}x - 5$.

Find the equation of the line perpendicular to the given line that passes through the given point.

7 $y = x - 5$, $(0, 1)$ **9** $y = -x + 4$, $(0, 3)$ **11** $y = 1 - 2x$, $(0, -2)$

8 $y = 2x + 1$, $(0, 2)$ **10** $y = 3 - x$, $(0, 7)$ **12** $y = \frac{2}{3}x - 2$, $(0, \frac{1}{3})$

13 State whether each of the following pairs of lines are parallel, perpendicular or neither.

   a $y = 1 + x$ **b** $y = 1 - 3x$ **c** $y = 5x + 3$
   $y = 3 - x$ $y = 5 + \frac{1}{3}x$ $y = 5x - 2$

**CONVERSION
GRAPHS**

Straight line graphs are useful in a variety of practical applications where the coordinates of points represent real quantities.

For example, if you go to Spain on holiday, it helps if you can convert a price given in pesetas to the equivalent price in sterling and vice-versa.

When the exchange rate is 210 pesetas (pta) to £1,
then, if n pta is equivalent to £N, we see that $n = 210N$,

Comparing this formula with $y = mx$

shows that plotting values of n against values of N will give a straight line through the origin.

To draw the graph we need points on the line; we can get these directly from the exchange rate, i.e. as £1 = 210 pta, £10 = 2100 pta, and so on.

We can put these values in a table, and then plot the points.

£ Sterling	1	10	100
Pesetas	210	2100	21 000

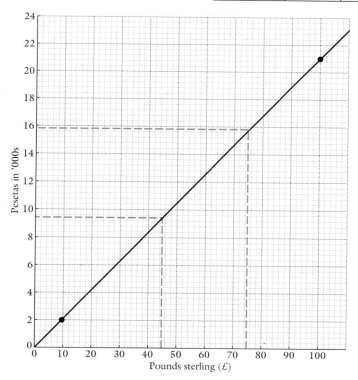

Pesetas in '000s

Pounds sterling (£)

From the graph we can immediately see that £45 ≈ 9400 pesetas and that 15 800 pesetas ≈ £75

The cost in pesetas is directly proportional to the cost in pounds sterling as the two costs are always in the ratio 210 : 1. The graph representing any two quantities that are directly proportional is like this one, that is a straight line through the origin

1 Use the conversion graph opposite for this question.

    a Why do you think this graph is drawn on graph paper rather than 5 mm squared paper?

    b The scales on the axes are not equal. Why do you think that 1 cm for 10 units is *not* used for both scales?

    c Can you find the cost in pounds sterling of a mountain bike priced at 100 000 pta from this graph?

    d To what values would the axes need to be scaled for this graph to be used to convert prices up to £500 into pesetas?

    e Why do you think the readings from the graph are given as approximations?

2 This graph can be used to convert values up to £100 into Norwegian kroner. Use the graph to find

    a the cost in pounds sterling of a pair of shoes priced at 760 kroner

    b how many kroner are equivalent to £46

    c the exchange rate on which this graph is based.

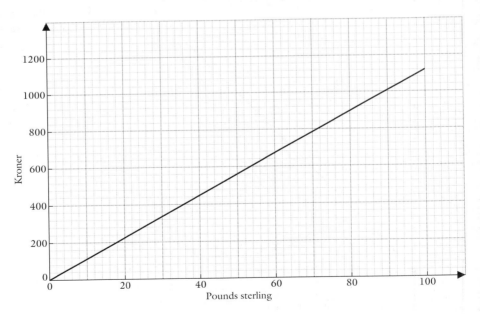

3 There are electronic gadgets that will convert from one currency into any other currency. The calculations can also be done using an ordinary calculator. Discuss the advantages and disadvantages that conversion graphs have over electronic aids.

Use 2 mm graph paper for the remaining questions in this exercise.

4 The table shows the conversion from US dollars to pounds sterling for various amounts of money.

US dollars	50	100	200
Pounds (£)	35	70	140

Plot these points on a graph and draw a straight line to pass through them. Let 5 cm represent 50 units on both axes.
Use your graph to convert

a 160 dollars into pounds (£)   **c** £122 into dollars

b 96 dollars into pounds (£)   **d** £76 into dollars

e What is the exchange rate between pounds (£) and dollars ($) that this graph is based on ?

5 Marks in an examination range from 0 to 65. Draw a graph which enables you to express the marks in percentages from 0 to 100.
Note that a mark of 0 is 0% while a mark of 65 is 100%.
Use your graph

a to express marks of 35 and 50 as percentages

b to find the original mark for percentages of 50% and 80%.

6 The table gives temperatures in degrees Fahrenheit (°F) and the equivalent values in degrees Celsius (°C).

Temperature in °F	50	104	158	194
Temperature in °C	10	40	70	90

Plot these points on a graph for Celsius values from 0 to 100 and Fahrenheit values from 0 to 220. Let 2 cm represent 20 units on each axis. Use your graph to convert

a 97 °F into °C   **c** 25 °C into °F

b 172 °F into °C   **d** 80 °C into °F

e Explain why this graph does not pass through the origin.

7 **a** Given that 20 mph = 32 km/h, find in km/h

   i 40 mph   **ii** 60 mph

b Use the values found in part **a** to draw a graph to convert from speeds in mph to speeds in km/h.

8 Given that 1 gallon is equivalent to 4.546 litres, draw a graph to convert from 0 to 100 litres into gallons.

Given an example of a situation where this conversion chart would be useful.

MIXED EXERCISE.

EXERCISE 13L

1 Write down the gradient of the line whose equation is $y = 3x - 2$.

2 Draw a sketch of the line whose equation is $y = -3x$.

3 What is the value of y where the line $y = 3x - 2$ cuts the y-axis?

4 The table gives the coordinates of three points on a line.
What is the equation of the line?

x	-2	0	2
y	-1	1	3

5 $A(3, a)$ and $B(b, 1)$ are points on the line $y = -4x + 2$.
Find a and b.

6 For the line shown find

a the gradient
b the y-intercept
c the equation.

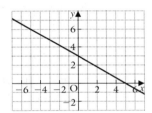

INVESTIGATIONS

1

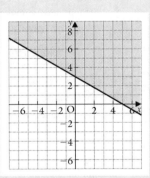

a What is the relationship between the coordinates of points on the line shown in this diagram?

b What can you say about the relationship between the coordinates of points in the shaded area?

c How can you describe the shaded area? Explain your reasoning.

2 The diagram shows a section through a swimming pool.

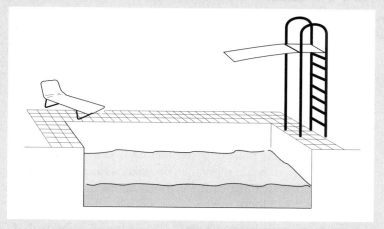

The pool is being filled with water at a constant rate of 50 litres per minute.

a Which of these graphs illustrates how the depth, d m, of water in the pool changes with time, t minutes?

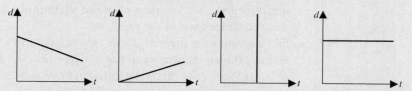

Give reasons for your choice and why you rejected the other graphs.

b Another pool has a cross-section like this.

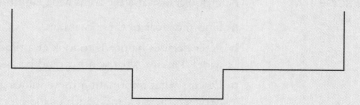

Sketch a graph showing how the depth of water in this pool varies with time as it is filled with water at a constant rate.

CURVED
GRAPHS

Peter is designing floor tiles. He has decided to make rectangular tiles that are twice as long as they are wide.

He must now consider what size to make the tiles and whether to make them in more than one size. One factor in the decision must be the floor area that each tile covers as this is clearly related to the lengths of the sides of the tile.

To get some idea of how the area varies with different lengths of the sides, Peter can

- find the area for some particular lengths; for example, for a tile measuring 5 cm by 10 cm, the area is $5 \times 10 = 50 \text{ cm}^2$.
  This is likely to lead to values being found in a haphazard order where it is difficult to see any pattern in the relationship between the areas and the dimensions of the tiles.
- find a formula for the area, $A \text{ cm}^2$, in terms of the width, x cm, of a tile and use this to tabulate values of A that correspond to some values of x. If Peter does this systematically, he may be able to get some sense of how the area varies with the length of the sides.
- try and get a 'picture' of how area varies with the dimensions.

EXERCISE 14A

1 Peter looks first at a tile 5 cm wide, and then at a tile 12 cm wide.

    a Find the areas of these two tiles.

    b Peter decides he needs to look at a tile between 5 cm and 12 cm wide. Discuss what width would be a sensible choice.

    c Discuss what information these values give about the relationship between the area and width of the tiles.

2 Peter decided to use a formula to work out some values for the areas of different width tiles.

The area of a rectangle is length × width.

Therefore $A = 2x \times x$

i.e. $A = 2x^2$

2x cm

A cm²

x cm

Peter used this formula to find the values in the table.

x	1	2	4	6	8	10	12
A	2	8	32	72	128	200	288

a Discuss what the results show about the relationship between the area covered by a tile and its width.

b Discuss whether it is possible to find easily the width of a tile that covers an area of $150\,\text{cm}^2$.

3 Discuss how the results in the table could be illustrated and if any of your suggestions help with the problems in question **2**, parts **a** and **b**.

GRAPHS
INVOLVING
CURVES

You may have discovered, from question **3** above, that plotting values of A against values of x on a graph gives a useful picture of how A varies when x varies.

When these values are plotted, we find that the points do not lie on a straight line, but they do lie on a smooth curve.

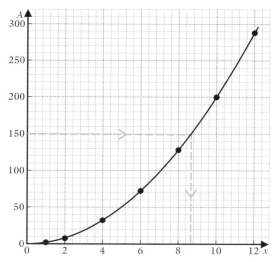

Even with the scales on the axes being far from equal, the graph shows that A increases at a much faster rate than x.

This graph can be used to 'read' the width of a tile whose area is $150\,\text{cm}^2$, or any other area.

When the area is $150\,\text{cm}^2$, i.e. when $A = 150$,
from the graph we see that x is about **8.7**.
So a tile whose width is estimated as $8.7\,\text{cm}$ will cover an area of $150\,\text{cm}^2$.

When two quantities that are related are plotted one against the other we often find that the points lie on a smooth curve. If they do, useful information can be found easily from the graph.

1 In the text before this exercise, we read the value of x when $A = 150$ and gave x as 'about **8.7**'. Discuss why we cannot give an exact value for x.

2 This graph shows the relationship between Jason's height and his age (the points shown give his heights on his birthdays).

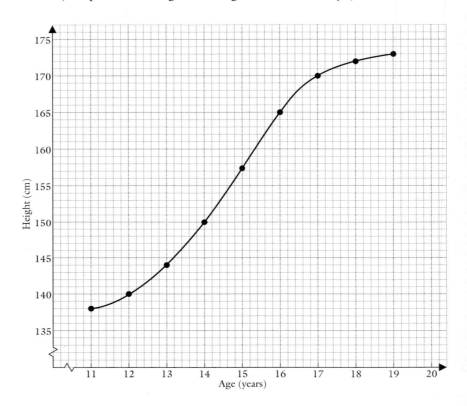

a Use the graph to estimate Jason's height when he was

   i $15\frac{1}{2}$ years old   **ii** $13\frac{1}{2}$ years old.

b Estimate how old Jason was when he was 171 cm tall.

c The flattest part of the curve is between 18 and 19 on the horizontal axis. How would you describe his growth rate for this part of the curve ?

d Between which two consecutive birthdays did he grow most ? How would you describe the part of the curve you used to find this ?

e Would you be able to get as much information from the graph if the vertical scale started at zero ? Explain your answer.

3 The weights of lead spheres of various diameters are shown in the table.

Diameter, D, in mm	4	5.2	6.4	7.2	7.9	8.8
Weight, W, in grams	380	840	1560	2230	2940	4070

Use 2 cm for 1 unit on the D-axis and 2 cm for 500 units on the W-axis to plot this information on a graph.
Use a pencil to draw a smooth curve through the points – you may need several attempts to do this. Most people find the best way to draw a curve is to keep their hand on the inside of the curve, like this:

It also helps to try and 'see' the curve before you start drawing it.
Use your graph to estimate

a the weight of a lead sphere of diameter 6 mm

b the diameter of a lead sphere of weight 2 kg.

4 Recorded speeds of a motor car at various times after starting from rest are shown in the table.

Time in seconds	0	5	10	15	20	25	30	35	40
Speed in km/h	0	62	112	148	172	187	196	199	200

Taking 2 cm for 5 sec and 1 cm for 10 km/h, plot these results and draw a smooth curve to pass through them.

Use your graph to estimate

a the time when the car reaches   **i** 100 km/h   **ii** 150 km/h

b its speed after   **i** 13 sec   **ii** 27 sec.

5 The weight of a puppy at different ages is given in the table.

Age, A, in days	10	20	40	60	80	100	120	140
Weight, W, in grams	50	100	225	425	750	875	950	988

Draw a graph to represent this data, taking 1 cm = 10 days on the A-axis and 1 cm = 50 g on the W-axis.
Hence estimate

a the weight of the puppy after   **i** 50 days   **ii** 114 days

b the age of the puppy when it weighs   **i** 500 g   **ii** 1000 g

c the weight it puts on between day 25 and day 55

d its birth weight

e the period of 10 days within which its weight increased most.

6 The cost of fuel, £ C, per nautical mile for a ship travelling at various speeds, v knots, is given in the table.
(1 knot means 1 nautical mile per hour).

v	12	14	16	18	20	22	24	26	28
C	18.15	17.16	16.67	16.5	16.5	16.67	16.94	17.36	17.82

Draw a graph to show how cost changes with speed.
Use 1 cm = 1 knot and 5 cm = £1. (Take 16 as the lowest value for C and 10 as the lowest value for v.)

Use your graph to estimate

a the most economical speed for the ship and the corresponding cost per nautical mile

b the speeds when the cost per nautical mile is £17

c the cost when the speed is   **i** 13 knots   **ii** 24.4 knots.

7 The time of sunset at Greenwich on different dates, each two weeks apart, is given in the table.

	May		June		July		Aug	
Date, D	15	29	12	26	10	24	7	21
Time, T	2045	2105	2118	2122	2116	2101	2039	2012

Using 1 cm for 1 week on the D-axis and 8 cm for 1 hour on the T-axis, plot these points on a graph and join them with a smooth curve. Take 1900 as the lowest value of T and start D at 15 May.

From your graph estimate

a the time of sunset on 17 July

b the date on which the sun sets at 2027.

8 A rectangle measuring l cm by b cm has an area of 24 cm^2. The table gives different values of l with the corresponding values of b.

l	1	2	3	4	6	8	12	16
b	24		8		4		2	1.5

Copy and complete the table and draw a graph to show this information. Join the points with a smooth curve.
Take 1 cm for 1 unit on the l-axis and 1 cm for 2 units on the b-axis.

Use your graph to estimate the value of

a l when b is   **i** 14   **ii** 2.4

b b when l is   **i** 10   **ii** 2.8

**CONSTRUCTING
A TABLE FROM A
FORMULA**

In the last exercise, the data was given in tables. In many problems we start with a formula and have to construct our own table.

On page 278, when Peter wanted to find how the width of his tiles affected the area they cover, he started with tiles twice as long as they were wide and this gave the formula

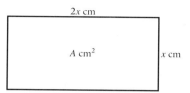

$$A = 2x^2$$

To draw a graph showing this relationship, we need some corresponding values of A and x.
We will take values of x from 0 to 4 at half-unit intervals.
(Remember that $2x^2$ means $2 \times x^2$, i.e. $2 \times x \times x$.)

When $x = 0$, $\quad A = 2 \times 0^2 \quad = 2 \times 0 \quad = 0$,
when $x = 0.5$, $\quad A = 2 \times 0.5^2 = 2 \times 0.25 = 0.5$, and so on.

We could continue with this list and then enter the corresponding values of A and x in a table, but it is more efficient to use a table from the start, i.e.

x	0	0.5	1	1.5	2	2.5	3	3.5	4
x^2	0	0.25	1	2.25	4	6.25	9	12.25	16
$A(= 2x^2)$	0	0.5	2	4.5	8	12.5	18	24.5	32

Using values of x on the horizontal axis and values of A up the vertical axis, we plot these points. Then we can draw a smooth curve through the points.

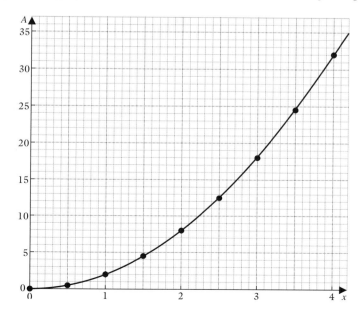

Sometimes a situation leads to a formula where the letters can stand for negative as well as positive numbers.

For example at various times after a stone is thrown up from the top of a cliff, the height of the stone relative to the cliff top may be positive or negative, i.e. the stone may be above or below the cliff top.

EXERCISE 14C

1 A wave generator in a pool causes the water level to rise and fall. There is a mark on the side wall of the pool showing the normal water level.

When the wave generator is on, the level, h cm, of the water above normal can be estimated from $h = 1 - t^2$ for values of t from -1.5 to 1.5, where t seconds is the time from where the water level is at its highest point.

a Copy and complete this table.

t	-1.5	-1	-0.5	0	0.5	1	1.5
t^2	2.25		0.25				
$h = 1 - t^2$	-1.25		0.75				

b Plot these points on a graph using 2 cm for one unit on both axes. Draw a smooth curve through the points.

c Where is the water level

i half a second before it reaches its highest point

ii half a second after it reaches its highest point?

d What is the value of t when $h = -1$? Interpret this value in the context of the problem.

e Where is the water level 1.2 seconds after it reaches its highest point?

2 A stone is dropped down a vertical mine shaft. After t seconds it has fallen s metres where $s = 5t^2$. (In this question the downward measurement is positive.)

a Construct a table to show the relationship between s and t for values of t from 0 to 6 at half-unit intervals.

b Use these values to draw a graph, using a scale of 1 cm to $\frac{1}{2}$ second on the horizontal axis and 1 cm to 25 metres on the vertical axis.

Use your graph to find

c how far the stone has fallen 3.4 seconds after it was dropped

d how long the stone takes to fall 100 metres.

**THE EQUATION
OF A CURVE**

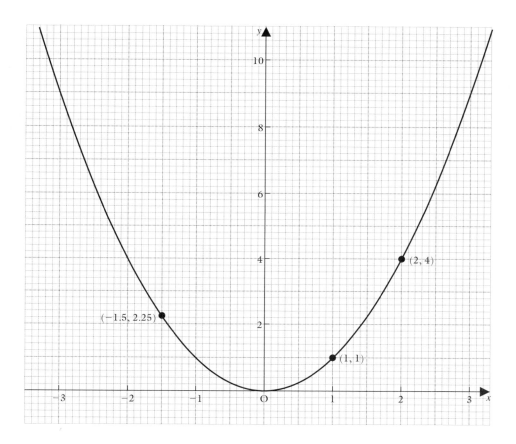

If we look at the relationship between the x- and y-coordinates of particular points on this curve, we find that the y-coordinate is always equal to the square of the x-coordinate.

This relationship between the coordinates is true for all the points on this curve, that is, for every point on the curve $y = x^2$.

This formula giving the y-coordinate of a point on the curve in terms of its x-coordinate is called the *equation of the curve*.

We refer briefly to *the curve* $y = x^2$.

To draw the graph of a curve we start by using the equation of the curve to make a table of corresponding values of x and y (in the same way as we used a formula earlier in this chapter).

When drawing a curve

- do not take too few points – about ten are usually needed
- to decide where to draw the y-axis, look at the range of x-values in your table
- to decide where to draw the x-axis, look at the range of y-values in your table
- to draw a smooth curve through the points, turn the graph paper to a position where your wrist is inside the curve
- be prepared to add more points to the table for any section of the curve where its shape is not clear. This is often necessary round where the curve turns.

EXERCISE 14D

1 **a** Make your own copy of the graph of $y = x^2$ using the graph on the previous page as a guide. Use 2 cm for 1 unit on the x-axis and 1 cm for 1 unit on the y-axis.

   b On your copy, draw the graph of $y = 3x^2$ using values of x from -1.5 to 1.5 to make your table.

   c Repeat part **b** for the graph of $y = \frac{1}{2}x^2$ using values of x from -3 to 3.

   d What can you say about the shape of the graph of $y = ax^2$ when a is any positive number?

   e Use a graphics calculator to investigate the shape of the curve $y = ax^2$ for some other positive values of a. Sketch the shapes of the graphs. How does the evidence you find affect your answer to part **d**?

2 **a** Make another copy of the graph on page 285 but this time extend the y-axis down to give values of y from -4.

   b On your copy draw the graphs of $y = x^2 + 4$ and $y = x^2 - 4$ using values of x from -2 to 2 to make the tables for both graphs.

   c What can you say about the shape of the graph of $y = x^2 + a$ when a is any number, positive or negative?

   d Use a graphics calculator to investigate the shape of the curve $y = x^2 + a$ for some other values of a. Sketch the shapes of the graphs. How does the evidence you find affect your answer to part **c**?

3 Use a graphics calculator to investigate the shape of the graphs of $y = -x^2$, $y = -2x^2$, $y = -3x^2$. How do the shapes of these curves compare with the shapes of the curves in question **1**?

Draw the graph of $y = x^2 + x - 6$ for whole number values of x from -4 to $+4$. Take 1 cm as 1 unit on the x-axis and 1 cm as 2 units on the y-axis. Use your graph to find

a the lowest value of $x^2 + x - 6$ and the corresponding value of x

b the values of x when $x^2 + x - 6$ is 4.

x	-4	-3	-2	-1	0	1	2	3	4	$-\frac{1}{2}$
x^2	16	9	4	1	0	1	4	9	16	$\frac{1}{4}$
x	-4	-3	-2	-1	0	1	2	3	4	$-\frac{1}{2}$
-6	-6	-6	-6	-6	-6	-6	-6	-6	-6	-6
$x^2 + x - 6$	6	0	-4	-6	-6	-4	0	6	14	$-6\frac{1}{4}$

It is clear that the middle of the curve is where $x = -\frac{1}{2}$.
To find this point accurately, we can add another column to the table.

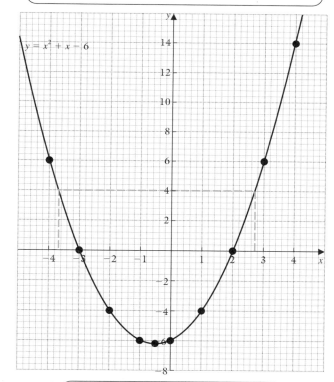

$y = x^2 + x - 6$

a

$x^2 + x - 6 = y$, so we want the lowest value of y.

From the graph, the lowest value of $x^2 + x - 6$ is $-6\frac{1}{4}$.
This occurs when $x = -\frac{1}{2}$

b

$x^2 + x - 6 = y$, so when $x^2 + x - 6$ is 4, y is 4.

The values of x when $x^2 + x - 6$ is 4 are -3.7 and 2.7.

4 Draw the graph of $y = x^2 - 2x - 3$ for whole number values of x in the range -3 to 5. Take 2 cm as 1 unit for x and 1 cm as 1 unit for y. Use your graph to find

    a the lowest value of $x^2 - 2x - 3$ and the corresponding value of x

    b the values of x when $x^2 - 2x - 3$ has a value of
        i 1   **ii** 8

5 Draw the graph of $y = 6 + x - x^2$ for whole number values of x from -3 to 4. Take 2 cm as 1 unit on both axes. Use your graph to find

    a the highest value of $6 + x - x^2$ and the corresponding value of x

    b the values of x when $6 + x - x^2$ has a value of
        i -2 **ii** 4

THE SHAPE OF THE CURVE $y = ax^2 + bx + c$

The equations of all the curves in the last exercise have an x^2 term and/or an x term and/or a number.

All these curves have the same basic shape.

When the x^2 term is positive, e.g. $y = x^2$, the curve is this way up.

If the x^2 term is negative, e.g. $y = -x^2 + x - 6$, the shape is upside down.

All curves with this shape are called *parabolas*.
The equation of a parabola has the form $y = ax^2 + bx + c$, where a, b and c are numbers, and a cannot be zero but b and/or c can be zero.

For example, $y = x^2$, $y = 3x^2$, $y = x^2 - 6$, $y = -x^2 + x - 6$

EXERCISE 14E

1 The equation of a curve is $y = 6x^2$. Which of these sketches shows the curve?

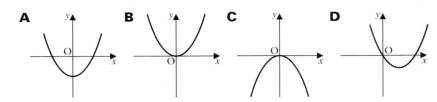

2 Explain why you rejected some of the curves in question **1**.

3 Draw a sketch of the curve $y = -5x^2$.

INVESTIGATION

The readings in the table were taken from the computer in a car that showed the petrol consumption for different speeds.

Speed, v mph	5	10	20	30	40	50	60	70
Petrol consumption, n miles per gallon	9	15	29	37	42	38	27	15

a Plot these points on a graph using 1 cm for 5 mph on the horizontal axis and 1 cm for 2 miles per gallon on the vertical axis.

b Now draw the curve whose equation is $n = 2v - \dfrac{v^2}{40}$.

c How well does the curve fit the points?

d Try adjusting the equation of the curve to get a better fit.

e Does your curve fit the points better for some values than others. Why do you think this is?

f Can you use your curve to predict the petrol consumption when the car is travelling at 100 mph? Explain your answer.

You can use a spreadsheet to work out a table of values to plot a graph. For example, when $y = x^2 + x + 1$, to find the values of y for values of $x = 0, 0.2, 0.4, \ldots 2$ enter 0 in cell A1, 0.2 in cell B1, 0.4 in cell C1 and so on up to 2 in cell K1.

Then in row 2 use the formula "+A1^2+A1+1" in the cell below 0 and then use "fill" to give the values of y, i.e.

x	0	0.2	0.4	0.6	0.8	1	1.2	1.4	1.6	1.8	2
y	1	1.24	1.56	1.96	2.44	3	3.64	4.36	5.16	6.04	7

You can also use the chart tool to draw the graph:

You need to use values of x that are close together because the chart program joins the points with straight lines.

Try using a spreadsheet to draw the graphs of some of the questions in Exercise 14D.

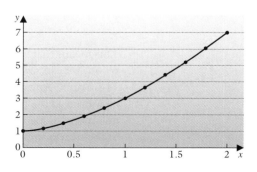

CONTINUOUS DATA

The length of a piece of wood is being measured

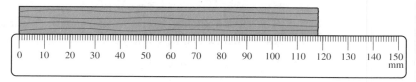

The length is close to 120 mm, so we can give the length as 120 mm to the nearest 10 millimetres.

If we magnify the scale where the end of the wood is, we can see that the length is 118 mm to the nearest millimetre.

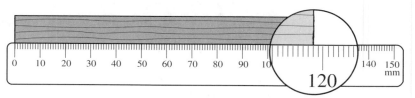

If we use stronger magnification, we can see that it becomes impossible to measure this length any more accurately because the edge is not smooth enough.

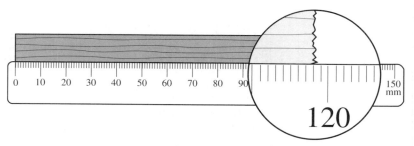

- This means that we cannot give an exact value for the length of this wood; we can say that the length lies between 117.5 mm and 118.5 mm. Any other piece of wood will have a length that can be measured to the nearest centimetre, or millimetre, or tenth of a millimetre, depending on the smoothness of the ends and the accuracy of the measuring scale. But it is *not* possible to give a length exactly. In general the length of a piece of wood can be anywhere on a scale.

- On the other hand if you count the number of pupils in your class, you can give an exact value. For example, 28 pupils *is* exact. If we show this number on a number line, it can be marked at only one point. If we mark another number of pupils on the number line, it will be at a separate and distinct point (we cannot have 1.5 pupils, or any other part of a pupil!)

EXERCISE 15A

1 Write down some possible values for

   a the number of people standing at a bus stop

   b the shoe size of a 13-year-old boy

   c a person's weight

   d the time it takes to run 100 metres.

2 Discuss with your group whether the values in question **1** are exact or can only be given correct to a number of significant figures.

3 Jane's height is 152 cm correct to the nearest centimetre. Roy's height is 152 cm correct to the nearest centimetre.
Discuss this statement: 'Jane and Roy are exactly the same height.'

4 You probably have been, or soon will be, involved in a conversation like this: Bus driver: 'How old are you?'
           Sam: 'I'm 13.'
Does Sam mean that he was 13 on his last birthday, or does he mean that he is 13 years old correct to the nearest year? (We will assume that Sam is telling the truth!)
Discuss the difference between these possibilities.

DISCRETE AND
CONTINUOUS
DATA

The examples in the last exercise show that there are two different types of values:

- those that are exact and distinct; for example, the possible number of people standing at a bus stop, possible shoe sizes.
Quantities like these are *discrete*.

- those that can only be given in a range on a continuous scale; for example, heights, weights, times.
Quantities like these are *continuous*.

EXERCISE 15B

State whether each of the following quantities is
a discrete   **b** continuous.

   1 The number of pupils in a school.

   2 The time it takes you to get to school.

   3 The number of peas in a pod.

   4 The length of a classroom.

5 Your friends' weight.

6 The volume of liquid in a bottle.

7 The number of words on a page of this book.

8 The time it takes for a train to travel from Manchester to Birmingham.

9 The number of passengers on a train.

10 The temperature at midday on the Met Office roof.

11 The hourly rate of pay for a person working in a supermarket.

12 The age of a pupil in your school.

13 The cost of a pair of shoes.

ROUNDING CONTINUOUS VALUES

The examples in **Exercise 15A** show that sometimes we round values to the nearest unit such as saying that a length is 118 mm to the nearest millimetre. This means that the length lies somewhere in the interval from half a millimetre less than 118 mm up to half a unit greater than 118 mm, as shown on the scale:

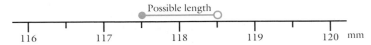

If the length is x mm, we can say that $117.5 \leqslant x < 118.5$.

Sometimes we round values down. For example, if you are asked to give your age, you will usually give the number of years reached on your last birthday; when Sam says he is 13 years old, he must be less than 14 but could be any age between 13 and 14.

i.e.

If Sam is x years old we could write $13 \leqslant x < 14$

There are occasions when we round up. If we spend 1 hour 20 minutes in this car-park, we need to round the time up to 2 hours to calculate the charge made.

Avenue Road Car Park
Tariff:
up to 1 hour – 50p
up to 2 hours – £1
up to 3 hours – £1.50
over 3 hours – £5

A time rounded up to 3 hours could be any time from just over 2 hours up to and including 3 hours.

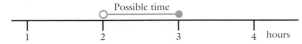

If t hours is the time we can write $2 < t \leqslant 3$.

EXERCISE 15C Copy this number line and mark the range in which the given times can lie.

1 3 hours to the nearest hour

2 5 hours rounded down to the nearest hour

3 2 hours rounded up to the next complete hour

We are told that an apple weighs 25 g to the nearest gram. Show, on a sketch the range in which the weight lies.

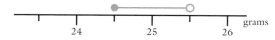

The weight is from 24.5 g up to 25.5 g.

Show on a sketch the range in which these given measurements lie.

4 12 m to the nearest metre

5 4 years to the nearest complete year

6 120 cm to the next complete centimetre

7 15 seconds to the nearest complete second

8 25 mm to the nearest complete millimetre

9 54 kg to the nearest kilogram

10 10 minutes to the next complete minute.

11 The weight of a tomato is 25 g correct to the nearest gram. The weight of another tomato is also 25 g correct to the nearest gram. Is it true to say that these two tomatoes have exactly the same weight? Explain your answer.

12 The lengths of two stakes are both given as 1.4 m correct to 1 decimal place? Are these two stakes the same length? Explain your answer.

This is a list of the heights of 55 children. Each height is rounded down to the nearest complete centimetre, so 141.2 cm and 141.8 cm are both rounded down to 141 cm.

The list has been extracted from a database which has sorted the heights into numerical order.

131	134	136	137	139	141	142	144	145	147	149
132	134	136	137	139	141	142	144	145	148	150
132	134	136	138	140	142	143	144	146	148	150
133	135	136	138	140	142	143	144	147	149	152
133	135	137	139	140	142	144	145	147	149	153

This information can be shown in the following frequency table.

Height (to the nearest complete cm)	131	132	133	134	135	136	137	138	139	140	141	142	143	144	145	146	147	148	149	150	151	152	153
Frequency	1	2	2	3	2	4	3	2	3	3	2	5	2	5	3	1	3	2	3	2	0	1	1

There are 2 children whose heights are recorded as 132 cm.
This means that each child's height is in the range 132 cm up to (but not including) 133 cm.

However the two children are unlikely to be exactly the same height.
This can be shown in the frequency table by giving the heights, x cm, as

$131 \leqslant x < 132, \quad 132 \leqslant x < 133,$ and so on.

Because of the space that this notation takes, it is sometimes simplified to 131–, 132–, and so on, where 131–, 132–, means a number from 131 up to, but not including 132, etc.

Height (cm)	131–	132–	133–	134–	135–	136–	137–	138–	139–	140–	141–	142–	143–	144–	145–	146–	147–	148–	149–	150–	151–	152–	153–
Frequency	1	2	2	3	2	4	3	2	3	3	2	5	2	5	3	1	3	2	3	2	0	1	1

Now we can see that *continuous values are already in groups.*

It is easier to understand this information if it is grouped into wider ranges, so we will start the first group at 130 cm, the second group at 135 cm, the third group at 140 cm, and so on. This will give us five groups which we can write as

$$130 \leqslant x < 135, \qquad 135 \leqslant x < 140, \qquad 140 \leqslant x < 145,$$

$$145 \leqslant x < 150, \qquad 150 \leqslant x < 155$$

Any height that is less than 135 cm belongs to the first group, but a height of 135 cm belongs to the second group.

Looking down the list of heights we can see that there are 8 children whose heights are in the first group, 14 children whose heights are in the second group, and so on.

We can write this information in a frequency table.

Height (cm)	Frequency
$130 \leqslant x < 135$	8
$135 \leqslant x < 140$	14
$140 \leqslant x < 145$	17
$145 \leqslant x < 150$	12
$150 \leqslant x < 155$	4
	Total: 55

EXERCISE 15D

1 Use the frequency table immediately above to answer the following questions.

a How many children had a height less than 135 cm?

b How many children had a height of at least 150 cm?

c In which group do the heights of most children lie?

d Two children were away when the survey was carried out. Their heights are 152 cm and 140 cm rounded down to the nearest complete centimetre.
In which group should we place
i 152 cm **ii** 140 cm?

2 This is a list of the weights of 100 adults. The weights are in kilograms rounded *up* to the next complete kilogram, e.g. 63.2 kg is rounded up to 64 kg. The list is in numerical order.

```
47  51  54  60  63  64  66  68  69  70  72  78  80  85  100
48  51  54  61  63  64  66  68  70  70  73  78  80  88  104
49  52  55  62  63  65  66  68  70  70  73  79  80  90  110
49  52  58  62  63  65  66  68  70  71  74  79  80  92  112
49  53  58  63  63  65  67  69  70  71  75  79  82  94  115
50  53  59  63  64  66  67  69  70  72  78  80  83  95  118
51  53  60  63  64  66  68  69  70  72
```

a What is the smallest weight?

b How many people have a weight that is 50 kg or less?

c Copy and complete this frequency table.

Weight, w kg	Frequency
$40 < w \leqslant 60$	
$60 < w \leqslant 80$	
$80 < w \leqslant 100$	
$100 < w \leqslant 120$	
	Total:

d How many people have a weight greater than 100 kg?

e How many people have a weight of 80 kg or less?

f Which group contains the largest number of weights?

3 For four weeks Emma kept a record of the time she had to wait for the bus to school each morning. The results are shown in this frequency table.

Time, t minutes	Frequency
$0 < t \leqslant 5$	7
$5 < t \leqslant 10$	9
$10 < t \leqslant 15$	3
$15 < t \leqslant 20$	1

a One morning Emma waited 8 minutes and 30 seconds. In which group should this time be placed?

b On how many mornings did Emma have to wait more than 15 minutes?

c How often did Emma wait no more than 5 minutes?

d On how many mornings did Emma record the length of her wait?

e Did Emma ever have to wait for 20 minutes? Explain your answer.

4 The pupils in Class 8T were given a list of the weights, in grams correct to the nearest gram, of apples gathered from one tree. Eddie made this frequency table to show the weights.

Weight, w grams	Frequency
$85.5 \leqslant w < 87.5$	13
$87.5 \leqslant w < 89.5$	20
$89.5 \leqslant w < 91.5$	45
$91.5 \leqslant w < 93.5$	31
$93.5 \leqslant w < 95.5$	15
$95.5 \leqslant w < 97.5$	9

a One weight in the list is 92 g. In which group should it be placed?

b How many apples were gathered from the tree?

c Can you tell how many apples weighed 90 g to the nearest gram? Explain your answer.

d Nigel made a different tally chart to group the weights.

Weight, w grams	Tally
$85 \leqslant w < 87$	
$87 \leqslant w < 89$	
$89 \leqslant w < 91$	
$91 \leqslant w < 93$	
$93 \leqslant w < 97$	

In which group would Nigel place an apple that weighed 90 g to the nearest gram?

e One apple weighs 89 g correct to the nearest gram. In what range does the weight of this apple lie?

f What problems would Nigel have in placing this weight of 89 g (to the nearest gram) in one of his groups? Would Eddie have the same problems?

5 The pupils in Class 8P were set 3 questions for homework and were asked to write down the number of minutes they spent on this homework.

This frequency table summarises the results.

Time, t minutes	Frequency
$0 < t \leqslant 5.5$	2
$5.5 < t \leqslant 10.5$	15
$10.5 < t \leqslant 15.5$	9
$15.5 < t \leqslant 20.5$	3
$20.5 < t \leqslant 25.5$	1

Use the information in the table to answer these questions. If you cannot give an answer, explain why.

a In which group does a time of 10 minutes, correct to the nearest minute, belong?

b How many pupils spent less than 10 minutes on their homework?

c How many pupils spent no time on their homework?

d How many pupils spent more than 15 minutes on their homework?

e What was the longest time spent by a pupil doing this homework?

BAR CHARTS FOR CONTINUOUS DATA

We can use the frequency table on page 295 to draw a bar chart.

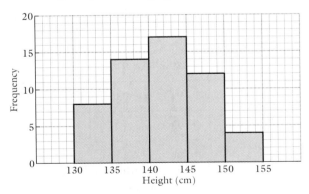

Notice that the horizontal axis gives the heights on a continuous scale, like part of a tape measure, so there are no gaps between the bars.

> A bar chart illustrating continuous data must have no gaps between the bars.

1 This bar chart summarises the times taken by all the pupils in Year 8 to complete a technology task.

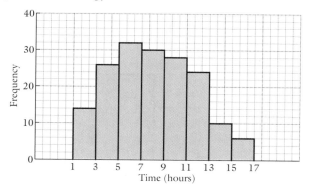

a How many pupils took more than 9 hours?

b How many pupils completed the task?

c There are 200 pupils in Year 8. How many of them did not complete the task?

d Jane was asked to estimate the number of pupils who spent less than 2 hours on this task. Which of these two answers do you think is the better estimate and why?   **i** 14   **ii** 7

e Is it true to say that most pupils took between 5 and 7 hours?

2 Here is a frequency table showing the times, in minutes, taken by the pupils in a class on their journeys from home to school on a particular morning.

Time, t minutes	Frequency
$0 \leqslant t < 10$	2
$10 \leqslant t < 20$	9
$20 \leqslant t < 30$	5
$30 \leqslant t < 40$	4
$40 \leqslant t < 50$	2
$50 \leqslant t < 60$	1

Copy and complete this bar chart, using the table.

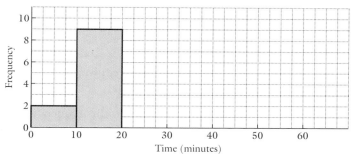

3 The heights of pupils belonging to a trampolining club were recorded; the bar chart summarises the information gathered.

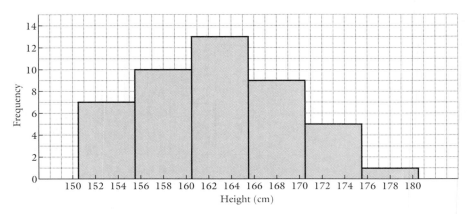

a How many heights were recorded?

b The heights were rounded when they were recorded. How do you think these heights were rounded and why do you think that?

c How many of these heights were more than 160 cm?

d In which group should a height recorded as 158 cm, to the nearest centimetre, be placed?

e How would you reply if asked to give the height of the shortest pupil in the club?

f Estimate the number of pupils who are taller than 168 cm, and explain how you get your estimate.

g What can you say about the number of pupils whose heights are 155 cm to 160 cm?

h Can you tell how many pupils are 162 cm tall?

4 At the health-centre, some babies were weighed one afternoon. Their weights, in kilograms, were recorded by the nurse as tally marks in this frequency table.

Weight, w kg	$4 \leqslant w < 8$	$8 \leqslant w < 12$	$12 \leqslant w < 16$
Tally	卌 //	卌 卌 /	卌 /
Frequency			

The next two babies were weighed at just under 12 kg and just over 12 kg.

a Add these weights to the frequency table and then complete the table.

b Draw a bar chart to illustrate this information.

5 Draw a bar chart to illustrate the data given in question **2**, **Exercise 15D**.

6 a Draw a bar chart to illustrate the data given in question **3**, **Exercise 15D**.

b Is it possible to say how the data has been rounded looking only at the bar chart?

7 This is a list of the weights in kilograms, rounded up to the nearest kg, of 30 fourteen-year-old boys.

$$
\begin{array}{cccccccccc}
50 & 56 & 60 & 62 & 65 & 65 & 67 & 67 & 68 & 69 \\
52 & 57 & 60 & 64 & 65 & 65 & 67 & 68 & 68 & 70 \\
55 & 57 & 61 & 64 & 65 & 66 & 67 & 68 & 69 & 75
\end{array}
$$

You are asked to draw a bar chart to illustrate this data.

a Decide on the groups that you will use and make a frequency table.

b Draw the bar chart.

c Which of your groups contains the largest number of weights?

8 These are two of the charts drawn to illustrate the data in question **7**.

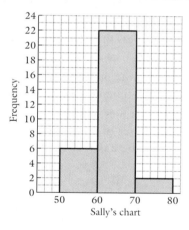
Sally's chart

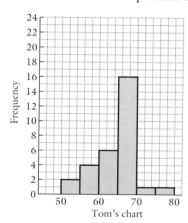
Tom's chart

a What groups did Sally use?

b What groups did Tom use?

c What are the advantages and disadvantages of Sally's choice of groups?

d What are the advantages and disadvantages of Tom's choice of groups?

e What are the advantages and disadvantages of your choice in question **7**?

9   The pupils in Class 8G were given two sets of five multiple choice questions to answer. They did the first set without any previous experience of this type of question. After the first set, they discussed the problems encountered and then they answered another set.

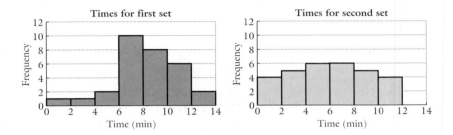

These bar charts illustrate the times taken by the pupils in Class 8G to answer each set of multiple choice questions.

a   Give two ways in which these distributions are different.

b   Jenny looked at these two bar charts and said, 'Practice helped us to do the second set more quickly.' Is she right?
Can you think of another reason why the times taken for the second test were, on average, less than those for the first test?

c   Pete took 13 minutes to answer the first set of questions. Is it true to say that he took less than 13 minutes to answer the second set? Justify your answer.

d   Stephen completed the first set in 8 minutes. What can you say, and why, about the time he took to complete the second set?

e   Erol thought these times would be easier to compare if they were drawn on the same set of axes. He used a chart drawing program and came up with these two charts.

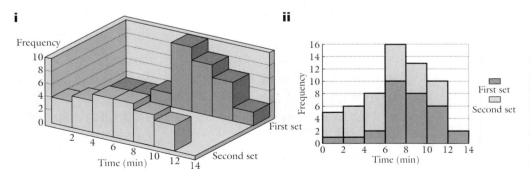

List some advantages and some disadvantages of each of these illustrations.

RANGE AND
MODAL GROUP

When values have been placed in groups we lose some of the information.

This bar chart, from page 298, illustrates the distribution of the heights of 55 children.

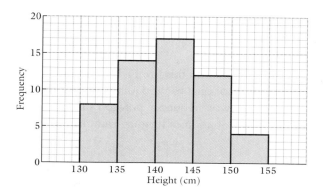

Using just this chart, we cannot tell how many times a height of 138 cm occurs. Also we cannot read from the bar chart either the smallest or the greatest height. This means that we cannot give the range of heights.

However we can *estimate* the range by assuming that the smallest height is at the lowest end of the first group, and that the largest height is at the top end of the last group,

i.e.   the smallest possible height is 130 cm

and   the largest possible height is 155 cm,

so the range is estimated as   155 cm − 130 cm = 25 cm.

> For a grouped frequency distribution
> the range is estimated as
> (the higher end of the last group) − (lower end of the first group)

Also, without looking at the original list, we cannot know which height occurs most often, so we cannot find the mode. However we can see that the group of heights from 140 cm to 145 cm contains more heights than any other group. We call this the *modal group*.

> For a grouped frequency distribution
> the modal group
> is the group with the highest number of items in it.

EXERCISE 15F

Give the modal group and estimate the range for the following grouped distribution from **Exercise 15E**.

1 Question **1**

3 Question **3**

2 Question **2**

4 **a** Question **8**, first set

    b Question **8**, second set

FREQUENCY POLYGONS

Another way to illustrate a grouped distribution is to draw a frequency polygon.

We can do this from a bar chart by marking a point in the middle of the top of each bar and joining these points with straight lines. This diagram shows a frequency polygon superimposed on the bar chart giving the distribution of heights from page 303.

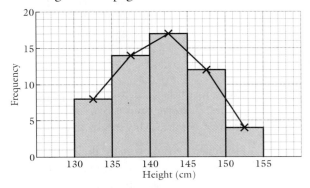

We can draw a frequency polygon directly from a frequency table without having to draw a bar chart first.

We do this by first identifying the middle value of each group (these are called the *mid-class values*).

For the distribution above, the mid-class value of the first group is halfway between 130 cm and 135 cm, i.e. $(130 + 135) \div 2$ cm $= 132.5$ cm.

The mid-class values of the other groups can be found in the same way.

It helps to add a column to the frequency table to show these mid-class values.

Height, x cm	Frequency	Mid-class value
$130 \leqslant x < 135$	8	132.5
$135 \leqslant x < 140$	14	137.5
$140 \leqslant x < 145$	17	142.5
$145 \leqslant x < 150$	12	147.5
$150 \leqslant x < 155$	4	152.5
	Total: 55	

Then, for each group, we plot a point on the graph above the mid-class value corresponding to the frequency, and then join all the points with straight lines.

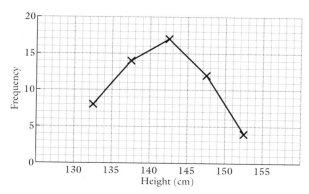

EXERCISE 15G   **1** Here is the frequency table constructed from the information in question **2** of **Exercise 15E**, showing the distribution of the times, in minutes, taken by the pupils in a class on their journeys from home to school on a particular morning.

Time, t minutes	Frequency	Mid-class value
$0 \leqslant t < 10$	2	5
$10 \leqslant t < 20$	9	15
$20 \leqslant t < 30$		
$30 \leqslant t < 40$		
$40 \leqslant t < 50$		
$50 \leqslant t < 60$		

a Copy and complete the table.

b Copy and complete the frequency polygon using your table.

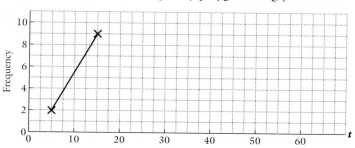

2 This frequency table gives the grouped weights of a sample of 100 eggs.

Weight (grams)	Frequency	Mid-class value
25–	25	27.5
30–	28	
35–	20	
40–	12	
45–	9	
50–55	6	

a Copy and complete the table.

b Illustrate the information with a frequency polygon.

c Give the modal group and estimate the range of these weights.

3 This frequency table gives the grouped weights of a different sample of 100 eggs.

Weight (grams)	Frequency
25–	22
30–	21
35–	23
40–	18
45–	8
50–	6
55–60	2

a Illustrate this information with a frequency polygon drawn on the same set of axes as the frequency polygon for question **2**.
(Use a different colour and make sure that you label your diagram to show which polygon illustrates which distribution.)

b Make one comparison of the distribution of weights in the two samples.

c What is the advantage of using frequency polygons rather than bar charts to illustrate these two distributions?

4 Draw a frequency polygon to illustrate this distribution of weights (given first in **Exercise 15E**, question **4**).

Weight, w kg	$4 \leqslant w < 8$	$8 \leqslant w < 12$	$12 \leqslant w < 16$
Tally	卌 //	卌 卌 /	卌 /
Frequency			

5 **a** Draw two frequency polygons on the same set of axes to illustrate the two distributions of times given in **Exercise 15E**, question **9**.

b The frequency polygons drawn for part **a** can be used to compare the two distributions. This comparison can also be made using the illustrations on page 302.
Discuss the advantages and disadvantages of each method.

6 Mrs Jones timed the pupils changing for games and this frequency polygon shows the results.

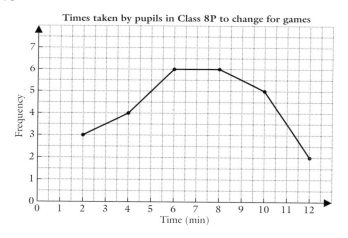

Times taken by pupils in Class 8P to change for games

a What are the mid-class values?

b Where does the first group start and where does it end?

c Where does the last group start and where does it end?

d Copy this diagram and superimpose a bar chart on your copy showing the same information.

e Make a frequency table showing this information. (Assume that the times have been rounded up to the nearest complete minute.)

f What is the range of times shown here?

g What is the modal group?

h How many pupils were timed?

7 After a lecture on the time being wasted by several members of Class 8P when changing for games, Mrs Jones repeated the check described in question **6**.

Times taken by pupils in Class 8P to change for games after lecture

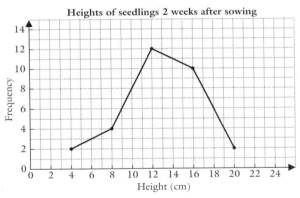

a Illustrate this distribution with a bar chart.

b What is the range and modal group of this distribution?

c How many pupils were timed?

d Say, with reasons, whether you think that the lecture had any effect on the times taken to change for games by the pupils in Class 8P.

8

Heights of seedlings 2 weeks after sowing

The diagram represents the results of a science experiment. Answer the following questions. If you cannot give an answer say why.

a How many seedlings were measured?

b How many seeds were planted?

c Estimate the range of heights.

d What was the most common height reached by the seedlings?

e Estimate the height which $\frac{2}{3}$ of these seedlings exceeded.

PRACTICAL
WORK

1 These are the heights (rounded up to the nearest complete centimetre) of 160 pupils in Year 7 at Headon Warren School. Each list is in numerical order.

GIRLS

127	143	147	148	152	155	157	158
134	143	147	148	153	155	157	159
134	144	147	148	153	155	157	160
134	144	147	149	154	155	157	160
134	145	147	150	154	155	157	160
137	145	147	150	154	156	157	161
138	145	147	151	154	156	157	161
139	146	147	151	155	156	157	161
140	146	148	152	155	156	158	163
141	146	148	152	155	157	158	165

BOYS

132	141	144	147	150	152	154	156
134	141	145	147	150	152	154	157
137	142	145	148	150	152	155	157
137	142	145	148	150	152	155	157
138	142	145	148	150	152	155	158
138	142	146	149	150	153	155	158
140	142	146	149	150	153	155	159
140	143	146	149	150	154	155	160
140	143	147	149	151	154	155	160
141	144	147	150	151	154	156	161

Compare the distribution of the heights of the girls with that of the boys.

In your report, you should explain what you have done with the data and why you have done it.

Do your results agree with what you expected?

2 If the data is available for your school, compare the heights of Year 7 pupils with those given above.

SIMULTANEOUS EQUATIONS

The Westbourne Hotel was chosen by Rackham Electronics for a weekend break for the staff. When Emma rang the hotel the receptionist told her that they had 134 rooms and could accommodate up to 253 in single or double rooms. She said that she wasn't certain how many single rooms this included, but promised to ring back as soon as she had checked. Emma told her not to bother as she could easily work that out from the information she had already been given.

Emma could find the number of single and double rooms

- either by trial and error
- or by forming equations and solving them.

EXERCISE 16A In the following problems, discuss what quantities are unknown and whether there is enough information to find them.

1 Henry takes delivery of 40 copies of a book knowing that there are 20 more with soft covers than with hard covers. He has an immediate unexpected order for 12 copies of the hardback. Does he have enough?

2 Wendy rings the box office to book tickets for a concert. There are only two ticket prices. She wants to book for a party of 18, some of whom would like the more expensive seats. The total cost is £405 and she wants to work out the two prices for the tickets. Should she be able to do so?

3 Stuart's grandfather, Wally, is 3 years younger than his grandmother, Eve. Their combined ages come to 181. Stuart wants to know the ages of his grandparents, but he doesn't think he has enough information. Has he?

FORMING EQUATIONS

Each of the problems above contains two unknown quantities.
Before problems similar to these can be solved, we must be able to form equations from given information. This means we have to identify the unknown quantities and then relate them to the information given.

EXERCISE 16B

> In Franconia it cost Henry 75 cents to send 1 postcard and 1 letter home. His wife, Isobel, sends 5 postcards and 1 letter which, in total, costs her 195 cents. Form two equations in words and symbols.
>
> > The unknown quantities are the cost of sending 1 postcard and the cost of sending 1 letter.
>
> Cost of sending 1 postcard + cost of sending 1 letter
> = 75 cents
> Cost of sending 5 postcards + cost of sending 1 letter
> = 195 cents

In questions **1** to **7** first identify the unknown quantities, then use the given information to write equations in words and symbols.

1 Olive Thorne frames pictures. For frames up to $40\,cm \times 30\,cm$ she charges one rate and for larger frames up to $80\,cm \times 60\,cm$ she charges a higher rate. One frame of each size together cost £65 whereas two large frames and four small frames cost £170.

2 At the garden centre each tray of bedding plants is marked at one of two prices. The cost of 3 trays at the lower price and 3 trays at the higher price is £8.25 and the cost of 3 trays at the cheaper rate and 5 trays at the dearer rate is £11.25.

3 Two numbers are such that their sum is 27 and their difference is 13.

4 Two numbers are such that their sum is 32 and the larger number is three times the smaller number.

5 The cost of one protractor and one set square is 120 p whereas the cost of one protractor and two set squares is 185 p.

6 It costs £30 for Mr Holder to take his son on a day-trip to Alton Towers. If he were to take his daughter as well it would cost £42.

7 When Owen buys one tabloid newspaper and one broadsheet newspaper he pays 60 p. He pays the newsagent on a Friday for all the papers he had received from Monday to Friday. One week, when he had collected the two papers every day except Thursday, he paid £2.65. His regular broadsheet newspaper had not been printed on Thursday so he had to make do with just the tabloid.

EQUATIONS WITH TWO UNKNOWN QUANTITIES

In previous chapters the equations we have solved have had one unknown quantity only, but in the equations formed in the exercise above there were two.

Using letters for unknown numbers, the equations formed from problems like those in **Exercise 16B** are of the type $2x + y = 8$.

Looking at the equation $2x + y = 8$, we can see that there are many possible values which will fit, for instance $x = 2$ and $y = 4$, or $x = 1$ and $y = 6$.

We could also have $x = -1$ and $y = 10$ or even $x = 1.681$ and $y = 4.638$.

Indeed, there is an infinite set of pairs of solutions.

In fact, if we rearrange the equation to $y = -2x + 8$, we know, from Chapter 13, that the pairs of values of x and y that satisfy this equation are the coordinates of points on a straight line.

If however we are *also* told that $x + y = 5$, then we shall find that not every pair of numbers which satisfies the first equation also satisfies the second. While $x = 2$, $y = 4$ satisfies the first equation, it does not satisfy $x + y = 5$. On the other hand $x = 3$, $y = 2$ satisfies both equations.

These two equations together form a pair of *simultaneous equations*. 'Simultaneous' means that the two equations are both satisfied by the same values of x and y; that is, when x has the same value in both equations then y has the same value in both.

There are several different methods for solving simultaneous equations. From Chapter 13 we know that one method is to draw the graphs of two straight lines and find where they intersect. We now look at an algebraic method that does not involve graphs.

ELIMINATION METHOD

Whenever we meet a new type of equation, we try to reorganise it so that it is similar to equations we have already met.

Previous equations have had only one unknown quantity, so we try to *eliminate* one of the two unknowns.

Consider the pair of equations

$$2x + y = 8 \qquad [1]$$
$$x + y = 5 \qquad [2]$$

In this case, if we try subtracting the second equation from the first we find that the y term disappears but the x term does not,

i.e. $[1] - [2]$ gives $\qquad\qquad x = 3$

Then, substituting 3 for x in equation [2], we see that $3 + y = 5$ so $y = 2$.

We can check that $x = 3$ and $y = 2$ also satisfy equation [1].

Notice that it is essential to number the equations and to say that you are subtracting them.

Sometimes it is easier to subtract the first equation from the second rather than the second equation from the first. (In this case we would write equation [1] again, underneath equation [2].)
Sometimes we can eliminate x rather than y.

EXERCISE 16C

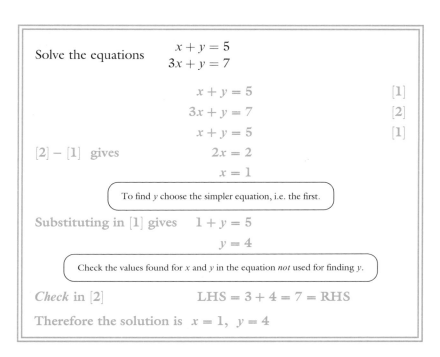

Solve the equations

$$x + y = 5$$
$$3x + y = 7$$

$$x + y = 5 \qquad [1]$$
$$3x + y = 7 \qquad [2]$$
$$x + y = 5 \qquad [1]$$

$[2] - [1]$ gives

$$2x = 2$$
$$x = 1$$

To find y choose the simpler equation, i.e. the first.

Substituting in [1] gives $\quad 1 + y = 5$
$$y = 4$$

Check the values found for x and y in the equation *not* used for finding y.

Check in [2] $\qquad$ LHS $= 3 + 4 = 7 =$ RHS

Therefore the solution is $x = 1$, $y = 4$

Solve the following pairs of equations.

1 $\quad x + y = 5$
$\quad 4x + y = 14$

2 $5x + y = 14$
$\quad 3x + y = 10$

3 $2a + b = 11$
$\quad 4a + b = 17$

4 $2x + 3y = 23$
$\quad x + 3y = 22$

5 $9c + 2d = 54$
$\quad c + 2d = 6$

6 $5x + 2y = 14$
$\quad 7x + 2y = 22$

7 $\quad x + 2y = 12$
$\quad x + y = 7$

8 $4p + 3q = -5$
$\quad 7p + 3q = -11$

9 $12x + 5y = 65$
$\quad 9x + 5y = 50$

10 $9x + 5y = 45$
$\quad 4x + 5y = 45$

Not all pairs of simultaneous equations can be solved by subtracting one from the other.

Consider $\qquad\qquad\qquad\qquad 4x + y = 6 \qquad\qquad\qquad\qquad$ [1]

$\qquad\qquad\qquad\qquad\qquad\qquad 2x - y = 0 \qquad\qquad\qquad\qquad$ [2]

If we subtract we get $\; 2x + 2y = 6 \;$ which is no improvement.

On the other hand, if we add we get $\; 6x = 6 \;$ which eliminates y.

If the signs in front of the letter to be eliminated are the same we should *subtract*; if the signs are different we should *add*.

EXERCISE 16D

Solve the equations $\qquad \begin{array}{l} x - 2y = 1 \\ 3x + 2y = 19 \end{array}$

$\qquad\qquad\qquad\qquad\qquad x - 2y = 1 \qquad\qquad$ [1]

$\qquad\qquad\qquad\qquad\qquad 3x + 2y = 19 \qquad\quad$ [2]

[1] + [2] gives $\qquad\qquad\qquad\quad 4x = 20$

$\qquad\qquad\qquad\qquad\qquad\qquad x = 5$

> It is easier to use the equation with the + sign to find y.

Substitute 5 for x in [2] $\qquad 15 + 2y = 19$

Take 15 from both sides $\qquad\qquad 2y = 4$

$\qquad\qquad\qquad\qquad\qquad\qquad y = 2$

Check in [1] $\qquad\qquad\qquad$ LHS $= 5 - 4 = 1 =$ RHS

Therefore the solution is $\; x = 5, \; y = 2$

Solve the following pairs of equations.

1 $\quad x - y = 2$
$\qquad 3x + y = 10$

2 $\quad 3a - b = 10$
$\qquad a + b = 2$

3 $\quad 6x + 2y = 19$
$\qquad x - 2y = 2$

4 $\quad 4x + y = 37$
$\qquad 2x - y = 17$

5 $\; 2x - y = 6$
$\qquad 3x + y = 14$

6 $\; 5p + 3q = 5$
$\qquad 4p - 3q = 4$

7 $\; 3x - 4y = -24$
$\qquad 5x + 4y = 24$

8 $\; 5x - 2y = 4$
$\qquad 3x + 2y = 12$

To solve the following equations, first decide whether to add or subtract.

9 $3x + 2y = 12$
    $x + 2y = 8$

13 $2x + 3y = 13$
     $2x + 5y = 21$

10 $x - 2y = 6$
     $4x + 2y = 14$

14 $5x - 2y = 24$
     $x + 2y = 0$

11 $x + 3y = 12$
     $x + y = 8$

15 $x + 3y = 0$
     $x - y = -4$

12 $4x + y = 19$
     $3x + y = 15$

16 $6a - b = 20$
     $6a + 5b = 8$

Solve the equations $4x - y = 10$
              $x - y = 1$

$$4x - y = 10 \qquad [1]$$
$$x - y = 1 \qquad [2]$$

The signs in front of the y terms are the same so we subtract. $-y - (-y) = -y + y = 0$

$[1] - [2]$ gives $3x = 9$
                    $x = 3$

Substitute 3 for x in $[2]$ $3 - y = 1$
Add y to both sides $3 = 1 + y$
Take 1 from both sides $2 = y$
                    $y = 2$

Check in $[1]$ $\text{LHS} = 12 - 2 = 10 = \text{RHS}$
Therefore the solution is $x = 3, \; y = 2$

Solve the following pairs of equations.

17 $2x - y = 4$
     $x - y = 1$

22 $6x - y = 7$
     $2x - y = 1$

18 $2p - 3q = -7$
     $4p - 3q = 1$

23 $5x - 2y = -19$
     $x - 2y = -7$

19 $x - y = 3$
     $3x - y = 9$

24 $2x - 3y = 14$
     $2x - y = 10$

20 $3p - 5q = -3$
     $4p - 5q = 1$

25 $3p - 5q = 35$
     $4p - 5q = 0$

21 $3p + 5q = 17$
     $4p + 5q = 16$

26 $3x + y = 10$
     $x + y = -2$

I think of two numbers. If I add three times the smaller number to the bigger number I get 14. If I subtract the bigger number from twice the smaller number I get 1. Find the two numbers.

> We can solve this problem by forming a pair of simultaneous equations. First we have to decide what the unknown quantities are, and then use letters to represent the unknown numbers.

Let the smaller number be x and the bigger number be y.

$$3x + y = 14 \qquad [1]$$
$$2x - y = 1 \qquad [2]$$

$[1] + [2]$ gives
$$5x = 15$$
$$x = 3$$

Substitute 3 for x in $[1]$ $9 + y = 14$
$$y = 5$$

Therefore, the two numbers are 3 and 5.

> Check by reading the original statements to see if the numbers fit.

Solve the following problems by letting the smaller number be x and the larger number y. Then form a pair of simultaneous equations.

27 The sum of two numbers is 20 and their difference is 4. Find the numbers.

28 The sum of two numbers is 16 and they differ by 6. Find the numbers.

29 If I add twice the smaller of two numbers to the larger number I get 14. If I subtract the larger number from five times the smaller number I get 7. Find the two numbers.

30 The sum of two numbers is 18. The larger number is twice the smaller number. What are they?

31 Find two numbers such that twice the first added to the second gives 25 whereas three times the first is 10 more than the second.

A shop sells bread rolls. Six brown rolls and six white rolls cost 132 p while six brown rolls and three white rolls cost 102 p. Find the cost of each type of roll.

> We do not know the cost of a brown roll or the cost of a white roll.

Let one brown roll cost x p and one white roll cost y p.

$$6x + 6y = 132 \qquad [1]$$
$$6x + 3y = 102 \qquad [2]$$

$[1] - [2]$ gives $\qquad\qquad 3y = 30$

Divide both sides by 3 $\qquad\quad y = 10$

Substitute 10 for y in $[2]$ $\quad 6x + 30 = 102$

Take 30 from both sides $\qquad\quad 6x = 72$

$$x = 12$$

Therefore one brown roll costs 12 p
and one white roll costs 10 p.

32 A cup costs x pence and a saucer costs y pence. Work in pence.

   a A cup and saucer cost £3.15. Write down an equation relating x and y.

   b A cup and two saucers cost £4.50. Write down another equation relating x and y. (If a saucer costs y pence how much do two saucers cost ?)

   c By subtracting the first equation from the second you can find the value of y. Hence find the cost of a cup and of a saucer.

33 In a test Harry scored x marks and Adam scored y marks.

   a Together Harry and Adam scored 42 marks. Write down an equation relating x and y.

   b Sam has twice as many marks as Adam, and the sum of Harry's and Sam's marks is 52. Write down another equation relating x and y.

   c Solve these two simultaneous equations by subtracting one equation from the other. What are the marks of each of the three boys ?

34 In a right-angled triangle the two angles apart from the right angle measure $x°$ and $y°$. The difference between x and y is 18 and x is bigger than y.

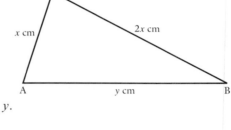

    a What other fact do you know about x and y?

    b Write down two equations relating x and y.

    c Find x and y.

35 In a triangle ABC, AC is x cm, BC is $2x$ cm and AB is y cm.

    a If AB is 2 cm longer than AC write down an equation relating x and y.

    b The perimeter of the triangle is 14 cm. Write down another equation relating x and y.

    c Find x and y.

If you attempted question **2** of **Exercise 16B** you will already have the equations necessary to solve question **36**. Use letters instead of words to represent the unknown numbers and remember to state what each letter represents.

36 At the garden centre each tray of bedding plants is marked at one of two prices. The cost of 3 trays at the lower price together with 3 trays at the higher price is £8.25 and the cost of 3 trays at the lower price and 5 trays at the higher price is £11.25. Find the two prices at which the garden centre sells trays of bedding plants.

37 The equation of a straight line is $y = mx + c$. When $x = 1$, $y = 6$ and when $x = 3$, $y = 10$. Form two equations for m and c and hence find the equation of the line.

38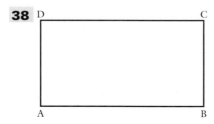

The perimeter of the rectangle is 31 cm. The difference between the lengths of AB and BC is $3\frac{1}{2}$ cm.

Find the lengths of AB and BC.

GRAPHICAL
SOLUTIONS OF
SIMULTANEOUS
EQUATIONS

We saw in a previous chapter that when we are given an equation we can draw a graph. Any of the equations in this chapter will give us a straight line. Two equations give us two straight lines which usually cross one another.

Consider the two equations $x + y = 4$
$$y = 1 + x$$

Suppose we know that the x-coordinate of the point of intersection is in the range $0 \leqslant x \leqslant 5$. We can then draw these lines for that range of values of x.

$x + y = 4$

x	0	4	5
y	4	0	-1

$y = 1 + x$

x	0	2	5
y	1	3	6

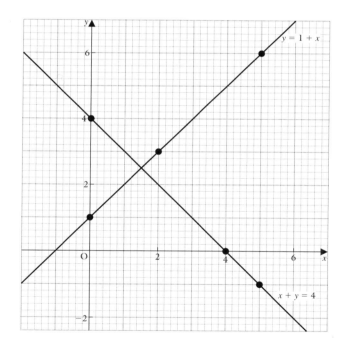

At the point where the two lines cross, the values of x and y are the same for both equations, so they are the solutions of the pair of equations.

From the graph we see that the solution is $x = 1.5$, $y = 2.5$

With this scale, it is only possible to give answers correct to 1 decimal place.

In questions **1** to **3** solve the equations graphically. In each case draw axes for x and y and use values in the ranges indicated, taking 2 cm to 1 unit. Give answers correct to 1 decimal place.

1 $x + y = 6$ $\qquad 0 \leqslant x \leqslant 6, \ 0 \leqslant y \leqslant 6$
$ \ y = 3 + x$

2 $x + y = 5$ $\qquad 0 \leqslant x \leqslant 6, \ 0 \leqslant y \leqslant 6$
$ \ y = 2x + 1$

3 $x + y = 1$ $\qquad -3 \leqslant x \leqslant 2, \ -2 \leqslant y \leqslant 4$
$ \ y = x + 2$

In question **4** rearrange each equation in the form $y = \ldots$, then use a graphics calculator to solve the equations.

4 $2x + y = 3$ $\qquad 0 \leqslant x \leqslant 3, \ -3 \leqslant y \leqslant 3$
$ \ x + y = 2\frac{1}{2}$

5 Allan and Nita are brother and sister. The sum of their ages is 30 and the difference between their ages is 6.
  a If Nita is x years old and her younger brother is y years old, form two equations in x and y.
  b Draw a pair of axes as shown, scaling both axes from 0 to 35. Use 2 cm to represent 5 units on both axes.

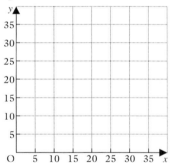

  c Draw graphs of the two equations and hence solve them.
  d **i** How old is Nita?   **ii** How old is Allan?
  e How accurate do you think your answers are? Give reasons for your answer.

6 A teaset includes one milk jug and six cups. The jug holds 130 ml more than a cup, but two cups hold 140 ml more than the jug.
  a If a cup holds x ml and the jug holds y ml, write down two simultaneous equations in x and y.
  b Use a graphics calculator to plot graphs of these two equations for the ranges $-200 \leqslant x \leqslant 500, \ 0 \leqslant y \leqslant 1000$.
  c Hence find the capacity of   **i** the jug   **ii** a cup.

7 At the beginning of the chapter a problem was posed concerning the number of single and double rooms at the Westbourne Hotel.

    a Use the information to form two simultaneous equations.

    b Draw suitable graphs to solve these equations and hence find the number of single rooms and the number of double rooms at the Westbourne Hotel.

    c Solve the equations algebraically. Which method do you find easier – graphical or algebraic? Discuss the advantages and disadvantages of each method.

MIXED EXERCISE

EXERCISE 16F

1 In a furniture store one chair plus the dining table costs £310 while a set of six chairs plus the dining table costs £660.
Use this information to form a pair of simultaneous equations, explaining what the letters that you use represent.

In questions **2** to **10** solve the simultaneous equations.

2 $x + y = 7$ $4x + y = 13$	**5** $5x - 2y = 14$ $3x + 2y = 18$	**8** $3x - y = 19$ $x - y = 3$
3 $x + y = 8$ $3x + y = 14$	**6** $5x - 3y = 3$ $4x + 3y = 24$	**9** $x - y = 1$ $12x + y = 25$
4 $4x - y = 18$ $x - y = 3$	**7** $3x - 2y = 13$ $5x + 2y = 43$	**10** $3x + 5y = 25$ $2x - 5y = 0$

11 The sum of two numbers is 25 and their difference is 9. Find the numbers.

12 The total mass of three large boxes and five small boxes is 12.7 kg, whereas the total mass of three large boxes and eight small boxes is 16.9 kg. Find the mass of each type of box.

13 a Solve the equations $y = 5 - x$, $y = 2 + x$ graphically, for $0 \leqslant x \leqslant 5$, $0 \leqslant y \leqslant 7$.

    b Check your answers by solving the equations algebraically.

14 Solve graphically, the equations $y = 3x - 2$, $y = 4 - 2x$ for $-1 \leqslant x \leqslant 3$, $-3 \leqslant y \leqslant 6$.
Give your answers correct to 1 decimal place.

INVESTIGATION

a Draw the graphs of $y = -x + 9$, $y = -x + 4$ for $0 \leqslant x \leqslant 9$, $0 \leqslant y \leqslant 9$.

b Why is it not possible to use these two graphs to solve the given pair of simultaneous equations? How are the two lines related?

c Repeat parts **a** and **b** for the equations $y = 2x + 3$, $y = 2x - 1$ for $0 \leqslant x \leqslant 4$, $-1 \leqslant y \leqslant 11$.

d Repeat parts **a** and **b** for the equations
$y = -2x + 3$, $y = -2x + \frac{7}{2}$ for $0 \leqslant x \leqslant 3$, $-3 \leqslant y \leqslant 4$.

e Find the gradient of each line given in part **c** and the gradient of each line given in part **d**.
What do you notice about the gradients of the lines in each pair? How would you describe the lines in each pair?

f Draw the graphs of $y = -x + 6$, $y = x - 4$ for $0 \leqslant x \leqslant 8$, $-5 \leqslant y \leqslant 7$. Hence solve the equations.
In what way are the graphs of these two equations related?

g Draw the graphs of $y = -\frac{1}{2}x + 2$, $y = 2x - 6$ for $0 \leqslant x \leqslant 6$, $-7 \leqslant y \leqslant 3$. In what special way are these two lines related?

h What conclusions can you draw that will determine whether the straight lines drawn to represent two linear equations are parallel, perpendicular or neither of these?

i Try solving the following pair of equations.
Comment on why the method breaks down.

$$y = 4 + 2x$$
$$y - 2x = 6$$

SOLVING EQUATIONS

Moira has designed a pendant earring to be made from sheet silver.

The basic shape is a triangle that is 5 mm higher than it is wide at the base.

There is enough money available to use 2400 mm² of sheet silver to make two of these pendants.

She needs to work out the dimensions of each triangle.

- One way in which this can be done is to try different heights and widths until values are found that give an area of 1200 mm² for each triangle.

- Alternatively Moira can form an equation using the information she has and then solve it.

EXERCISE 17A   Work in a group for these questions.

1 Ranjit wants to make a square that has an area of 500 mm² from sheet silver. Discuss how Ranjit can find the length of a side of his square by trying different lengths. Include in your discussion

   a what you think is the most efficient way to find this number

   b how accurate the answer should be; for example, is an answer to the nearest 10 mm good enough for this problem

   c whether your responses to parts **a** and **b** would be different if the area of the square was 100 mm².

2 Draw a diagram showing the square that Ranjit wants to make. Mark on it the area and, using a letter, the length of a side.

   a Discuss the relationship between the area and the length of a side, and hence form an equation.

   b Does this equation then help to organise the search for the length of the side?

3 Rita has a different problem. She wants to make an open box with a volume of $300\,\text{cm}^3$ from a rectangular sheet of cardboard measuring 50 cm by 25 cm. Discuss this problem and ways in which Rita can try to solve it. (You are not expected to solve Rita's problem, but you can if you want to!)

FORMING EQUATIONS

By discussing the problems in the last exercise, you may have discovered a way through to a solution. One way is to start by identifying unknown quantities and then use known information to find a relationship between these quantities, that is, by forming an equation.

You will already be familiar with forming and solving equations such as $3x - 8 = 25$. In equations like these the terms other than numbers contain only a number of xs; they are called *linear equations*.

In this chapter we look at problems which lead to equations containing terms involving powers of x, for example, $x^2 = 500$, $x^3 + x = 12$. These are called *polynomial equations*.

Consider Moira's problem again.

She has $2400\,\text{mm}^2$ of silver to make two triangular pendants, each 5 mm higher than it is wide at the base.

We can put this information on a diagram.

Using x mm as the width of the base, the height is $(x + 5)$ mm.

The area of a triangle is

$\frac{1}{2}$ base $\times$ height

So the area of one triangle is

$\frac{1}{2} \times x \times (x + 5)\,\text{mm}^2$

and the area of two triangles is

$2 \times \frac{1}{2} \times x \times (x + 5)$

Now $\quad 2 \times \frac{1}{2} = 1,$

so the expression simplifies to $\quad x \times (x + 5)$

$\therefore$ leaving out the multiplication sign, $\quad x(x + 5) = 2400$

Having formed the equation, Moira now needs to solve it. It will help if we can get rid of the bracket.

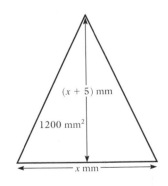
$(x + 5)$ mm

$1200\,\text{mm}^2$

x mm

SIMPLIFYING
EXPRESSIONS
WITH BRACKETS

We know that $2(3x-4)$ means multiply each term in the bracket by 2.

i.e. $2(3x-4) = 2 \times 3x - 2 \times 4$.

In the same way, $x(x+5)$ means $x \times x + x \times 5$.

Now $x \times x$ can be written as x^2
and $x \times 5$ can be written as $5x$

Therefore $x(x+5) = x^2 + 5x$

This is called multiplying out the bracket.

We cannot simplify this further because x^2 and x are unlike terms.

EXERCISE 17B

> Simplify **a** $4x \div (-2y)$ **b** $2p \times q$
>
> **a** $4x \div (-2y) = \dfrac{{}^{2}4x}{{}_{1}-2y}$ **b** $2p \times q = 2pq$ (Leaving out the $\times$ sign.)
>
> $\qquad\qquad\quad = -\dfrac{2x}{y}$

Simplify where possible

1 $x \times y$

2 $a \times a$

3 $2s \times 3s$

4 $4x \times 3x$

5 $u \div v$

6 $a \div (-b)$

7 $a \div a$

8 $6b \div 2c$

9 $s + t$

10 $m - n$

11 $m \times (-n)$

12 $3s - 2t$

13 $a - 3a$

14 $p \times 4p^2$

15 $8u \div 4w$

16 $2z - 3y$

17 $3s \times 2t$

18 $p \times 2p$

19 $q - 5q$

20 $r - (-4s)$

21 $8p \div 2q$

22 $(-3s) \times (-2t)$

23 $(b) \times (-2b)$

24 $(-x) \div (-y)$

Simplify **a** $a(a-b)-(b-a)$ **b** $3(a-b)-2(a+b)$

a Remember that $-(b-a)$ means minus b and minus $(-a)$.

$$a(a-b)-(b-a)=a^2-ab-b-(-a)$$
$$=a^2-ab-b+a$$

b $3(a-b)-2(a+b)=3a-3b-2a-2b$
$$=a-5b$$

Simplify

25 $a-3(a-b)$

26 $2a+a(a-3)$

27 $2(a-b)-(b-a)$

28 $4(a-c)+2(a-b)$

29 $y-2(y-z)$

30 $4(x+y)+2(x+z)$

31 $3(p+q)-2(p+r)$

32 $2x-(x+y)$

33 $2(p+q)-3(p-q)$

34 $a(a+b)-2(a-b)$

35 $x(x-y)+y(y-x)$

36 $4(b-c)-2(b+c)$

37 $4(p-q)+2(q-p)$

38 $w(w+x)-x(w-x)$

39 $3(m+n)-5(m-n)$

40 $5(b-c)-3(b+c)$

41 A rectangle is to be made with the following properties. The area is 45 cm^2 and the length is twice the width. If the width is x cm, copy the diagram and label it with all the information known. Hence form an equation and simplify it.

42 Another rectangle is to be made with these properties.

The area is 35 cm^2.
The length is 5 cm longer than the width.
If the width is x cm, copy the diagram and label it with all the information known. Hence form an equation and simplify it.

43 Ann is four years younger than her sister Sally. The product of their ages is 60.

   a If Sally is x years old, find Ann's age in terms of x.

   b Find the product of Ann's and Sally's ages in terms of x.

   c Form an equation in x and simplify it.

44 The volume of a cube is 80 cm^3.
If the length of one side of the cube is x cm,
form an equation in x.

45 The volume of the cuboid shown in the
diagram is 140 cm^3.
Find an expression for the volume in
terms of x. Hence form an equation in x.

FINDING SQUARE
ROOTS

If Beth wants to draw a square whose area is 16 cm^2, she must first find
the lengths of the sides.

If the sides are each x cm long, Beth can form the equation $x^2 = 16$.

Now x^2 means $x \times x$, so to solve this equation she needs to find a
number which, when multiplied by itself, gives 16. This number, as we
know from Chapter 7, is called the *square root* of 16.

Now $4 \times 4 = 16$, so 4 is a square root of 16,

But $(-4) \times (-4) = 16$, therefore -4 is also a square root of 16.

So when $x^2 = 16$, we have $x = 4$ or -4, written $x = \pm 4$.

This is true for the square root of any number,

i.e. any number has two square roots, one positive and one negative.

We use the symbol $\sqrt{}$ to mean the *positive square root*,

so when $x^2 = 16$ we can write $x = \pm \sqrt{16}$

Therefore the equation $x^2 = 16$ has *two* solutions, $x = 4$ *and* $x = -4$.

The side of a square cannot have a negative length so $x = 4$ is the only
solution that has meaning in this context, therefore Beth's square has
sides 4 cm long.

EXERCISE 17C

Use your calculator to solve the equation $x^2 = 725$, giving
solutions correct to 3 significant figures.

$x^2 = 725$ | Press | **7** **2** **5** **√** | The display shows 26.9258 . . .

$\therefore \quad x = \pm\sqrt{725} = \pm 26.92\ldots$

$= \pm 26.9$ (correct to 3 s.f.)

Check: $26.9^2 = 723.6\ldots$ | Press | **2** **6** **.** **9** **x²**

Use your calculator to solve the equations, giving solutions correct to
3 significant figures.

1 $x^2 = 38$ **2** $x^2 = 19$ **3** $x^2 = 420$

4 $x^2 = 42$ **8** $x^2 = 0.47$ **12** $x^2 = 63$

5 $x^2 = 4.2$ **9** $x^2 = 3.8$ **13** $x^2 = 122$

6 $x^2 = 680$ **10** $x^2 = 18$ **14** $x^2 = 39$

7 $x^2 = 68$ **11** $x^2 = 10\,700$ **15** $x^2 = 0.29$

The rectangle in the diagram has an area of $31\,\text{cm}^2$.
Form an equation in x and solve it
to find the width of the rectangle.

x cm

$3x$ cm

The area of the rectangle $= 4x \times x\,\text{cm}^2$
therefore $\qquad\qquad 4x^2 = 31$

$\qquad\qquad\qquad x^2 = 7.75$

$\boxed{\text{Dividing both sides by 4.}}$

$\therefore \qquad\qquad\quad x = \pm\sqrt{7.75}$

$\qquad\qquad\qquad\quad = \pm 2.783\ldots$

$\boxed{\text{The width of a rectangle cannot be a negative number, so}\\\text{only the positive square root gives a solution to the problem.}}$

The width of the rectangle is $2.78\,\text{cm}$ correct to 3 s.f.

---

16 The area of the square in the diagram is $50\,\text{mm}^2$.

   a Form an equation in x.

   b Solve the equation to find the
      length of a side.

x cm

17 The area of the rectangle in the diagram is $120\,\text{cm}^2$.

   a Form an equation in x.

   b Solve the equation to find the
      width of the rectangle.

x cm

$3x$ cm

18 The area of this triangle is $290\,\text{cm}^2$.
Form an equation and solve it to
find the height of the triangle.

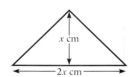

x cm

$2x$ cm

19 James is twice as old as Ali. The product of their ages is 32.

   a If Ali is x years old, write down James's age in terms of x.

   b Write down the product of the boys' ages in terms of x.

   c Form an equation in x and solve it to find each boy's age.

Consider Moira's problem again;
she needs to know the dimensions
of this triangular pendant.

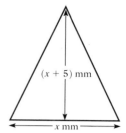

On page 324 we found the equation

$$x^2 + 5x = 2400$$

and now we need to solve it.

As this equation is not in the form '$x^2 = $ a number' we cannot use
square roots to find x, so we will try some values for x and see if we can
find one that fits.

If we try numbers for x at random, we could spend hours before finding
one that fits, so an organised approach is needed.

Firstly we give x a value that looks reasonable and try it in the equation.

Try $x = 50$: $50^2 + 5 \times 50 = 2750$; so 50 is too big.

The obvious next step is try a number smaller than 50;

Try $x = 40$: $40^2 + 5 \times 40 = 1800$; so 40 is too small.

Now it is clear that the value of x is between 40 and 50, so we try a
number between 40 and 50 to get an improvement.

Try $x = 45$: $45^2 + 5 \times 45 = 2250$; so 45 is too small.

If we use a number line to illustrate where the results of our tries are in
relation to the result we want, we get a picture of how close we are and
where to go next.

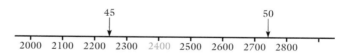

This shows that 45 is nearer than 50 to the result we want. It is sensible
to try next a number between 45 and 50, but nearer to 45.

Try $x = 47$: $47^2 + 5 \times 47 = 2444$; so 47 is too big, but only just.

We can continue placing the trials on the number line and use the
information to judge what number to try next.

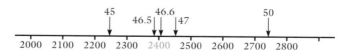

Try $x = 46.5$: $46.5^2 + 5 \times 46.5 = 2394.75$; so 46.5 is just too small.

Try $x = 46.6$: $46.6^2 + 5 \times 46.6 = 2404.56$; so 46.6 is just too big.

Now we can see that x lies between 46.5 and 46.6.

Now x cm is the width of the base of a triangle that can be made from 1200 mm² of silver. In this context, a measurement that gives slightly under 1200 mm² for the area is sensible. Also it is probably impossible to cut sheet silver more accurately than to the nearest tenth of a millimetre. Therefore we do not need to look any further for an acceptable solution to the problem, that is taking x as **46.5** gives a triangle whose measurements fit the requirements.

If the context from which the equation arose needed a more accurate solution, we could draw another number line for the range 2392 to 2408,

i.e.

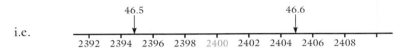

The next try could then be **46.55**.

EXERCISE 17D

Find, by trial and improvement, a positive solution of the equation $x^3 - 4x = 7$, giving the answer correct to 1 decimal place.

> If we try 1 as a trial solution, we get a negative result, so we will start with 2.

Try 2: $2^3 - 4 \times 2 = 0$;   too small
Try 3: $3^3 - 4 \times 3 = 15$;   too big
Try 2.5: $2.5^3 - 4 \times 2.5 = 5.6 \ldots$ too small
Try 2.6: $2.6^3 - 4 \times 2.6 = 7.1 \ldots$ too big

> Now we can see that the solution is nearer 2.6 than 2.5.

$x = 2.6$ (correct to 1 d.p.)

Find a positive solution of each equation by trial and improvement, giving answers correct to 1 decimal place.

1 $x^2 + 7x = 12$ (Start by trying $x = 1$)

2 $x^2 + x = 4$ (Start by trying $x = 1$)

3 $x^3 - 2x = 10$ (Start by trying $x = 2$)

4 $x^2 + 3x = 7$ (Start by trying $x = 1$)

5 $2x^2 - 3x = 10$ (Start by trying $x = 2$)

6 $x^3 + 2x = 500$ (Start by trying $x = 10$)

7 $x^3 + x^2 + x = 50$ (Start by trying $x = 3$)

8 $x^3 - x = 12$ **12** $8x^2 - x^3 = 12$

9 $2x^2 + 5x = 20$ **13** $x^4 - 3x^3 = 10$

10 $5x^2 - 7x = 4$ **14** $3x^3 + 2x^2 = 7$

11 $3x^3 + x = 8$ **15** $x^2 + 3 = 5x$

USING A SPREADSHEET

A spreadsheet takes the work out of the calculations needed for each trial. This leaves you free to concentrate on the choice of numbers to try. The diagram shows how a solution to $x^3 - 4x = 7$ can be found correct to 2 decimal places.

	A	B	C	
1	x^3 − 4x = 7			
2				
3		Try x =	x^3 − 4x	
4		1	− 3	Enter formula B4^3 − 4*B4 in this cell.
5		2	0	
6		3	15	
7		2.5	5.625	
8		2.6	7.176	
9		2.55	6.381375	Use the fill command for these cells.
10		2.57	6.694593	
11		2.59	7.013979	
12		2.58	6.853512	
13			0	
14			0	

From the entries in cells B11, C11 and B12, C12 we can see that $x = 2.59$ gives a result nearer to 7 than does $x = 2.58$, so we deduce that $x = 2.59$ (correct to 2 d.p.)

EXERCISE 17E

Use a spreadsheet to find solutions to the equations in **Exercise 17D**, correct to 2 decimal places.

USING GRAPHS
TO SOLVE
EQUATIONS

Solving equations by trial and improvement can be time consuming, particularly when we need a solution correct to more than 2 significant figures.

Using a graph is another method for finding solutions of equations.

Consider again the equation $x^3 - 4x = 7$.

We can start by rearranging this equation so that the right-hand side is zero; subtracting 7 from each side gives $x^3 - 4x - 7 = 0$

Draw the graph of $y =$ the left-hand side i.e. $y = x^3 - 4x - 7$

x	-1	-0.5	0	1	1.5	2	2.5	3
y	-4	-5.125	-7	-10	-9.625	-7	-1.375	8

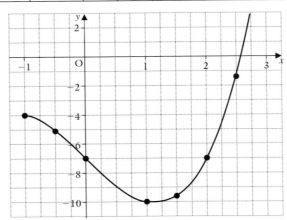

We want the value of x when $x^3 - 4 - 7 = 0$, i.e. when $y = 0$.
$y = 0$ where the graph crosses the x-axis,
and this is where $x = 2.6$ (correct to 1 d.p.)

EXERCISE 17F

1 a Copy and complete the table of values for $y = 3x^2 - 4x - 2$

x	0	0.5	1	1.5	2	2.5
y	-2	-3.25				6.75

b Using 2 cm for 1 unit on the y-axis and 4 cm for 1 unit on the x-axis, draw the graph of $y = 3x^2 - 4x - 2$

c Hence find the positive solution of the equation $3x^2 - 4x - 2 = 0$

d How accurate is your answer? Give reasons.

e If the x-axis were extended in the negative direction, would the curve cross the x-axis at another point? Explain your answer.

f What is the implication of your answer to part **e** for the solution of the equation $3x^2 - 4x - 2 = 0$?

g If this equation comes from a problem where x cm is a length, what implication does this have for the solution of the problem?

2 **a** Copy and complete the table where $y = 5x^2 - 2x - 1$

x	-1	-0.5	0	0.5	1	1.5
y		1.25				

b Using 2 cm for 1 unit on the y-axis and 4 cm for 1 unit on the x-axis, draw the graph of
$$y = 5x^2 - 2x - 1$$

c How many solutions are there to the equation $5x^2 - 2x - 1 = 0$?

d Give these solutions, stating how accurate you judge your answers to be.

e If this equation comes from a problem where x seconds is the time taken for a ball thrown in the air to reach ground level again, which solution has a meaning?

f If the equation comes from a problem where x metres is the height of a ball above the top of a fence, which solution has a meaning?

3 **a** What graph would you draw to find solutions to the equation
$$x^2 + 2x = 1?$$

b Draw the graph for values of x from -3 to 1 at intervals of half a unit.

c Give the solutions to the equation.

d Which of these solutions has meaning when the equation comes from a problem where x hours is the time after midnight at which the moon is at a given height above the horizon?

4 **a** Write down the graph that you would draw to find the solutions to the equation $3x^2 = 7$.

b *Sketch* this graph.

c What does your sketch show you about the number of solutions to the equation $3x^2 = 7$?

d Use your calculator to find these solutions correct to 4 decimal places.

USING A
GRAPHICS
CALCULATOR

A graphics calculator takes the hard work out of graph plotting because there is no need to work out a table of values.

To plot the graph of $y = x^3 - 4x - 7$, set the range to

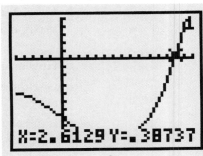

x min: -1
x max: 3
scale: 0.2
y min: -10
y max: 5
scale: 1

Then, enter $(y =) \ x^3 - 4x - 7$

and **GRAPH** the function.

The diagram shows the display when the TRACE button is used.

From the display it is clear that the curve crosses the x-axis where $x = 2.6$ and that this value is correct to 1 decimal place. This can be confirmed by moving the cursor to where the graph crosses the x-axis.

If a more accurate answer is needed, the ZOOM function can be used to draw a small section near where the curve crosses the x-axis,

e.g.

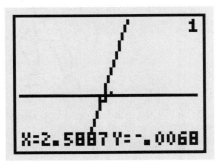

Moving the cursor around the point, we see that x lies between $2.588 \ldots$ and $2.590 \ldots$, i.e. $x = 2.59$ correct to 2 d.p.

EXERCISE 17G

Use a graphics calculator to find solutions to the equations given in **Exercise 17F**, giving answers correct to 2 decimal places.

EXERCISE 17H

Solve the problems by forming an equation from the information given and then solving the equation by any method you choose.

1 A rectangular carpet has an area of $15 \, \text{m}^2$ and is twice as long as it is wide. Find the length and width of this carpet.

2 Jeff is x years old. Steve is 5 years older than Jeff and the product of their ages is 204. How old is Steve?

3 An open box has a volume of $870\,\text{cm}^3$.
It has a square cross-section and it is
5 cm longer than it is deep.
Find the dimensions of the box.

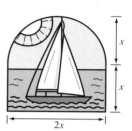

x cm

x cm

$(x + 5)$ cm

4 The area of this window is $2.5\,\text{m}^2$.
Find the height of the window.

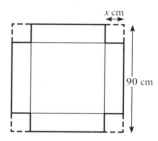

x

x

$2x$

5 An open box with a volume of
$32\,000\,\text{cm}^3$ is made by cutting
squares from the corners of a
square sheet of cardboard whose
edges are 90 cm long. The sides
are folded up and fixed together.
Find the length of the sides of the
squares cut out of the cardboard.

x cm

90 cm

INVESTIGATION

a What do you notice when you use your calculator to find
$\sqrt{0.04}$ $\sqrt{0.4}$ $\sqrt{4}$ $\sqrt{40}$ $\sqrt{400}$ $\sqrt{4000}$?
b What do you notice when you use your calculator to find
$\sqrt{0.06}$ $\sqrt{0.6}$ $\sqrt{6}$ $\sqrt{60}$ $\sqrt{600}$ $\sqrt{6000}$?
c Repeat part **a** for the sequence
$\sqrt{0.012}$ $\sqrt{0.12}$ $\sqrt{1.2}$ $\sqrt{12}$ $\sqrt{120}$ $\sqrt{1200}$
Try some different numbers.
d Using the results from part **a** and, *without* using your calculator,
write down the first significant figure of $\sqrt{400\,000}$.
e *Without* using your calculator, find the first significant figure of
$\sqrt{50}$, $\sqrt{5000}$, $\sqrt{0.5}$.
Is this the same as the first significant figure of $\sqrt{5}$?
f Try to explain how to find the first significant figure of the square
root of a number without using a calculator. Test your explanation
on some other numbers.

SUMMARY 4

Multiplication and division of algebraic fractions

The same rules apply to fractions with letter terms as to fractions with numbers only,

i.e. to multiply fractions, multiply together the numerators and multiply together the denominators,

e.g. $\quad \dfrac{2x}{3} \times \dfrac{5x}{7} = \dfrac{2x \times 5x}{3 \times 7} = \dfrac{10x^2}{21}$

and to divide by a fraction, turn it upside down and multiply,

e.g. $\quad \dfrac{x}{6} \div \dfrac{2x}{3} = \dfrac{\cancel{x}^1}{\cancel{6}_2} \times \dfrac{\cancel{3}^1}{2\cancel{x}_1} = \dfrac{1}{4}$

Simplification of brackets

$x(2x - 3)$ means $x \times 2x + x \times (-3)$

Therefore $x(2x - 3) = 2x^2 - 3x$

Linear equations

When an equation contains brackets, first multiply out the brackets,

e.g. $\quad 3x - 2(3 - x) = 6$

gives $\quad 3x - 6 + 2x = 6 \quad$ which can be solved easily.

When an equation contains fractions, multiply each term in the equation by the lowest number that all denominators divide into exactly. This will eliminate all fractions from the equation,

e.g. if $\quad \dfrac{x}{2} + 1 = \dfrac{2}{3}$

multiplying each of these three terms by 6 gives

$$\dfrac{6}{1} \times \dfrac{x}{2} + 6 \times 1 = \dfrac{6}{1} \times \dfrac{2}{3} \quad \text{which simplifies to} \quad 3x + 6 = 4$$

which can be solved easily.

Simultaneous equations

Two equations with two unknown quantities are called simultaneous equations.

A pair of simultaneous equations can be solved algebraically by eliminating one of the letters; when the signs of this letter are different, we add the equations, when the signs are the same we subtract the equations.

For example, to eliminate y from $\quad 2x + y = 5 \qquad [1]$

$$\text{and} \quad 3x - y = 7 \qquad [2]$$

we add [1] and [2] to give $\qquad\qquad 5x = 12 \qquad [3]$

The value of x can be found from [3]. This value is then substituted for x in [1] or [2] to find y.

To eliminate y from $\qquad\qquad 4x + y = 9 \qquad [1]$

$$\text{and} \quad 3x + y = 7 \qquad [2]$$

we work out $[1] - [2]$ to give $\qquad x = 2$

Now we can substitute 2 for x either in [1] or in [2] to find y.

Polynomial equations

Polynomial equations in one unknown contain terms involving powers of x, e.g. $x^3 - 2x = 4$ and $2x^2 = 5$ are polynomial equations.

Equations containing only an x^2 term and a number can be solved by *finding square roots*. Now the square root of 9 can be 3, but it can also be -3, that is, a positive number has two square roots, one positive and one negative.
For example, if $x^2 = 9$ then $x = \pm\sqrt{9}$, so $x = +3$ and $x = -3$ are solutions.

More complex equations can be solved by *trial and improvement,* that is, by trying possible values for x until we find a value that fits the equation.
Equations can also be solved by *drawing a graph,*
for example to solve $x^3 - x = 10$, we can start by arranging the equation as $x^3 - x - 10 = 0$ and then drawing the graph of
$y = x^3 - x - 10$. The solutions are the values of x where this graph crosses the x-axis.

Inequalities

An inequality remains true when the same number is added to, or subtracted from, both sides,

e.g. $\quad$ if $x > 5$ then $x + 2 > 5 + 2$

$$\text{and} \quad x - 2 > 5 - 2$$

An inequality also remains true when both sides are multiplied, or divided, by the same *positive* number,

e.g. $\quad$ if $x > 5$ then $2x > 10$

$$\text{and} \quad \frac{x}{2} > \frac{5}{2}$$

However, multiplication or division by a negative number must be avoided because this destroys the inequality.

GRAPHS

The equation of a line or curve gives the y-coordinate of a point in terms of its x-coordinate. This relationship between the coordinates is true only for points on the line or curve.

Straight lines

The *gradient* of a straight line can be found from two points, P and Q, on the line, by calculating

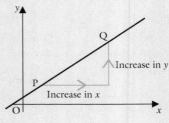

$$\frac{\text{increase in } y \text{ in moving from P to Q}}{\text{increase in } x \text{ in moving from P to Q}}$$

$$= \frac{y\text{-coordinate of } Q - y\text{-coordinate of P}}{x\text{-coordinate of } Q - x\text{-coordinate of P}}$$

When the gradient is positive, the line slopes uphill when moving from left to right.
When the gradient is negative, the line slopes downhill when moving from left to right.

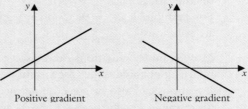

Positive gradient   Negative gradient

The equation of a straight line is of the form $y = mx + c$

where m is the gradient of the line

and c is the y-intercept,

e.g.   the line whose equation is $y = 2x - 3$

has gradient 2 and y-intercept -3.

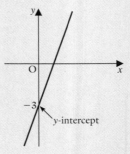

An equation of the form $y = c$ gives a line parallel to the x-axis.

An equation of the form $x = b$ gives a line parallel to the y-axis.

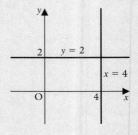

Two simultaneous equations can be solved graphically by drawing two straight lines and finding the coordinates of their point of intersection.

for example, to solve $2x - y = 7$ [1]

and $3x + y = 9$ [2]

we first rearrange the equations so they are each in the form $y = \dots$

i.e.   to $y = 2x - 7$ [3]

and $y = -3x + 9$ [4]

We then plot these lines:
As accurately as we can read from the graph, the point of intersection is
(3.2, −0.6).
So the solution is
$x = 3.2$ and $y = -0.6$

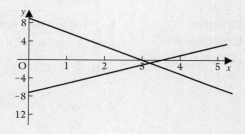

Curved graphs

A *parabola* is a curve whose equation is in the form $y = ax^2 + bx + c$.

The shape of this curve looks like this:

When the x^2 term is positive it is the way up shown above.

When the x^2 term is negative the curve is upside down.

This, for example, is the graph of $y = x^2 - 2x$:

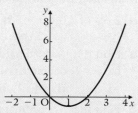

whereas the graph of $y = 2x - x^2$ looks like this:

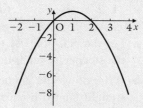

STATISTICS

Discrete values are exact and distinct, e.g. the number of people in a queue, shoe size.

Continuous values can only be given in a range on a continuous scale, e.g. the length of a piece of wood.

For a grouped frequency distribution,
the *range* is estimated as

the highest end of the last group − the lowest end of the first group

the *modal group* is the group with the largest number of items in it.

REVISION EXERCISE 4.1 (Chapters 12 to 14)

1 Solve the equations

a $5 + 4(x + 3) = 25$

c $0.84x = 4$

b $\dfrac{x}{7} = 4$

d $\dfrac{x}{6} + \dfrac{x}{3} = 4$

2 Simplify

a $\dfrac{3x}{5} \times \dfrac{15}{x}$

c $4(7 - 2x)$

b $\dfrac{5x}{9} \div \dfrac{2}{3}$

d $3(5 - 2x) - 4(2 - 3x)$

3 a Solve the following inequalities and illustrate each solution on a number line.

i $x + 9 < 0$ **ii** $9 - x > 3$ **iii** $3x - 5 \leqslant 4$ **iv** $4 > 5 - 2x$

b Solve the pair of inequalities $5 + x \geqslant 4$ and $5 - x \geqslant 1$, and find the range of values of x that satisfy both of them.

4 Solve the following equations, giving your answers correct to 3 significant figures where necessary.

a $2.6x = 10.4$ **b** $4.2x + 3 = 7.6$ **c** $0.9x = 10.4$

5 a Find the y-coordinates of points on the line $y = -2x$ which have x-coordinates of

i 3 **ii** -4 **iii** $\frac{1}{2}$ **iv** $-\frac{3}{4}$

b Find the x-coordinates of points on the line $y = 3x$ which have y-coordinates of

i 2 **ii** -5 **iii** $2\frac{1}{2}$ **iv** 3.6

6 a For the pair of lines $y = 4x$ and $y = \frac{1}{4}x$, state which line is the steeper. Sketch both lines on the same diagram.

b Write down the gradient, m, and the y-intercept, c, for the straight line with equation   **i** $y = 7x - 5$ **ii** $y = 4 - 5x$

7 Draw x- and y-axes on graph paper. Scale the x-axis from -8 to 8 and the y-axis from -10 to 10, using $1\,\text{cm}$ as 1 unit on both axes. Draw the graphs of $y = \frac{1}{2}x + 2$ and $y = -x + 8$. Hence find the value of x and the value of y that satisfies both equations simultaneously.

8 The equation of a curve is $y = 4x^2$. Which of these sketches could be the curve?

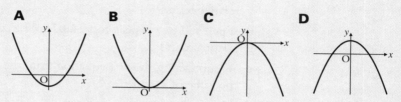

A   **B**   **C**   **D**

Give reasons why you rejected three of the sketches.

9 Draw a sketch of the curve with equation

a $y = \frac{1}{2}x^2$ **b** $y = -\frac{1}{2}x^2$ **c** $y = \frac{1}{2}x^2 + 2$

10 A manufacturer produces a set of jugs, identical in shape but different in size. The capacities of some of the jugs are given in the table.

Height, $h\,\text{cm}$	7.2	9	10.8	12.6	14.4	16.2	18
Capacity, $C\,\text{cm}^3$	64	125	216	343	512	729	1000

Draw a graph to show how capacity changes with height. Use $1\,\text{cm}$ as 1 unit on the h-axis and as 50 units on the C-axis. Take 6 as the lowest value for h and 0 as the lowest value for C.
Use your graph to estimate

a the capacity of a jug $17\,\text{cm}$ high

b the height of a jug that holds $650\,\text{cm}^3$.

**REVISION
EXERCISE 4.2
(Chapters 15
to 17)**

1 State whether the possible values of each quantity are discrete or continuous.

a The number of passengers on a bus.

b The time a bus takes to travel from one bus stop to the next one.

c The amount of milk in a carton.

d The cost of a cup of coffee.

e Your best friend's height.

2 Show, on a sketch, the range in which these measurements lie.

   a 62 kg to the nearest kilogram

   b 14 years rounded down to the nearest year

   c 140 cm rounded up to the nearest centimetre

3 A ball point pen costs x pence and a pencil costs y pence.

   a A pen and a pencil together cost £1.51. Write down an equation relating x and y.

   b A pen and two pencils cost £2.07. Write down another equation relating x and y.

   c By subtracting the first equation from the second equation find the value of y.

   d What is the cost of   **i** a pencil   **ii** a ball point pen?

4 At a veterinary clinic all the dogs seen in one day were weighed. Their weights, in kg, were recorded in this observation chart..

Weight, w kg	$0 < w \leqslant 10$	$10 < w \leqslant 20$	$20 < w \leqslant 30$	$30 < w \leqslant 40$
Tally	卌	卌 卌 ////	卌 卌 卌 ///	///
Frequency				

   a The vet saw two dogs after-hours. One weighed just over 20 kg and the other just under 10 kg. Copy the table, add tally marks for these two weights and then complete the table.

   b How many of the dogs weighed that day were

   i more than 20 kg   **ii** 30 kg or less?

   c How many dogs were there altogether?

   d Draw a bar chart to illustrate this information.

5 Solve the simultaneous equations

   a $x + y = 5$
   $3x + y = 9$

   b $5a - b = 27$
   $5a + 3b = 19$

6 The sum of two numbers, x and y, is 22 and their difference is 4. If the larger number is x, form a pair of simultaneous equations in x and y and solve them. What are the two numbers?

7 Solve graphically the equations $x + y = 6$
$$y = 3x - 1$$
Draw axes for $0 \leqslant x \leqslant 6$ and $0 \leqslant y \leqslant 6$, taking 2 cm as 1 unit. Give your answers correct to 1 decimal place.

8 Simplify

    a $3p \times 4q$ **b** $3(a+b)-(b-a)$ **c** $p(p+q)-q(p-q)$

9 The area of a square is $50\,\text{cm}^2$. If the length of one side of the square is x cm, form an equation in x and solve it. What is the length of one side?

10 Find, by trial and improvement, a positive solution of the equation $x^2 + 7x = 140$, giving your answer correct to 1 decimal place. (Start by trying $x = 8$.)

REVISION EXERCISE 4.3
(Chapters 12 to 17)

1 Solve the equations

    a $\dfrac{3x}{4} = 21$ **c** $5.7x = 3$

    b $\dfrac{x}{4} - \dfrac{x}{12} = \dfrac{1}{2}$ **d** $5x = 3x - (x - 18)$

2 a Simplify   **i** $\dfrac{4a}{7} \times \dfrac{a}{12}$ **ii** $\dfrac{15b}{7} \div \dfrac{27b}{14}$

    b Solve the following inequalities, illustrating each solution on a number line.

        i $x - 6 \leqslant -4$ **ii** $5 - x > -3$ **iii** $5x - 2 > 7$

3 a Determine whether each of the following straight lines makes an acute angle or an obtuse angle with the positive x-axis.

        i $y = -5x$ **iii** $y = -\tfrac{1}{2}x - 4$

        ii $y = 0.4x$ **iv** $y = 3 + 4x$

    b **i** Find the gradient of this line.

       ii Where does the line cross the y-axis?

      iii Find the equation of the line.

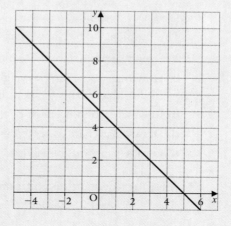

4 The table gives some lengths in inches and the equivalent lengths in centimetres.

Length (inches)	0	10	50
Length (centimetres)	0	25.4	127

a Choose your own scales and draw a graph to convert lengths up to 100 inches into centimetres.

b Use your graph to convert
i 36 inches into centimetres   **iii** 200 centimetres into inches
ii 73 inches into centimetres   **iv** 120 centimetres into inches.

5 The equation of a curve is $y = -2x^2$. Which of these sketches could be the curve?

A   **B**   **C**   **D**

  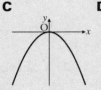

Explain why you rejected three of the curves.

6

Weight, w kg	Frequency
$40 \leqslant w < 50$	5
$50 \leqslant w < 60$	13
$60 \leqslant w < 70$	18
$70 \leqslant w < 80$	7
$80 \leqslant w < 90$	7
$90 \leqslant w < 100$	4

This frequency table shows the weights of a group of teenage boys.

a How many boys are there in the group?

b How many boys have a weight that is
i less than 50 kg   **ii** at least 70 kg?

c In which group do most of the weights lie?

d Draw a bar chart to represent this information.

e On a separate diagram illustrate this data with a frequency polygon.

7 Solve the simultaneous equations $7p + 3q = 33$
$4p - 3q = 0$

8 The sizes of the three angles in a triangle are $105°$, $x°$ and $y°$ where x is larger than y. The difference between x and y is 12.

   a What other fact do you know about x and y?

   b Write down two equations relating x and y.

   c Find x and y.

9 The area of a rectangle is $30\,\text{cm}^2$. The width of the rectangle is $x\,$cm and the rectangle is 6 cm longer than it is wide. Form an equation in x and simplify it.

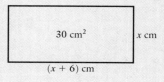

10 Find, by trial and improvement, a positive solution of the equation $x^3 - 2x = 50$, giving your answer correct to 1 decimal place. (Start by trying $x = 3$.)

REVISION
EXERCISE 4.4
(Chapters 1 to 17)

1 **a** Give 0.09273 correct to

     i 2 decimal places

     ii 3 significant figures

     iii 1 significant figure

  b Use a calculator, giving your answers correct to 3 significant figures, to find

     i 0.846×12.55

     ii $16.87 \div 24.66$

2 **a** Calculate   **i** $3\frac{1}{7} \times 21$ **ii** $8\frac{1}{4} \times \frac{5}{11}$ **iii** $3\frac{3}{5} \div 2\frac{7}{10}$

  b Which is the larger and by how much, $\frac{8}{9}$ of $2\frac{1}{4}$ or $\frac{7}{12} \div \frac{3}{10}$?

3 **a** Express 46% as a fraction in its lowest terms.

  b Give $60\,\text{mm}^2$ as a fraction of $3\,\text{cm}^2$.

  c The product of two numbers is 12. If one of the numbers is $2\frac{1}{4}$ find the other.

4 **a** If $4 : 7 = 16 : x$, find x.

  b A carpet is made from wool and nylon in the ratio $4 : 1$ by weight. How much wool is there in a carpet that weighs 55 kg?

5 Find the area of each shape

a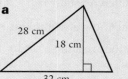

28 cm 18 cm 32 cm

b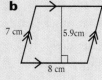

7 cm 5.9cm 8 cm

c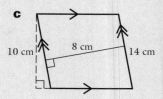

10 cm 8 cm 14 cm

6 a If $v = u + at$ find v when $u = 4$, $a = -3$ and $t = 2$.

  b The volume of this cuboid is $V \, \text{cm}^3$.
  Form a formula for V.

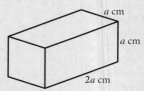

a cm a cm $2a$ cm

7

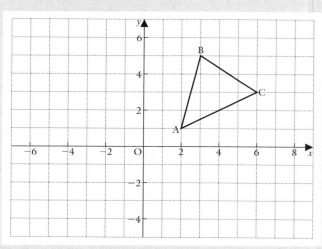

  a Copy the diagram and draw the image $A'B'C'$ when ABC is rotated through $180°$ about the point $(0, 1)$.

  b Draw the image $A''B''C''$ when ABC is rotated $90°$ clockwise about the point $(2, 0)$.

8 Solve the equations

  a $0.4x = 5.6$

  c $\dfrac{3x}{4} - \dfrac{x}{6} = 14$

  b $\dfrac{x}{3} - \dfrac{3}{7} = 3$

  d $10x - 3(x + 7) = 0$

9 Using squared paper and 1 square to one unit on both axes, plot the points $(-3, 6), (-2, 3), (1, -6)$ and $(2, -9)$. Find the equation of the straight line that passes through these points.

10 Given below is a list of the heights, in centimetres rounded up to the nearest centimetre, of 60 girls. The list is in numerical order.

```
132   132   133   134   134   134   135   136   136   136
136   137   137   137   138   138   138   139   139   139
139   139   139   140   140   140   140   140   140   140
140   141   142   142   143   144   144   145   145   146
146   146   146   146   147   147   147   147   147   147
148   148   148   149   149   150   152   152   153   154
```

a What is the height of

   i the tallest person   **ii** the shortest person?

b Copy and complete this frequency table.

Height, h cm	Frequency
$130 < h \leqslant 135$	
$135 < h \leqslant 140$	
$140 < h \leqslant 145$	
$145 < h \leqslant 150$	
$150 < h \leqslant 155$	

c How many girls have a height that is

   i greater than 140 cm   **ii** 145 cm or less?

d Which group is the modal group?

e Draw

   i a bar chart

  ii a frequency polygon, to illustrate this information.

**REVISION
EXERCISE 4.5
(Chapter 1 to 17)**

1 One letter is chosen at random from the letters in the word WEDNESDAY. What is the probability that it is

  a the letter D   **b** not a vowel   **c** not the letter E?

2 Find

  a $6\frac{3}{4} + 1\frac{7}{9}$ **b** $5\frac{1}{3} - 1\frac{3}{5}$ **c** $6\frac{3}{4} \times 1\frac{7}{9}$ **d** $5\frac{1}{3} \div 1\frac{3}{5}$

3 a Find   **i** 5.5% of 16 cm   **ii** $2\frac{3}{4}$% of 930 mm

  b A company employs 200 people. Next year it expects to increase the number of employees by 33%. How many employees does it expect to have next year?

4 a Is it possible for each exterior angle of a regular polygon to be 36°?

b Is it possible for each interior angle of a regular polygon to be 156°?

In each case, where it is possible, give the number of sides.

5 The table shows the prices charged by a publishing company for books of different lengths.

Number of pages	320	512	390	200	450	280
Sale price (£)	5.75	9.50	6.75	3.99	8.50	5.25

a Show this information on a graph. Use 2 mm graph paper, a horizontal scale of 2 cm for 50 pages and mark the axis from 200 to 600. Use a vertical scale of 2 cm for £1 and mark the axis from 3 to 10. (These scales give a small division on the horizontal scale for 5 pages and a small division on the vertical scale for 10 pence.)

b The company is due to publish a new book with 475 pages. About how much should it cost?

c Another new book is priced at £7.25. About how many pages should it have?

6 a Calculate

  i $(+5) \times (2)$ **iii** $7 \times (-4)$

  ii $(-7) \times (-5)$ **iv** $2(-5)$

b Simplify

  i $11x - 3(x + 4)$

  ii $2(x + 5) - (x - 6)$

  iii $5(x + 5) - 4(x + 3)$

c If $P = 2x + 3y$ find P when

  i $x = 3$ and $y = 4$ **ii** $x = 6$ and $y = -3$

7 a Write down the gradient, m, and y-intercept, c, for the straight line with equation

  i $y = -4x + 6$ **ii** $y = 3 - x$

b Sketch the straight line with equation

  i $y = 2x - 5$ **ii** $y = 6 - \frac{1}{2}x$

8 The sketch shows the graph of $y = x^2$.

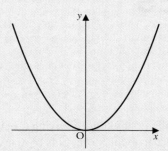

Draw a similar sketch of your own and on it add sketches

of $y = x^2 + 4$ (mark it A)

and $y = x^2 - 4$ (mark it B).

9 Solve the simultaneous equations

a $7x + 3y = 2$ **b** $2x + 5y = 17$
   $5x + 3y = 4$ $4x - 5y = 19$

10 The area of a rectangle is $100 \, \text{cm}^2$.
The rectangle is $4x \, \text{cm}$ long and $3x \, \text{cm}$ wide.

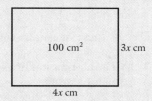

a Form an equation in x.

b Solve the equation.

c Find   **i** the length of the rectangle   **ii** its width.

VOLUMES

Sally is thinking of buying a new pedal-bin for the kitchen. At the Department Store, bins of many different shapes and sizes are on display. Her final choice is between two bins, each priced at £9.99; one is circular in cross-section while the other is square.
Both bins measure 240 mm across the top and both are 38 cm high.

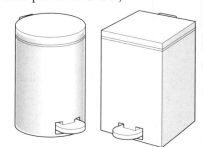

- Which pedal-bin has the larger capacity?

- How much more will it hold than the other bin?

We all have to answer these and similar questions at one time or another, and the answers involve knowing how to calculate volumes.

EXERCISE 18A   Discuss these questions with members of your group.

1 What information does a builder need before he can order the blocks to build a wall?

2 A manufacturer of reinforced steels joists (RSJs) can supply joists to order with different cross-sections. The cost, for each RSJ of a given length, depends on the amount of material used. Given below are sketches, drawn to scale, of several different RSJs. All are the same length.

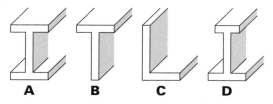

a Which RSJ do you think would be **i** the cheapest **ii** the dearest?

b What information do you need to find the exact cost of each one?
(Assume that if you use twice the amount of material you will double the cost.)

3 Alex wants to put a new tiled roof on his garage. He measures the area of the roof and sets off for the builders' merchant. After choosing a rectangular tile he calculates its area, then divides his answer into the area of the roof. He thinks this gives him the number of tiles he should buy.

a Explain why this calculation will not give him the number of tiles he needs.

b What measurements should he make and how should he use them to get a more accurate value of the number of tiles he needs?

c Why should he order more tiles than he found in part **b**?

d He wants to transport the tiles home using his pick-up truck. How can he calculate the number of journeys he needs to make if the tiles must not be loaded above the height of the sides of the truck which is 50 cm high?

e Can you think of a reason why he should stop loading the tiles before they are 50 cm high?

VOLUME OF A CUBOID

Remember that we find the volume of a cuboid (that is, a rectangular block) by multiplying length by width by height,

i.e.   volume = length × width × height

or $V = l \times w \times h$

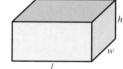

Remember also that the measurements must all be in the same unit before they are multiplied together.

EXERCISE 18B

> Find the volume of a wooden cuboid of length 10 cm, width 66 mm and height 7 cm.
>
> Width = 66 mm = 6.6 cm, $V = l \times w \times h$
>
> = 10 × 6.6 × 7 = 462
>
> Volume = 462 cm³

1 Find the volume of a cuboid of length 9 cm, width 6 cm and height 4 cm.

2 Find the volume of a cuboid of length 12 in, width 8 in and height 4.5 in.

3 Find the volume of a rectangular block of metal 300 mm long, 20 mm wide and 30 mm high.

4 Find the volume of a cuboid of length 6.2 cm, width 3.4 cm and height 5 cm.

Find the volume of the following cuboids, changing the units first if necessary. Do *not* draw a diagram.

	Length	Width	Height	Volume units
5	3.2 cm	5 mm	10 mm	mm^3
6	4 cm	$3\frac{1}{4}$ cm	$4\frac{1}{2}$ cm	cm^3
7	9.2 m	300 cm	1.8 m	m^3
8	6.2 m	32 mm	20 cm	cm^3

Some solids can be made by putting two or more cuboids together.

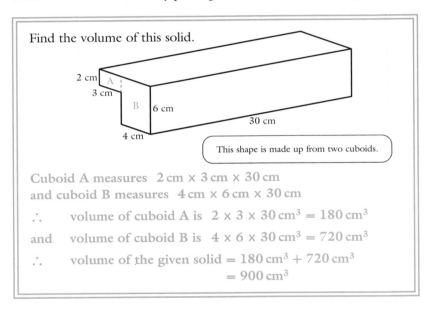

Find the volume of this solid.

2 cm   A   3 cm   B   6 cm   30 cm   4 cm

This shape is made up from two cuboids.

Cuboid A measures 2 cm × 3 cm × 30 cm
and cuboid B measures 4 cm × 6 cm × 30 cm

∴   volume of cuboid A is $2 \times 3 \times 30 \, cm^3 = 180 \, cm^3$

and   volume of cuboid B is $4 \times 6 \times 30 \, cm^3 = 720 \, cm^3$

∴   volume of the given solid $= 180 \, cm^3 + 720 \, cm^3$
                        $= 900 \, cm^3$

In questions **9** to **12** find the volume of each solid assuming that it has been made from two or more cuboids.

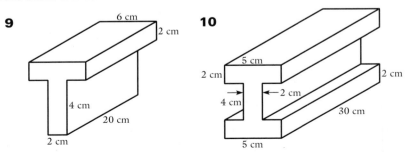

9   6 cm   2 cm   4 cm   20 cm   2 cm

10   5 cm   2 cm   2 cm   4 cm   2 cm   30 cm   5 cm

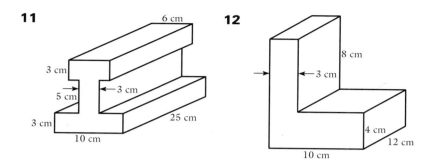

11

6 cm

3 cm

5 cm

3 cm

3 cm

10 cm

25 cm

12

8 cm

3 cm

4 cm

12 cm

10 cm

In questions **13** and **14** find the capacity of each container.

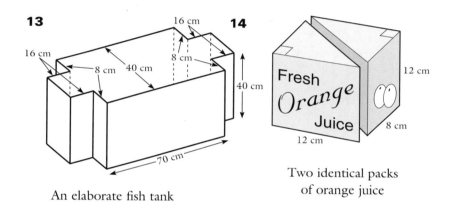

13

16 cm

16 cm

8 cm

40 cm

8 cm

70 cm

40 cm

An elaborate fish tank

14

Fresh
Orange
Juice

12 cm

12 cm

8 cm

Two identical packs
of orange juice

**VOLUMES OF
SOLIDS WITH
UNIFORM
CROSS-
SECTIONS**

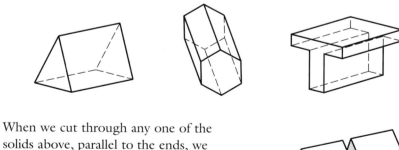

When we cut through any one of the
solids above, parallel to the ends, we
always get the same shape as the end.
This shape is called the *cross-section*.

As the cross-section is the same shape and size wherever the solid is cut,
the cross-section is said to be *uniform* or *constant*. These solids are also
called *prisms* and we can find the volumes of some of them.

First consider a cuboid (which can also be thought of as a rectangular prism).

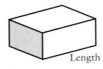

Volume = length × width × height
= (width × height) × length
= area of shaded end × length
= area of cross-section × length

Now consider a triangular prism. If we enclose it in a cuboid we can see that its volume is half the volume of the cuboid.

Volume = ($\frac{1}{2}$ × width × height) × length
= area of shaded triangle × length
= area of cross-section × length

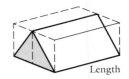

This is true of any prism so that

> Volume of a prism = area of cross-section × length

EXERCISE 18C

This solid represents a plastic block from a child's building set.
Find its volume.

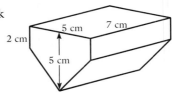

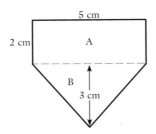

We draw the cross-section only, not the solid. The block can be thought of as a cuboid plus a triangular prism.

Area of A = 2 × 5 cm² = 10 cm²
Area of B = $\frac{1}{2}$ × 5 × 3 cm² = 7.5 cm²
Area of cross-section = 17.5 cm²

Volume = area × length
= 17.5 × 7 cm³
= 122.5 cm³ = 123 cm³ (correct to 3 s.f.)

Find the volumes of the following prisms. Draw a diagram of the cross-section but do *not* draw a picture of the solid.

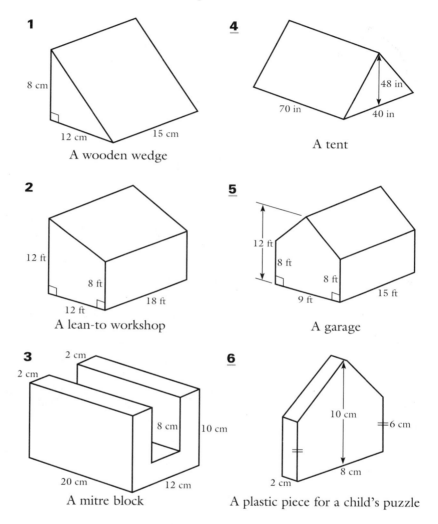

1

8 cm

12 cm   15 cm

A wooden wedge

4

48 in

70 in   40 in

A tent

2

12 ft

8 ft

12 ft   18 ft

A lean-to workshop

5

12 ft

8 ft   8 ft

9 ft   15 ft

A garage

3

2 cm

2 cm

8 cm   10 cm

20 cm   12 cm

A mitre block

6

10 cm   6 cm

8 cm

2 cm

A plastic piece for a child's puzzle

The following two solids are standing on their ends so the vertical measurement is the length.

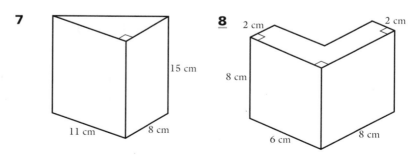

7

15 cm

11 cm   8 cm

8

2 cm   2 cm

8 cm

6 cm   8 cm

In questions **9** to **12**, the cross-sections of the prisms and their lengths are given. Find their volumes.

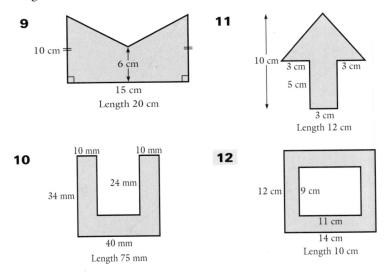

9

10 cm

6 cm

15 cm

Length 20 cm

11

10 cm

3 cm   3 cm

5 cm

3 cm

Length 12 cm

10

10 mm   10 mm

24 mm

34 mm

40 mm

Length 75 mm

12

12 cm   9 cm

11 cm

14 cm

Length 10 cm

In questions **13** and **14** find the volume of the solid shown.

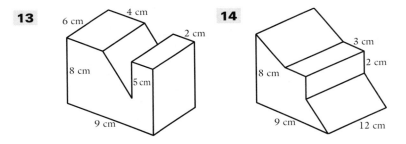

13

6 cm   4 cm

2 cm

8 cm

5 cm

9 cm

14

3 cm

2 cm

8 cm

9 cm   12 cm

15 A tent is in the shape of a triangular prism. Its length is 2.4 m, its height is 1.8 m and the width of the triangular end is 2.4 m. Find the volume enclosed by the tent.

16 A trench 15 m long is dug. Its cross-section, which is uniform, is in the shape of a trapezium with its parallel sides horizontal. Its top is 2 m wide, its base is 1.6 m wide and it is 0.8 m deep. How much earth is removed in digging the trench?

0.2 m ← 1.6 m → 0.2 m

0.8 m

1.6 m

17

The area of the cross-section of the given solid is 42 cm^2 and the length is 32 cm. Find its volume.

18

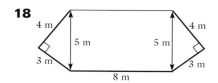

A solid of uniform cross-section is 12 m long. Its cross-section is shown in the diagram. Find its volume.

19

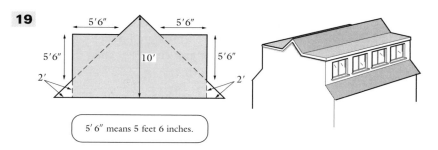

5′ 6″ means 5 feet 6 inches.

The diagram shows the cross-section through the roof space inside a chalet bungalow. There are dormer windows along the whole of both sides.

a Use the measurements given on the diagram to find the area of cross-section of the usable space (shown shaded).

b If the bungalow is 40 ft long find the total volume of the roof space that can be used.

20

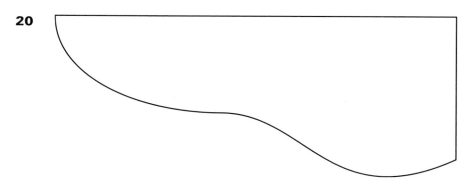

The diagram shows the cross-section of a gutter which carries away the rain water that falls on the roof of a house. The total length of gutter along the front of the house is 12 m.

a Trace this cross-section and place your tracing over a 1 cm grid. By counting squares, estimate the area of cross-section in square centimetres.

b Hence find the maximum volume of water that the gutter can hold. Give your answer in cm^3 correct to 2 significant figures.

c What is the capacity, in litres, of the gutter?

VOLUME OF A CYLINDER

A cylinder can be thought of as a circular prism so its volume can be found using

$$\text{volume} = \text{area of cross-section} \times \text{length}$$
$$= \text{area of circular end} \times \text{length}$$

From this we can find a formula for the volume.

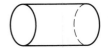

We usually think of a cylinder as standing upright so that its length is represented by h (for height). If the radius of the end circle is r, then the area of the cross-section is πr^2

$\therefore \qquad \text{volume} = \pi r^2 \times h$

$$V = \pi r^2 h$$

EXERCISE 18D

Find the volume inside a cylindrical mug of internal diameter 8 cm and height 6 cm. Use the π button on your calculator.

As the diameter is 8 cm, the radius is 4 cm.

The volume is given by the formula
$V = \pi r^2 h$
$\quad = \pi \times 4 \times 4 \times 6 = 301.59\ldots$

Therefore volume of mug is 302 cm³ (correct to 3 s.f.)

An alternative method would be:

Area of cross-section $= \pi r^2$
$\qquad\qquad\qquad = \pi \times 4 \times 4\,\text{cm}^2 = 50.265\ldots\text{cm}^2$

Volume $=$ area of cross-section $\times$ length
$\qquad\quad = (\,50.265\ldots \times 6\,)\,\text{cm}^3$
$\qquad\quad = 301.59\ldots\text{cm}^3$
$\qquad\quad = 302\,\text{cm}^3$ (correct to 3 s.f.)

Find the volumes of the following cylinders. Give all your answers correct to 3 significant figures.

1 Radius 2 cm, height 10 cm

2 Radius 3 cm, height 4 cm

3 Radius 3 in, height $2\frac{1}{2}$ in

4 Diameter 2 cm, height 1 cm

5 Radius 1 cm, height 4.8 cm

6 Radius 4 in, height 3 in

7 Radius 12 cm, height 1.8 cm

8 Radius 7 cm, height 9 cm

9 Radius 3.2 cm, height 10 cm **13** Diameter 2.4 cm, height 6.2 cm

10 Diameter 10 cm, height 42 cm **14** Radius 6 cm, height 3.6 cm

11 Radius 4.8 mm, height 13 mm **15** Diameter 16.2 cm, height 4 cm

12 Radius 8 cm, height 44 mm **16** Diameter 16 mm, height 5.2 cm

Find the volumes of the following compound shapes. Give your answers correct to 3 significant figures.
Draw diagrams of the cross-sections but do not draw pictures of the solids.

17
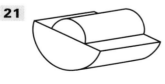
A half-cylinder of length 16 cm and radius 4 cm.

18
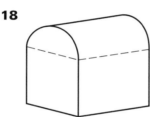
A solid of length 6.2 cm, whose cross-section consists of a square of side 2 cm surmounted by a semicircle.

19
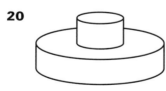
A disc of radius 9 cm and thickness 0.8 cm.

20

A solid made of two cylinders each of height 5 cm.
The radius of the smaller one is 2 cm and of the larger one is 6 cm.

21
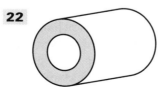
A solid made of two half-cylinders each of length 11 cm. The radius of the larger one is 10 cm and the radius of the smaller one is 5 cm.

22

A tube of length 20 cm. The inner radius is 3 cm and the outer radius is 5 cm.

23 The diagram shows a cylindrical hole, diameter 42 mm, in a concrete block which is buried in the ground to support a clothes line.

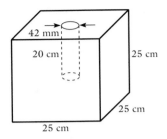

a What volume of water would fill the hole?

b What volume of concrete is used to make the block?

24 The walls of a cylindrical wooden egg cup are 3 mm thick and the base is 6 mm thick.

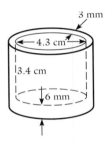

a Calculate the volume of wood in the egg cup.

b How many millilitres of water will the egg cup hold?

25 At the beginning of the chapter Sally was thinking about buying a pedal-bin for the kitchen. Which one has the larger capacity?

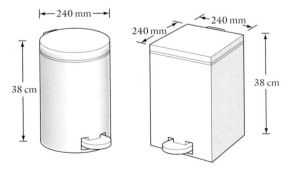

Find the difference between the capacities of the two bins.

26 Having decided to buy the cylindrical bin, Sally needs to buy some plastic bin liners. She has a choice of three packs: in pack A the capacity of the liners is 5 litres, in pack B the capacity is 10 litres, and in pack C the capacity is 25 litres. Which pack should she buy and why?

DENSITY

It is often important to know the mass of one unit of volume of the material from which an object is made.
This is sometimes called the *density* of the material,

i.e.

$$\text{density} = \frac{\text{mass}}{\text{volume}}$$

A jeweller, for example, might need to know that the mass of $1\,\text{cm}^3$ of silver is $10.5\,\text{g}$, that is, that the density of silver is $10.5\,\text{g/cm}^3$. (g/cm^3 means grams per cubic centimetre.)
She could then find the volume of silver she has if she knows its mass or she could calculate what mass is needed to give a certain volume.

EXERCISE 18E

A gold ingot is a cuboid measuring $4\,\text{cm} \times 4\,\text{cm} \times 2\,\text{cm}$. Given that the density of gold is $19.3\,\text{g/cm}^3$, find the mass of the ingot.

Volume of ingot $= 4 \times 4 \times 2\,\text{cm}^3$
$\qquad = 32\,\text{cm}^3$
$\qquad$ Mass $= 32 \times 19.3\,\text{g}$
$\qquad\qquad = 617.6\,\text{g}$
$\qquad\qquad = 618\,\text{g}$ (correct to 3 s.f.)

> Density $= 19.3\,\text{g per cm}^3$, so $32\,\text{cm}^3$ weighs $32 \times 19.3\,\text{g}$

> For a substance as valuable as gold the mass would probably be found more accurately, but to do this we need to know the density more accurately than correct to 1 d.p.

Give all answers correct to 3 significant figures.

1 The density of oak is $0.8\,\text{g/cm}^3$. Find the mass of a piece of oak that has a volume of $120\,\text{cm}^3$.

2 The density of copper is $8.9\,\text{g/cm}^3$. Find the mass of a length of copper pipe if $260\,\text{cm}^3$ of copper has been used to make it.

3 A block of brass has a volume of $14.3\,\text{cm}^3$. Given that the density of brass is $8.5\,\text{g/cm}^3$, find the mass of the block.

4 Find the mass of 1 litre of milk given that the density of milk is $0.98\,\text{g/cm}^3$.

5 A rectangular block of wood measures $6\,\text{cm}$ by $8\,\text{cm}$ by $24\,\text{cm}$. The density of the wood is $0.85\,\text{g/cm}^3$. Find the mass of the block.

6 The diagram shows a block of paraffin wax cast into the shape of a triangular prism. The density of the wax is $0.9 \, \text{g/cm}^3$. Find the mass of the wax.

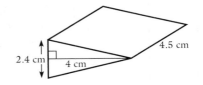

4.5 cm

2.4 cm

4 cm

7 The diagram shows the cross-section of a water trough, 1.5 m long, which is full of ice.

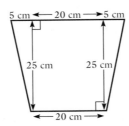

5 cm ← 20 cm → 5 cm

25 cm   25 cm

← 20 cm →

Using the measurements given on the diagram find

a the area of cross-section in cm^2

b the mass of the ice in the trough given that the density of ice is $0.917 \, \text{g/cm}^3$.

8 A hollow cylinder has an internal diameter of 1.2 cm and is 3 cm deep. It is filled with mercury which has a density of $13.6 \, \text{g/cm}^3$. Find the mass of the mercury.

MIXED EXERCISE

EXERCISE 18F

1 Find the volume of a concrete block measuring 16 cm by 24 cm by 1.4 m.

In questions **2** and **3**, the cross-sections of the prisms and their lengths are given. Find their volumes.

2

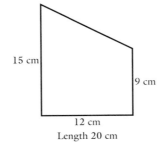

15 cm

9 cm

12 cm

Length 20 cm

3

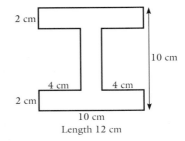

2 cm

10 cm

4 cm   4 cm

2 cm

10 cm

Length 12 cm

4 Find the capacity of a cylinder with

a radius 3 cm and height 6.5 cm

b diameter 8 in and height 4 in.

(Give your answers correct to 3 significant figures.)

5 The diagram shows a solid model of a
house which has been made from wood.

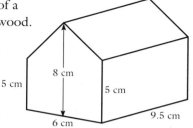

a Use the measurements given on
the diagram to find
i the area of the uniform
cross-section
ii the volume of wood used to
made the model.

b If the density of the wood is 8.5 g/cm^3, find the mass of the model.

INVESTIGATION

1 This investigation develops the investigation given in Chapter 20
of Book 7A.

A rectangular piece of steel sheet measuring 14 cm by 20 cm is to be
used to make an open rectangular box. The diagram shows one way
of doing this. The four corner squares are removed and the sides
folded up. The four vertical seams at the corners are then sealed.

a Copy and complete the following table which gives the
measurements of the base and the capacity of the box,
when squares of different sizes
are removed from the corners.
Continue to add numbers to
the first column in the table as
long as it is reasonable to do so.
These numbers should follow
the pattern indicated.

Length of edge of square (cm)	Measurements of base (cm)	Capacity of box (cm^3)
0.5	13 × 19	0.5 × 13 × 19 = 123.5
1	12 × 18	1 × 12 × 18 = 216
1.5		
2		
2.5		
3		

b What is the last number you entered in the first column of the
table? Justify your choice.

c What size of square should be removed to give the largest
capacity recorded in the table?

d Investigate whether you can find a number that you have not
already entered in the first column that gives a larger capacity
than any value you have found so far.

2 a Betty has a bag of identical triangular wooden blocks.

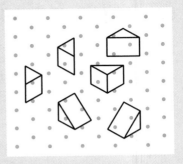

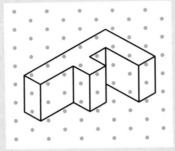

She uses some of them to make the letter F.
How many blocks does she need?

b Betty also has a bag of triangular blocks identical to each other but with a different shape and size from those in the first bag.

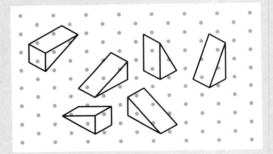

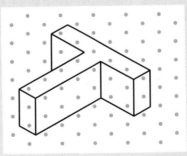

How many of these does she need to make the letter T?

c How does the volume of one of the larger blocks compare with the volume of one of the smaller blocks?

d Which letter uses the more wood, the F or the T?

e How many cubical blocks of side 1 unit would Betty need to make **i** the F **ii** the T?

f Which other letters in the alphabet can be drawn on isometric paper using cubes? Draw at least two of them. Can you draw additional letters if you use blocks like those shown in part **a**?

ENLARGEMENT

Melanie is designing some cut-out models. She designs them on squared paper, but to turn them into saleable products she needs to draw them much larger. She could use a photocopier or computer to do this but she does not have easy access to either. How can she overcome this problem of enlarging a given shape to a more acceptable size?

ENLARGEMENTS

All the transformations we have used so far (reflections, translations and rotations) have moved the object or turned it over to produce the image, but its shape and size have not changed. In each case, the image and the object are congruent. Now we consider a different transformation, an enlargement, that keeps the shape but alters the size.

EXERCISE 19A

Discuss which of the following use enlargement in their work.

1 photographer   **3** architect   **5** social worker

2 doctor   **4** automotive engineer   **6** travel agent

Think of the picture thrown on the screen when a slide projector is used.

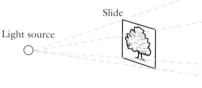

The picture on the screen is the same as that on the slide but it is very much bigger.

We can use the same idea to enlarge any shape.

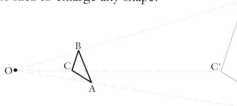

$\triangle A'B'C'$ is the image of $\triangle ABC$ under an *enlargement, centre O.*
O is the *centre of enlargement.*
We call the dotted lines *guidelines.*

CENTRE OF
ENLARGEMENT

In the next exercise we investigate how to find the centre of enlargement when we have the original shape and its image.

EXERCISE 19B   In all these questions, one shape is an enlargement of the other.

1 Copy the diagram using 1 cm to 1 unit. Draw A′A, B′B, C′C, and D′D and continue all four lines until they meet.
The point where the lines meet is called the centre of enlargement.
Give the coordinates of the centre of enlargement.
Consider the sides and angles of the two squares. What do you notice ?

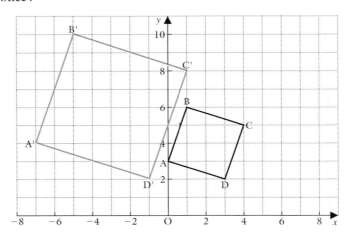

2 Repeat question **1** using this diagram.

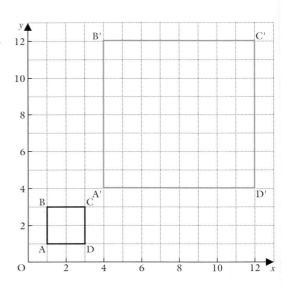

3 Copy the diagram using 1 cm to 1 unit. Draw P′P, Q′Q and R′R and continue all three lines until they meet.

Give the coordinates of the centre of enlargement.

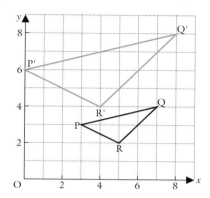

Repeat question **3** using the diagrams in questions **4** and **5**.

4

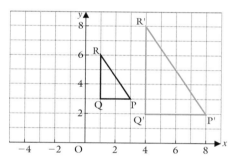

5

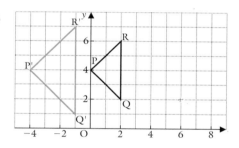

6 In questions **3** to **5** name pairs of lines that are parallel.

7 Draw axes for x and y from 0 to 9 using 1 cm as 1 unit.
Draw $\triangle ABC$: A(2, 3), B(4, 1), C(5, 4)
Draw $\triangle A′B′C′$: A′(2, 5), B′(6, 1), C′(8, 7).
Draw A′A, B′B and C′C and extend these lines until they meet.

a Give the coordinates of the centre of enlargement.

b Measure the sides and angles of the two triangles. What do you notice?

8 Draw axes for x and y from 0 to 10 using 1 cm as 1 unit.
Draw $\triangle XYZ$ with X(8, 2), Y(6, 6) and Z(5, 3)
and $\triangle X′Y′Z′$ with X′(6, 2), Y′(2, 10) and Z′(0, 4).
Find the centre of enlargement and label it P.
Measure PX, PX′, PY, PY′, PZ, PZ′. What do you notice?

The centre of enlargement can be anywhere, including a point inside the object or a point on the object.

The centres of enlargement in the diagrams below are marked with a cross.

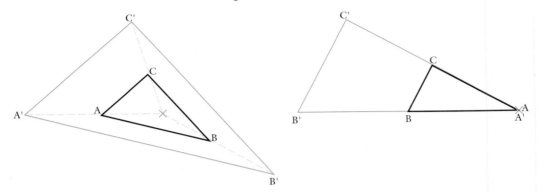

9 Copy the diagram using 1 cm as 1 unit. Draw A′A, B′B, C′C and D′D and extend the lines until they meet. Give the coordinates of the centre of enlargement.

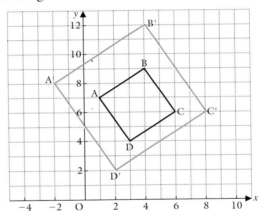

10 In the diagram below which point is the centre of enlargement?

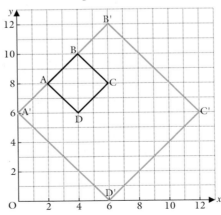

11 Copy the diagram using 1 cm as 1 unit. Draw A'A, B'B and C'C and extend the lines until they meet. Give the coordinates of the centre of enlargement.

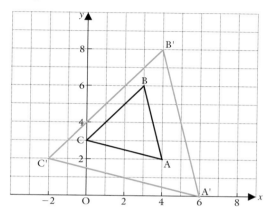

12 In the diagram below which point is the centre of enlargement?

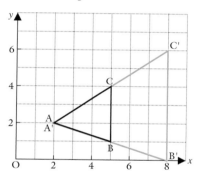

13 Draw axes for x and y from -3 to 10 using 1 cm as 1 unit.
Draw $\triangle ABC$ with $A(4, 0)$, $B(4, 4)$ and $C(0, 2)$.
Draw $\triangle A'B'C'$ with $A'(5, -2)$, $B'(5, 6)$ and $C'(-3, 2)$.
Find the coordinates of the centre of enlargement.

SCALE FACTORS   If we measure the lengths of the sides of the two triangles PQR and P'Q'R' and compare them, we find that the lengths of the sides of $\triangle P'Q'R'$ are three times those of $\triangle PQR$.

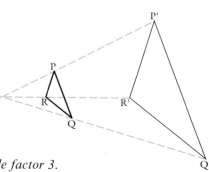

We say that $\triangle P'Q'R'$ is the image of $\triangle PQR$ under an enlargement, centre O, with *scale factor 3*.

FINDING AN
IMAGE UNDER
ENLARGEMENT

If we measure OR and OR′ in the diagram at the bottom of the previous page, we find OR′ is three times OR. This enables us to work out a method for enlarging an object with a given centre of enlargement (say O) and a given scale factor (say **3**).

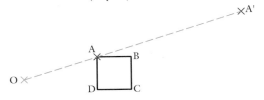

Measure OA. Multiply it by 3. Mark A′ on the guideline three times as far from O as A is,

i.e. $$OA' = 3 \times OA$$

Repeat for B and the other vertices of ABCD.

Then A′B′C′D′ is the image of ABCD. To check, measure A′B′ and AB. A′B′ should be three times as large as AB.

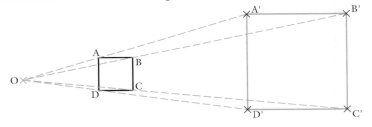

EXERCISE 19C

1 Find the scale factor used for each of questions **1** to **5** in **Exercise 19B**.

2 Copy the diagram using 1 cm as 1 unit. P is the centre of enlargement. Draw the image of the square ABCD under an enlargement, scale factor 2.

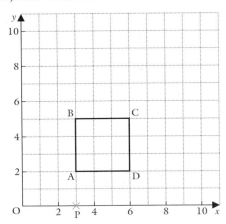

3 Copy the diagram using 1 cm as 1 unit. P is the centre of enlargement. Draw the image of △ ABC under an enlargement, scale factor 2.

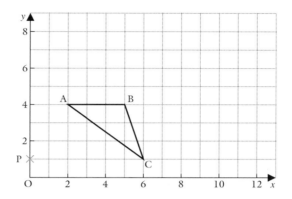

4 Repeat question **3** using this diagram.

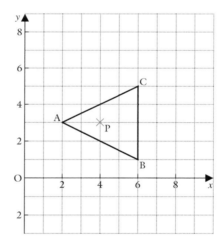

In questions **5** to **7** draw axes for x and y from 0 to 10, using 1 cm as 1 unit. In each case, find the image A′B′C′ of △ABC using the given enlargement. Check by measuring the lengths of the sides of the two triangles.

5 △ABC: A(3, 3), B(6, 2), C(5, 6).
Enlargement with centre (5, 4), and scale factor 2.

6 △ABC: A(2, 1), B(4, 1), C(3, 4).
Enlargement with centre (1, 1), and scale factor 3.

7 △ABC: A(1, 2), B(7, 2), C(1, 6).
Enlargement with centre (1, 2), and scale factor $1\frac{1}{2}$.

8 On plain paper, mark a point P near the left-hand edge. Draw a small object (a pin man perhaps, or a square house) between P and the middle of the page. Using the method of enlargement, draw the image of the object with centre P and scale factor 2.

9 Repeat question **8** with other objects and other scale factors. Think carefully about the space you will need for the image.

10 Draw axes for x and y from 0 to 10 using 1 cm as 1 unit. Draw the square ABCD with A$(5, 7)$, B$(6, 9)$, C$(8, 8)$ and D$(7, 6)$. Taking the point P$(10, 9)$ as the centre of enlargement and a scale factor of 2, draw the image of ABCD by counting squares and without drawing guidelines.

11 Draw axes for x and y from 0 to 10 using 1 cm as 1 unit. Draw $\triangle$ABC with A$(2, 2)$, B$(5, 1)$ and C$(3, 4)$. Taking the origin as the centre of enlargement and a scale factor of 2, draw the image of $\triangle$ABC by counting squares and without drawing guidelines.

12 Draw axes for x and y. Scale the x-axis from 0 to 16 and the y-axis from 0 to 24, using 5 mm as 1 unit. Plot the following points and join them in order by straight lines:
$(3, 0), (3, 3), (0, 3), (2, 6), (1, 6), (3, 9), (2, 9), (4, 12),$
$(6, 9), (5, 9), (7, 6), (6, 6), (8, 3), (5, 3)$ and $(5, 0)$.
What shape have you drawn?
Take the origin as the centre of enlargement and a scale factor of 2, draw the image of this shape by counting squares. Do not draw guidelines.
How are the coordinates of the image related to the coordinates of the object?

FRACTIONAL SCALE FACTORS

We can reverse the process of enlargement and shrink or reduce the object, producing a smaller image. If the lengths of the image are one-third of the lengths of the object then the scale factor is $\frac{1}{3}$.

There is no satisfactory word to cover both enlargement and shrinking (some people use 'dilation' and some 'scaling') so *enlargement* tends to be used for both. You can tell one from the other by looking at the size of the scale factor. A scale factor smaller than 1 gives a smaller image while a scale factor greater than 1 gives a larger image.

EXERCISE 19D

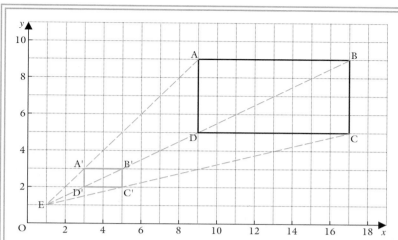

A′B′C′D′ is the image of the rectangle ABCD, and E is the centre of enlargement.

a Write down the coordinates of the centre of enlargement.

b Find the scale factor.

a The coordinates of E, the centre of enlargement are (1, 1).

b From the diagram we can see that EA = 4EA′,
EB = 4EB′, EC = 4EC′ and ED = 4ED′.

The scale factor is given by

$$\frac{\text{Distance from the centre of enlargement to a point on the image}}{\text{Distance from the centre of enlargement to the corresponding point on the original rectangle}}$$

Scale factor = $\dfrac{\text{EA}'}{\text{EA}}$ = $\dfrac{1}{4}$

i.e. the scale factor is $\frac{1}{4}$.

1 In this question the square A′B′C′D′ is the image of the square ABCD.
Find the centre of enlargement and the scale factor.

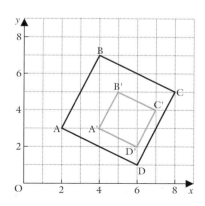

In questions **2** to **5** △A′B′C′ is the image of △ABC. Give the centre of enlargement and the scale factor.

2

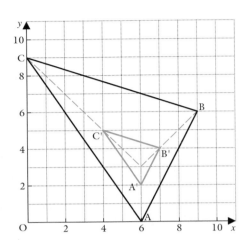

3

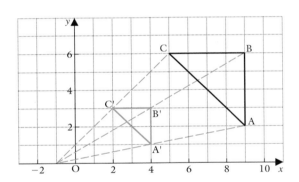

4 Draw axes for x and y from -2 to 8 using 1 cm as 1 unit.
   Draw △ABC with A($-1, 4$), B($5, 1$) and C($5, 7$),
   and △A′B′C′ with A′($2, 4$), B′($4, 3$) and C′($4, 5$).

5 Draw axes for x and y from 0 to 9 using 1 cm as 1 unit.
   Draw △ABC with A($1, 2$), B($9, 2$) and C($9, 6$),
   and △A′B′C′ with A′($1, 2$), B′($5, 2$) and C′($5, 4$).

6 Draw axes for x and y from -1 to 11 using 1 cm as 1 unit.
   Draw △ABC with A($4, 0$), B($10, 9$) and C($1, 6$).
   With the point ($4, 3$) as the centre of enlargement and using a scale factor of $\frac{1}{3}$ draw the image △A′B′C′ of △ABC.

**PRACTICAL
WORK**

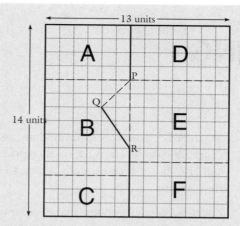

Daniel designs packaging. The sketch shows a design that will fold to make a package suitable for sending books of various sizes by post. The design can be produced in several different sizes if it is enlarged by a given scale factor.

The basic rectangle is divided into 6 smaller rectangles, marked A, B, C, D, E and F on the sketch. It is cut along the heavy lines and fold marks are pressed into it along the broken lines.

By folding along PQ and PR, you will find that rectangle B now fits over rectangle E and this acts as the base of the package. The remaining 4 rectangles fold up and over allowing books of different thicknesses to be packed.

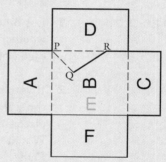

a On an A4 sheet of 5 mm squared paper, draw an enlargement of the diagram with a scale factor of 2.

b Cut it out and fold it to show how it is used.

c If you have a much larger sheet of paper or card, repeat parts **a** and **b** using a larger scale factor, for example, 4.

d **i** Can the fold PQ be drawn at any angle?
   ii Can the cut QR be made at any angle?
   Give reasons for your answers.

e What is the ratio of the length of the original rectangle to its breadth? If you alter this ratio, can you still make a satisfactory package? Investigate.

f Can you improve on the design?

SCALE DRAWING

An architect makes an accurate drawing of a building before it is built. Everything is shown much smaller than it will be in the completed building, but it is all carefully drawn to scale, that is, all the proportions are correct.

Likewise an engineer also makes accurate drawings when he designs small parts for a new machine. However, these drawings usually show small components much larger than they will be when they are manufactured.

In these, as in many other occupations, scale drawings are essential to high quality products and services.

EXERCISE 20A

1 List other occupations where accurate drawings are used. Say whether each occupation tends to make scale drawings that show each object larger or smaller than the finished product.

2 If you had a map of the area within 20 miles of where you live, and the scale was not given, would you be able to decide how far one place was from another?
Would you give the same answer if the map was for a small area of France where you would like to take a holiday? Justify your answer.

3 What information do you think is needed on any scale drawing?

CONSTRUCTING TRIANGLES AND QUADRILATERALS

In Book 7A, Chapter 12, we saw that we could construct a triangle provided we were given enough information.

The necessary information is

either   one side and two angles

or   two sides and the angle between those two sides

or   three sides

Some quadrilaterals can also be constructed using similar methods.

EXERCISE 20B

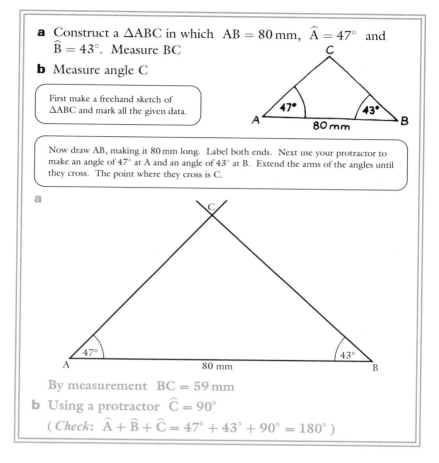

a Construct a $\triangle ABC$ in which $AB = 80$ mm, $\hat{A} = 47°$ and $\hat{B} = 43°$. Measure BC

b Measure angle C

> First make a freehand sketch of $\triangle ABC$ and mark all the given data.

> Now draw AB, making it 80 mm long. Label both ends. Next use your protractor to make an angle of 47° at A and an angle of 43° at B. Extend the arms of the angles until they cross. The point where they cross is C.

a

By measurement BC = 59 mm

b Using a protractor $\hat{C} = 90°$

(*Check:* $\hat{A} + \hat{B} + \hat{C} = 47° + 43° + 90° = 180°$)

In questions **1** to **4** use the given information to construct each triangle. Remember to draw a freehand diagram first. It may be necessary to calculate the third angle before you can begin the construction.

1 A metal plate $\triangle ABC$ in which $AB = 65$ mm, $\hat{A} = 50°$, $\hat{B} = 45°$. Measure AC.

2 $\triangle XYZ$ in which $\hat{Y} = 67°$, $XY = 3.7$ cm, $YZ = 6.8$ cm. Measure $\hat{X}$ and $\hat{Z}$.

3 A triangular plastic marker DEF in which $DE = 115$ mm, $EF = 92$ mm, $DF = 69$ mm. Measure $\hat{D}$, $\hat{E}$ and $\hat{F}$. (Remember to use compasses for this construction.)

4 Construct a quadrilateral ABCD in which $AB = 47$ mm, $AD = 80$ mm, $BD = 87$ mm, $A\hat{B}C = 120°$, $A\hat{D}C = 85°$. Measure BC, DC and $\hat{C}$. (When you have drawn the rough sketch think of it as made up of two triangles.)

ACCURATE
DRAWING WITH
SCALED-DOWN
MEASUREMENTS

If you are asked to draw a car-park which is a rectangle measuring 50 m by 25 m, you obviously cannot draw it full size. To fit it on to your page you will have to scale down the measurements. In this case you could use 1 cm to represent 5 m on the car-park. This is called the *scale*; it is usually written as 1 cm ≡ 5 m, and must *always* be stated on any scale drawing.

EXERCISE 20C

Start by making a freehand drawing of the object you are asked to draw to scale. Mark all the full-size measurements on your sketch. Next draw another sketch and put the scaled measurements on this one. Then do the accurate scale drawing.

The end wall of a building is a rectangle with a triangular top. The rectangle measures 12 m wide by 6 m high. The base of the triangle is 12 m and the sloping sides are 8 m long. Using a scale of 1 cm to 2 m, make a scale drawing of this wall. Use your drawing to find, to the nearest tenth of a metre, the distance from the ground to the ridge of the roof.

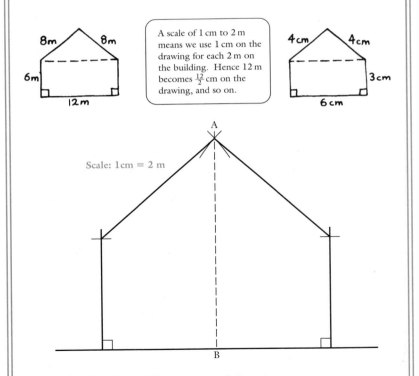

A scale of 1 cm to 2 m means we use 1 cm on the drawing for each 2 m on the building. Hence 12 m becomes $\frac{12}{2}$ cm on the drawing, and so on.

Scale: 1cm ≡ 2 m

From the drawing, AB measures 5.7 cm.
So the height of the wall is 5.7 × 2 m = 11.4 m

In questions **1** to **5** use the scale given in brackets to make a scale drawing of the given object.

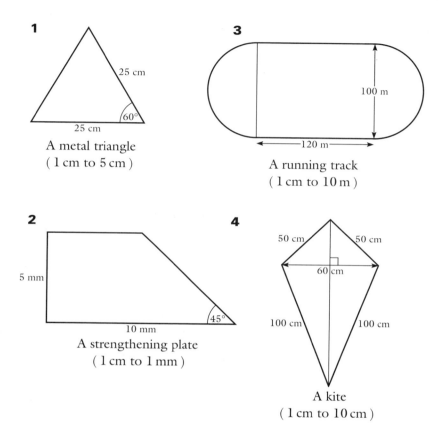

1

25 cm

25 cm

60°

A metal triangle
(1 cm to 5 cm)

3

100 m

120 m

A running track
(1 cm to 10 m)

2

5 mm

10 mm

45°

A strengthening plate
(1 cm to 1 mm)

4

50 cm

50 cm

60 cm

100 cm

100 cm

A kite
(1 cm to 10 cm)

5 A field is rectangular in shape. It measures 300 m by 400 m. A land drain goes in a straight line from one corner of the field to the opposite corner. Using a scale of 1 cm to 50 m, make a scale drawing of the field and use it to find the length of the land drain.

6 The end wall of a ridge tent is a triangle. The base is 2 m and the sloping edges are each 2.5 m. Using a scale of 1 cm to 0.5 m, make a scale drawing of the triangular end of the tent and use it to find the height of the tent.

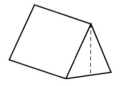

7 The surface of a swimming pool is a rectangle measuring 25 m by 10 m. Choose your own scale and make a scale drawing of the pool. Now compare and discuss your drawing with those of other pupils.

In questions **8** to **11** choose your own scale.

Choose a scale that gives lines that are long enough to draw easily; in general, the lines on your drawing should be at least 5 cm long. Avoid scales that give lengths involving awkward fractions of a centimetre, such as thirds; $\frac{1}{3}$ cm cannot be read from your ruler.

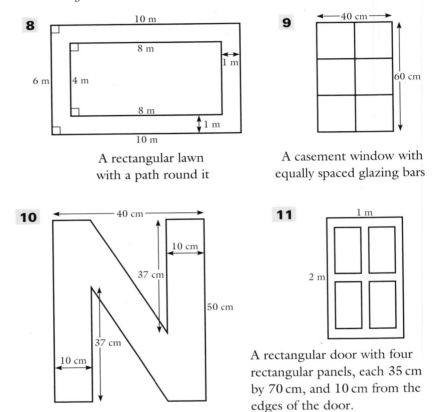

8

10 m

8 m

1 m

6 m 4 m

8 m

1 m

10 m

A rectangular lawn
with a path round it

9

←— 40 cm —→

60 cm

A casement window with
equally spaced glazing bars

10

←——— 40 cm ———→

10 cm

37 cm

50 cm

37 cm

10 cm

11

1 m

2 m

A rectangular door with four
rectangular panels, each 35 cm
by 70 cm, and 10 cm from the
edges of the door.

**ANGLES OF
ELEVATION**

If you are standing on level ground and can see a tall building, you will have to look up to see the top of that building.

If you start by looking straight ahead and then look up to the top of the building, the angle through which you raise your eyes is called the *angle of elevation* of the top of the building.

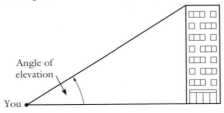

Angle of
elevation

You

There are instruments for measuring angles of elevation. A simple one can be made from a large card protractor and a piece of string with a weight on the end.

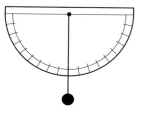

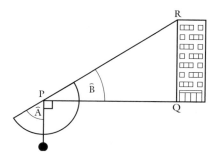

You can read the size of $\widehat{A}$.
Then the angle of elevation $\widehat{B}$ is given by $\widehat{B} = 90° - \widehat{A}$.
(This method is not very accurate.)

If your distance from the foot of the building and the angle of elevation of the top are both known, you can make a scale drawing of $\triangle$PQR.
(Note that *only* $\triangle$PQR needs to be drawn.)
This drawing can then be used to work out the height of the building.

EXERCISE 20D

From a point A on the ground, which is 50 m from the base of a tree, the angle of elevation of the top of the tree is 22°. Using a scale of 1 cm ≡ 5 m, make a scale drawing and use it to find the height of the tree.

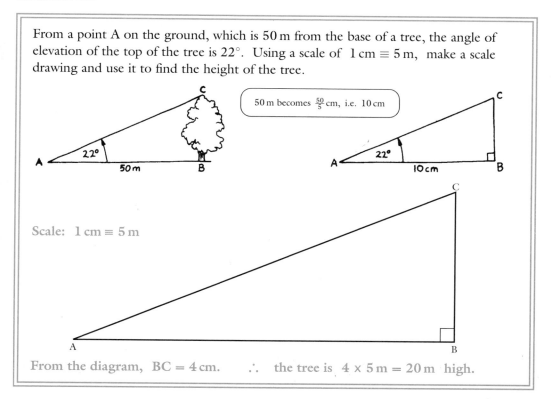

50 m becomes $\frac{50}{5}$ cm, i.e. 10 cm

Scale: 1 cm ≡ 5 m

From the diagram, BC = 4 cm.   ∴   the tree is 4 × 5 m = 20 m high.

In questions **1** to **4** A is a point on the ground, Â is the angle of elevation of C, the top of the side wall BC of a building. Using a scale of 1 cm ≡ 5 m, make a scale drawing and use it to find the height of the wall BC.

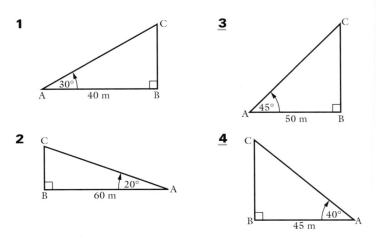

1

3

2

4

In questions **5** to **9** make a scale drawing of the triangle only; do not add unnecessary details such as buildings.

5

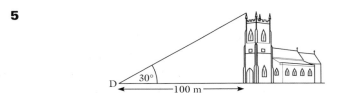

From a point D on the ground which is 100 m from the foot of a church tower, the angle of elevation of the top of the tower is 30°. Use a scale of 1 cm to 10 m to make a scale drawing. Use your drawing to find the height of the tower.

6

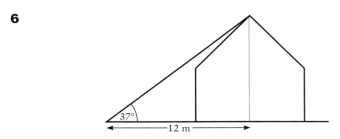

From the opposite side of the road, the angle of elevation of the top of the roof of my house is 37°. The horizontal distance from the point where I measured the angle to the middle of the house is 12 m. Make a scale drawing, using a scale of 1 cm to 1 m, and use it to find the height of the top of the roof.

7

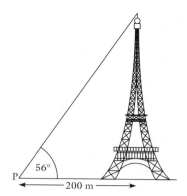

56°

P

— 200 m —

From a point P on the ground which is 200 m from the middle of the base of the Eiffel Tower, the angle of elevation of the top is 56°. Use a scale of 1 cm to 20 m to make a scale diagram and find the height of the Eiffel Tower.

8 From a point on the ground which is 300 m from the base of the National Westminster Tower, the angle of elevation of the top of the tower is 31°. Using a scale of 1 cm to 50 m , make a scale drawing and find the height of the National Westminster Tower. (This is a high office building in the City of London.).

9 The top of a radio mast is 76 m from the ground. From a point P on the ground, the angle of elevation of the top of the mast is 40°. Use a scale of 1 cm to 10 m to make a scale drawing to find how far away P is from the bottom of the mast.
(You will need to do some calculation before you can do the scale drawing.)

ANGLES OF DEPRESSION

An *angle of depression* is the angle between the line looking straight ahead and the line looking *down* at an object below you.

If, for example, you are standing on a cliff looking out to sea, the diagram shows the angle of depression of a boat.

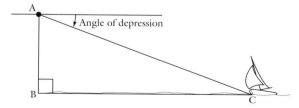

A

Angle of depression

B

C

If the angle of depression and the height of the cliff are both known, you can make a scale drawing of △ABC. Then you can work out the distance of the boat from the foot of the cliff.

EXERCISE 20E

From the top of a cliff 20 m high, the angle of depression of a boat out at sea is 24°.

Using a scale of 1 cm to 5 m, make a scale drawing to find the distance of the boat from the foot of the cliff.

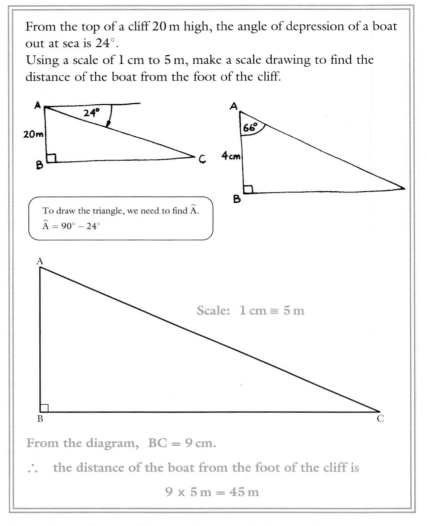

To draw the triangle, we need to find $\hat{A}$.

$\hat{A} = 90° - 24°$

Scale: 1 cm ≡ 5 m

From the diagram, BC = 9 cm.

∴   the distance of the boat from the foot of the cliff is

$$9 \times 5\,\text{m} = 45\,\text{m}$$

Draw simple diagrams like those in the worked example; do not draw buildings, trees, etc.

In questions **1** to **4** use a scale of 1 cm ≡ 10 m.

1

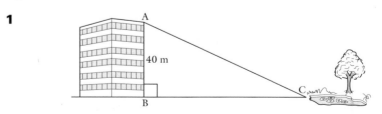

From the top A, of a building the angle of depression of the front edge of an ornamental pond C, is 25°. Find BC, the distance of the pond from the base of the building.

2

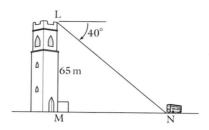

From the top L, of a tower the angle of depression of a bus N on the road below, is 40°. Find MN, the distance of the bus from the foot of the tower.

3

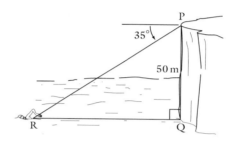

From the top P, of a cliff the angle of depression of a swimmer R out at sea, is 35°. Find RQ, the distance of the swimmer from the foot of the vertical cliff.

4

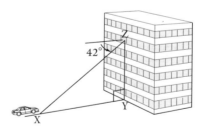

From Z, the position of my window in a multi-storey office block, the angle of depression of my car X, in the car park is 42°. Given that Z is 90 m above ground level, find XY, the distance of my car from the point on the ground immediately beneath my window.

5 From the top of the Eiffel Tower, which is 300 m high, the angle of depression of a house is 20°. Use a scale of 1 cm to 50 m to make a scale drawing and find the distance of the house from the base of the tower.

6

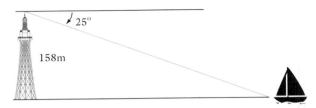

From the top of Blackpool Tower, which is 158 m high, the angle of depression of a yacht at sea is 25°. Use a scale of 1 cm to 20 m to make a scale drawing to find the distance of the yacht from the base of the tower.

7 From an aircraft flying at a height of 300 m, the angle of depression of the end of the runway is noted as 18°. Using a scale of 1 cm to 100 m, make a scale diagram to find the horizontal distance of the aircraft from the runway.

8 The Sears Tower in Chicago is an office building and it is 443 m high. From the top of this tower, the angle of depression of a ship on a lake is 40°. How far away from the base of the building is the ship? Use a scale of 1 cm to 50 m to make your scale drawing.

For the remaining questions in this exercise, make a scale drawing choosing your own scale.

9 From a point on the ground 60 m away, the angle of elevation of the top of a factory chimney is 42°. Find the height of the chimney.

10 From the top of a hill, which is 400 m above sea-level, the angle of depression of a boat house is 20°. The boat house is at sea-level. Find the distance of the boat house from the top of the hill.

11 The pilot of an aircraft, flying at 5000 m, reads the angle of depression of a point on the coast as 30°. At the moment that the angle is read from the cockpit instrument, how much further has the plane to fly before passing over the coast line?

12 An automatic lightship is stationed 500 m from a point A on the coast. There are high cliffs at A and from the top of these cliffs, the angle of depression of the lightship is 15°. How high are the cliffs?

13 An airport controller measures the angle of elevation of an approaching aircraft as 20°. If the aircraft is then 1.6 km from the control building, at what height is it flying?

THREE-FIGURE
BEARINGS

A bearing is a compass direction.

If you are standing at a point A and looking at a tree B in the distance, as shown in the diagram below, then using compass directions you could say that

from A, the bearing of B is SE

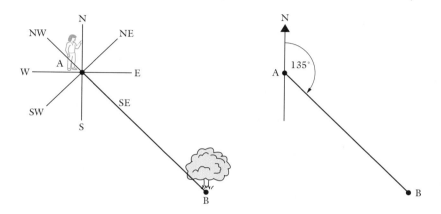

Using a *three-figure bearing* we first look north and then turn clockwise until we are looking at B. The angle turned through is the three-figure bearing.

In this case

from A, the bearing of B is 135°

A three-figure bearing is a clockwise angle measured from north.

If the angle is less than 100°, it is made into a three-figure angle by putting zero in front, so 20° becomes 020°.

EXERCISE 20F

Draw a freehand sketch to illustrate that the bearing of a lighthouse B from a ship, A, is 060°. Mark the angle in your sketch.

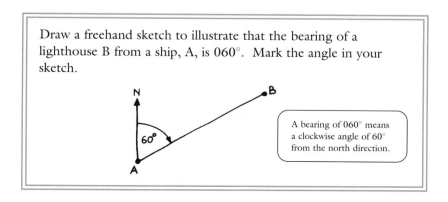

A bearing of 060° means a clockwise angle of 60° from the north direction.

From a ship C the bearing of a ship D is 290°. Make a freehand sketch and mark the angle.

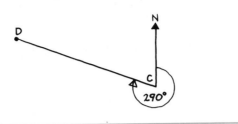

The bearing of an object from an aeroplane, a ship or a control tower is usually found by taking readings from a radar screen. Other bearings can be found by using a hand-held compass. In questions **1** to **15** draw a freehand sketch to illustrate each bearing. Start by drawing a north-pointing line from where the bearing is being measured. Mark the angle in your sketch.

1 From a ship P the bearing of a yacht Q is 045°.

2 From a control tower F the bearing of an aeroplane A is 090°.

3 From a point A the bearing of a radio mast M is 120°.

4 From a town T the bearing of another town S is 180°.

5 From a point H the bearing of a church C is 210°.

6 From a ship R the bearing of a port P is 300°.

7 From an aircraft A the bearing of an airport L is 320°.

8 From a town D the bearing of another town E is 260°.

9 From a helicopter G the bearing of a landing pad P is 080°.

10 From a point L the bearing of a tree T is 270°.

11 The bearing of a ship A from the pier P is 225°.

12 The bearing of a radio mast S from a point O is 140°.

13 The bearing of a yacht Y from a tanker T is 075°.

14 The bearing of a town Q from a town R is 250°.

15 The bearing of a tree X from a hill top Y is 025°.

In questions **16** to **19** make a freehand copy of the diagram. For part **b** add a north pointing line at the position from where the bearing is to be found.

16

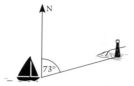

What is the bearing of

a the lighthouse from the ship

b the ship from the lighthouse?

18

What is the bearing of

a the church from the castle

b the castle from the church?

17

What is the bearing of

a the farm from the gate

b the gate from the farm?

19

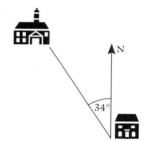

What is the bearing of

a the Town Hall from the Post Office

b the Post Office from the Town Hall?

20

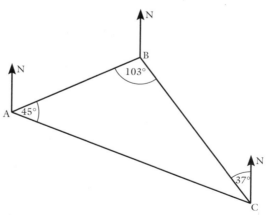

The sketch shows the positions of three villages A, B and C. Use the information given on the diagram to find the bearing of

a B from C   **c** A from B   **e** C from A

b C from B   **d** B from A   **f** A from C.

21

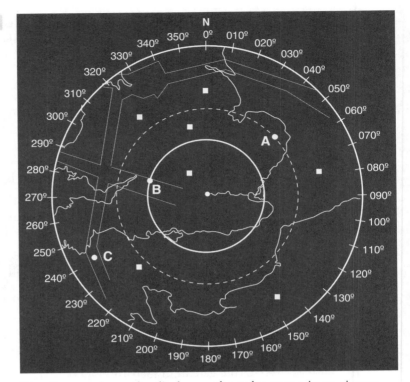

The diagram shows the display on the radar screen in an airport control tower. The centre of the screen shows the position of the control tower. Planes are shown at A, B and C.

a What is the bearing of A from the control tower?

b What is the bearing of B from the control tower?

c Find the bearing of the control tower from C.

d Use tracing paper to mark the positions of B and C. Find the bearing of B from C.

22

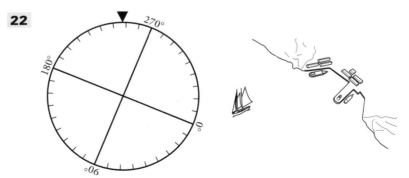

The compass shows the bearing on which a boat, which has left port, is heading. Draw a diagram to show the path of the boat, marking clearly an angle to show its direction in relation to north.

USING BEARINGS TO FIND DISTANCES   If we measure the bearing of a distant object from two different positions and make a scale diagram, we can use this diagram to find the distance of that object from one or other of the positions.

EXERCISE 20G

From one end A of a road the bearing of a building L is 015°. The other end B of the road is 300 m due east of A. From B the bearing of the building is 320°. Using a scale of 1 cm to 50 m, make a scale diagram to find the distance of the building from A.

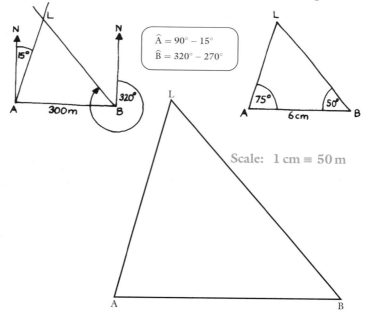

$\hat{A} = 90° - 15°$

$\hat{B} = 320° - 270°$

Scale: 1 cm ≡ 50 m

From the diagram, LA = 5.6 cm.

∴ the distance of the building from A is 5.6 × 50 m = 280 m

1

From a gate A the bearing of a tree C is 060°. From a second gate B, which is 100 m due east of A, the bearing of the tree is 330°.

a Draw a sketch to show the positions of A, B and C. Show the North direction and calculate the angles of △ABC. Show them in your sketch.

b Use a scale of 1 cm to 10 m to make a scale diagram and find the distance of the tree from gate A.

2

From a ship C the bearing of a lighthouse A is 320°. A rock B, which is 200 m due east of A, is on a bearing of 030° from the ship.

  a Draw a sketch to show the positions of A, B and C. Calculate the angles in △ABC and show them in your sketch.

  b Use a scale of 1 cm to 20 m to make a scale diagram and use it to find the distance of the ship from the rock B.

3 From a point A, the bearing of a tower T, is 030°. From a second point B, which is 400 m due north of A, the bearing of the tower is 140°.

  a Draw a sketch to show the positions of A, B and T. Calculate the angles in △ABT and show them in your sketch.

  b Using a scale of 1 cm to 50 m, make a scale drawing and use it to find the distance of the tower from A.

4 From a control point A, the bearing of a radar mast M is 060°. From a second control point B, which is 40 m due east of A, the bearing of the radar mast is 010°.

  a Draw a sketch to represent this information. Calculate the angles in △ABM and show them in your sketch.

  b Use a scale of 1 cm to 5 m and make a scale drawing to find the distance of the radar mast from the control point A.

5 From a ship P the bearing of a submarine S is 020°. From a second ship Q, which is 1000 m due north of P, the bearing of the submarine is 070°. Using a scale of 1 cm to 50 m, make a scale drawing to find the distance of the submarine from P.

6

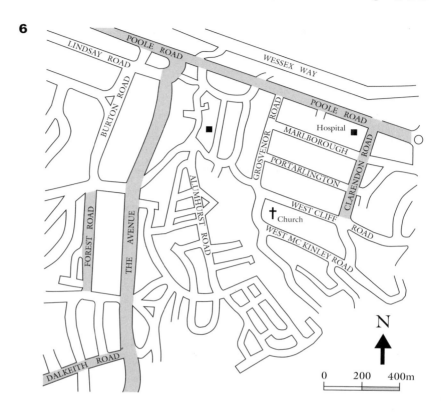

The map shows part of the seaside resort of Bournemouth. Use this map to answer the following questions. Do not draw on this map; if you need to draw, use tracing paper to copy the section of the map that you want to use.

a How long, approximately, is

   i Forest Road

   ii The Avenue from Dalkeith Road to Poole Road?

b What is the approximate bearing of

   i the north end of Clarendon Road from its south end?

   ii the south end of Clarendon Road from its north end?

   iii the church (**†**) on West Cliff Road from the Hospital (■) on Poole Road?

MIXED EXERCISE

EXERCISE 20H

1 Construct $\triangle ABC$ in which $AB = 64$ mm, $BC = 55$ mm and $\widehat{B} = 53°$. Measure and record the length of AC.

2

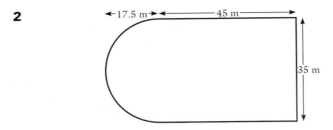

Make a scale drawing of this lawn using a scale of 1 cm to 5 m.

3 From the top of a tower, which is 150 m tall, the angle of depression of a house is 17°. Make a freehand sketch to show this information.

4

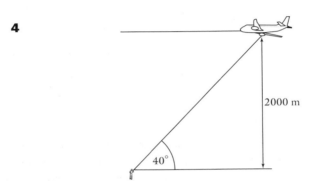

An aircraft is flying at a height of 2000 m, and from a point on the ground an observer measures the angle of elevation of the aircraft as 40°. Make a scale drawing, choosing your own scale, to find the distance from the observer to the point on the ground immediately beneath the aircraft.

5 From a farm F, a cottage C is 850 m due east. From the farm the bearing of a barn B is 063° and from the cottage the bearing of the barn is 330°.

    a Draw a sketch to show this information. Find the angles in △BCF and show them in your sketch.

    b Using a scale of 1 cm to 100 m make a scale diagram and use it to find the distance of the barn

       i from the farm

       ii from the cottage.

1 The whole class working together can collect the information for this exercise.

Measure your classroom and make a freehand sketch of the floor plan. Mark the position and width of doors and windows.

Choosing a suitable scale, make an accurate scale drawing of the floor plan of your classroom.

If you would like a bigger challenge show the position of all the furniture in the room.

2 The sketch shows the measurements of Ken Barker's bathroom. There is only one outside wall and the bottom of the window is 120 cm above the level of the floor. Draw an accurate diagram of the floor, using a scale of 1 cm to 10 cm.

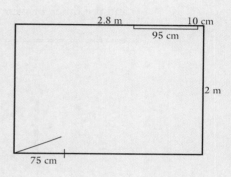

Ken wants to have a new bathroom suite. The units he would like to install, together with their measurements, are:

bath 170 cm × 75 cm
handbasin 60 cm × 42 cm, the longest edge against a wall
shower tray 80 cm square
toilet 70 cm × 50 cm, the shorter measurement against a wall
bidet 55 cm × 35 cm, the shorter measurement against a wall.

Using the same scale make accurate drawings of the plans of these units, cut them out and see if you can place them on your plan in acceptable positions.

If they will not all fit into the room, which unit(s) would you be prepared to do without? Give reasons for your answer.

Is it possible to arrange your chosen units so that all the plumbing is against

a the outside wall

b not more than two walls at right angles, one of which is the outside wall? Illustrate your answer with a diagram.

PYTHAGORAS' THEOREM

Thousands of years ago the Egyptians used knotted ropes to mark out land and buildings. They used a rope with 13 knots to make sure that there were right angles at the corners. One knot was made at each end and the other 11 were equally spaced along its length.

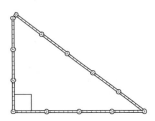

The rope was arranged in a triangle, as shown, so that there were 3 equal lengths along one side, 4 equal lengths along a second side and 5 equal lengths along the third side. The ends of the rope met and each section was pulled taut. The largest angle in this triangle is a right angle.

With such a simple system the corners of buildings like the great pyramids could be made square.

While the 3,4,5 triangle is the simplest triangle that contains a right angle, many other triangles were discovered that will do the same job.

The formal relationship between the lengths of the three sides is attributed to Pythagoras, a Greek philosopher and mathematician who lived about 2500 years ago.

RIGHT-ANGLED TRIANGLES

Any triangle whose largest angle is $90°$ is called a *right-angled triangle*.

The side opposite the right-angle is called the *hypotenuse*.

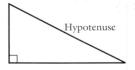

EXERCISE 21A

In this exercise we investigate the relationship between the lengths of the three sides. First we will collect some evidence. Bear in mind that, however accurate your drawing may be, it is not perfect.

Construct the triangles in questions **1** to **6** and in each case measure the third side, the hypotenuse.

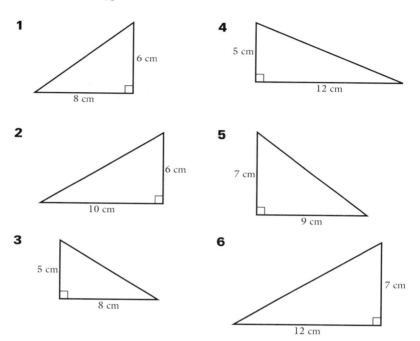

1

6 cm

8 cm

4

5 cm

12 cm

2

6 cm

10 cm

5

7 cm

9 cm

3

5 cm

8 cm

6

7 cm

12 cm

7 In each question from **1** to **6** find the squares of the lengths of the three sides. Copy and complete the following table.

	(Shortest side)^2	(Middle side)^2	(Side opposite right angle)^2
1			
2			
3			
4			
5			
6			

Can you see a relation between the numbers in the first two columns and the number in the third column ?

PYTHAGORAS'
THEOREM

If your drawings are reasonably accurate you will find that by adding the squares of the two shorter sides you get the square of the hypotenuse.

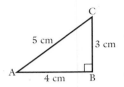

$$AB^2 = 16$$
$$BC^2 = 9$$
$$AC^2 = 25$$
$$25 = 16 + 9$$
so $$AC^2 = AB^2 + BC^2$$

This result is called Pythagoras' theorem, which states that

> in a right-angled triangle the square of the hypotenuse is equal to the sum of the squares of the other two sides.

EXERCISE 21B

Give your answers correct to 3 significant figures.

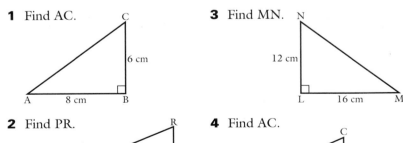

The diagram shows a rope arranged to form a triangle PQR. If $R = 90°$, $PR = 7\,m$ and $QR = 6\,m$, find the length of PQ.

$$PQ^2 = PR^2 + QR^2 \quad (\text{Pythagoras' theorem})$$
$$= 7^2 + 6^2$$
$$= 49 + 36 = 85$$
$$PQ = \sqrt{85} = 9.219\ldots$$
Length of PQ = 9.22 m correct to 3 s.f.

In the following right-angled triangles find the required lengths.

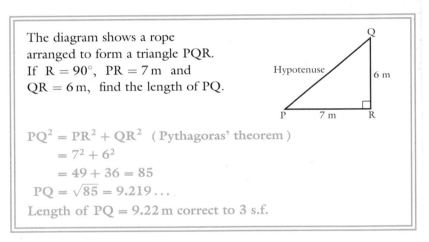

1 Find AC.

3 Find MN.

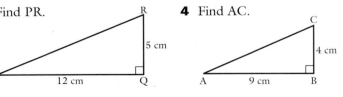

2 Find PR.

4 Find AC.

5 Find AC.

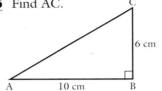

6 Find QR.

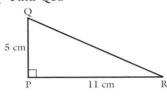

7

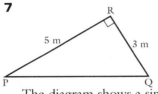

The diagram shows a simple roof support. What distance does PQ span?

8 Find QR.

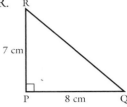

9 Find EF.

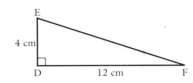

10

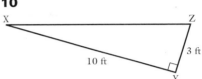

The diagram shows the cross-section of a drainage ditch. How wide is the ditch at the surface?

A manufacturer wants to make a quantity of plastic triangles like the one shown in the diagram. Use the given information to find the length of XY.

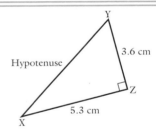

$XY^2 = XZ^2 + ZY^2$ (Pythagoras' theorem)
$= 5.3^2 + 3.6^2$
$= 28.09 + 12.96$
$= 41.05$
$XY = \sqrt{41.05} = 6.407\ldots$
Length of XY = 6.41 cm correct to 3 s.f.

$5.3^2 \approx 5 \times 5 = 25$
$3.6^2 \approx 4 \times 4 = 16$

11 Find AC.

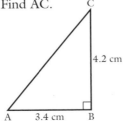

12 Find XY.

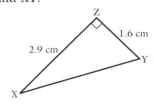

13 Find AC.

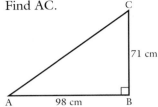

14 Find PR.

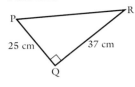

15

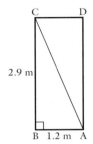

The sketch shows a door ABCD.
Find the length of the diagonal AC.

16

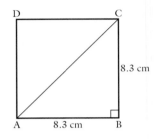

A small square table mat has an edge of **8.3** cm. How far is it from one corner to the opposite corner?

17 In $\triangle$ABC, $\widehat{B} = 90°$, AB $= 7.9$ cm, BC $= 3.5$ cm. Find AC.

18 In $\triangle$PQR, $\widehat{Q} = 90°$, PQ $= 11.4$ m, QR $= 13.2$ m. Find PR.

A man starts from A and walks 4 km due north to B, then 6 km due west to C. Find how far C is from A.

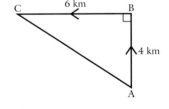

$$AC^2 = BC^2 + AB^2$$
$$\text{(Pythagoras' theorem)}$$
$$= 6^2 + 4^2$$
$$= 36 + 16$$
$$= 52$$
$$AC = \sqrt{52}$$
$$AC = 7.211\ldots$$
$$= 7.21 \text{ (correct to 3 s.f.)}$$

The distance of C from A is 7.21 km correct to 3 s.f.

---

19 A hockey pitch measures 55 m by 90 m. Find the length of a diagonal of the pitch.

20

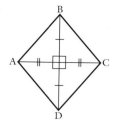

ABCD is a rhombus. AC = 10 cm and BD = 12 cm.

Find the length of a side of the rhombus.

21 A carpenter is making a teak front door that is to be 6 ft 4 inches high and 32 inches wide. He checks that it is square (i.e. that all the corners are 90°) by measuring both diagonals. How long should each one be when the door is finished?

22

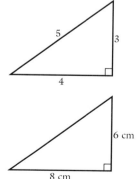

The diagram shows a nest of 4 squares set one within another. The side of the outer square is 20 cm. The midpoints of the sides are joined to give a second square and the process repeated to give the third and fourth squares. Find the length of a side of the smallest square.

THE 3,4,5, TRIANGLE

You will have noticed that, in most cases when two sides of a right-angled triangle are given and the third side is calculated using Pythagoras' theorem, the answer is not an exact number. There are a few special cases where all three sides are exact numbers.

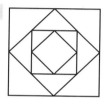

The simplest one is the 3,4,5 triangle. Any triangle that is an enlargement of this one has sides in the ratio 3 : 4 : 5 so whenever you spot this case you can find the missing side very easily.

For instance, in this triangle, $6 = 2 \times 3$ and $8 = 2 \times 4$. The triangle is an enlargement, by a factor of 2, of the 3, 4, 5 triangle, so the hypotenuse is 2×5 cm, that is, 10 cm.

The Eqyptians' knotted rope is based on this triangle.

Another right-angled triangle with exact sides which might be useful is the 5,12,13 triangle.

EXERCISE 21C

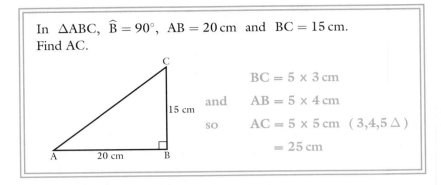

In △ABC, $\widehat{B} = 90°$, AB = 20 cm and BC = 15 cm.
Find AC.

and

so

BC = 5 × 3 cm

AB = 5 × 4 cm

AC = 5 × 5 cm (3,4,5 △)

= 25 cm

In each of the following questions decide whether the triangle is an enlargement of the 3,4,5 triangle or of the 5,12,13 triangle or of neither.

Find the hypotenuse, using the method you think is easiest.

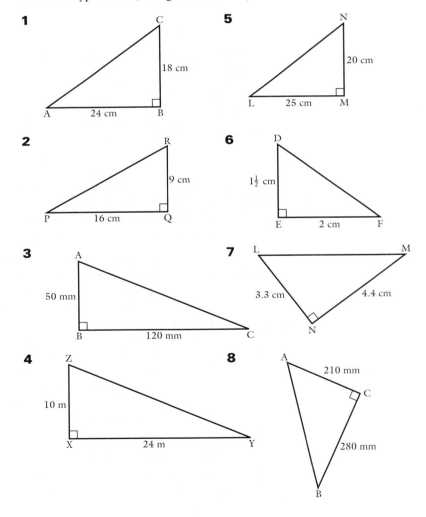

FINDING ONE OF
THE SHORTER
SIDES

If we are given the hypotenuse and one other side we can find the third side.

EXERCISE 21D

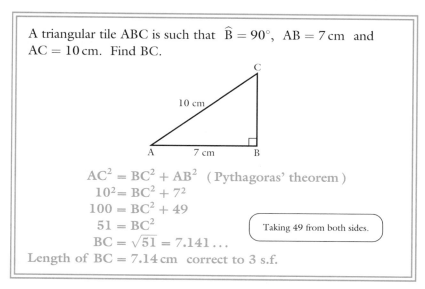

A triangular tile ABC is such that $\widehat{B} = 90°$, AB = 7 cm and AC = 10 cm. Find BC.

$$AC^2 = BC^2 + AB^2 \quad (\text{Pythagoras' theorem})$$
$$10^2 = BC^2 + 7^2$$
$$100 = BC^2 + 49$$
$$51 = BC^2$$
$$BC = \sqrt{51} = 7.141\ldots$$

Taking 49 from both sides.

Length of BC = 7.14 cm correct to 3 s.f.

Give all answers that are not exact to 3 significant figures.

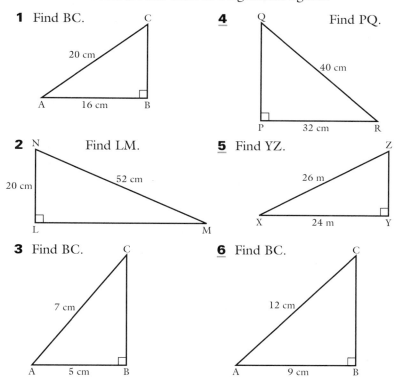

1 Find BC.

20 cm
A   16 cm   B

2 Find LM.

N
20 cm
52 cm
L   M

3 Find BC.

7 cm
A   5 cm   B

4 Find PQ.

Q
40 cm
P   32 cm   R

5 Find YZ.

Z
26 m
X   24 m   Y

6 Find BC.

C
12 cm
A   9 cm   B

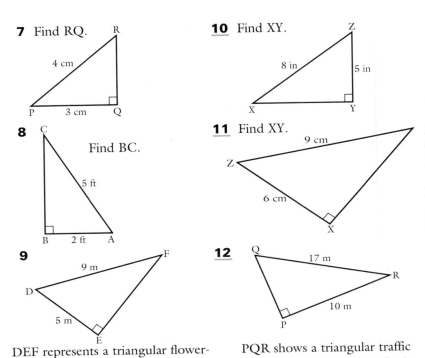

7 Find RQ.

R

4 cm

P   3 cm   Q

10 Find XY.

Z

8 in   5 in

X   Y

8 C

Find BC.

5 ft

B   2 ft   A

11 Find XY.

9 cm

Z

6 cm

X

9

F

9 m

D

5 m

E

DEF represents a triangular flower-bed. Find the length of EF.

12 Q

17 m

R

10 m

P

PQR shows a triangular traffic island. Find the length of PQ.

13 A wire stay 11 m long is attached to a telegraph pole at a point A, 8 m up from the ground. The other end of the stay is fixed to a point B, on the ground. Draw a suitable diagram showing this information. How far is B from the foot of the telegraph pole?

14 A diagonal of a football pitch is 130 m long and the long side measures 100 m. Find the length of the short side of the pitch.

15

15 cm

5 cm

The slant height of a cone is 15 cm and the base radius is 5 cm.
Find the height of the cone.

16 The diagram shows the points A(1, 5) and B(3, −1).
On your own copy of this diagram, mark the point C so that $A\hat{C}B = 90°$.
Use $\triangle ABC$ to find the length of AB.

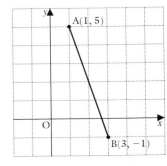

17 Repeat question 19 with the points A(−1, 2) and B(4, 6).

18 A rubbish skip is in the shape of a prism.

When viewed from the side,
the cross-section of the skip
is an isosceles trapezium ABCD,
where AB = 3.4 m, DC = 2 m
and AD = BC = 1.4 m

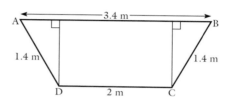

a Find the depth of the skip.

b Find the area of cross-section of the skip.

c The skip is 1.65 m wide.

   Calculate the capacity of the skip in

   i cubic metres   **ii** litres.

19 a Draw a triangle ABC in which AB = 8 cm, BC = 6 cm and
   AC = 10 cm.
   (Remember that in this triangle $A\widehat{B}C$ is 90°.) For this triangle
   we know that $AC^2 = AB^2 + BC^2$

b Now draw a triangle ABC in which AB = 8 cm, BC = 6 cm
   and $A\widehat{B}C$ has any value less than 90°. Measure AC.
   Find $AB^2 + BC^2$ and AC^2.
   Is $AB^2 + BC^2$ greater than AC^2 or less than AC^2 ?

c Next draw a triangle ABC in which AB = 8 cm, BC = 6 cm
   and $A\widehat{B}C$ has any value greater than 90°. Measure AC.
   Find $AB^2 + BC^2$ and AC^2.
   Is $AB^2 + BC^2$ greater or less than AC^2 ?

d Draw any other right-angled triangle ABC in which AC is the
   hypotenuse. Find $AB^2 + BC^2$ and AC^2. What conclusion can
   you draw about the angle B if

   i $AB^2 + BC^2 > AC^2$

   ii $AB^2 + BC^2 < AC^2$?

20 The sketch shows an Inn sign measuring 1.2 m by 0.8 m which is supported by a frame attached to the wall. The sloping edge of the frame is 1.8 m long and at its upper end is attached to the top of the wall. The bottom of the sign is 3 m above ground level.
How high is the wall?

PERIGAL'S DISSECTION

Perigal's dissection is a demonstration of Pythagoras' theorem. On squared paper, and using 1 cm to 1 unit, copy the left-hand diagram. Make sure that you draw an accurate square on the hypotenuse either by counting the squares or by using a protractor and a ruler. D is the centre of the square on AB. Draw DE parallel to AC. You will see that $DE = \frac{1}{2} AC$.

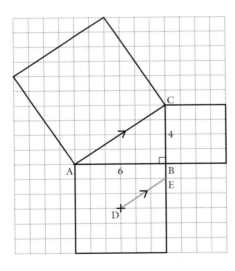
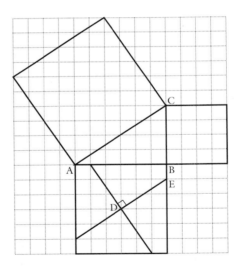

Now complete the drawing as in the right-hand diagram. Make sure that the angles at D are right angles.

Cut out the smallest square and the four pieces from the middle-sized square. These five pieces can be fitted exactly, like a jigsaw, into the outline of the biggest square.

Now try it yourself with a different right-angled triangle.

FINDING LENGTHS IN AN ISOSCELES TRIANGLE

An isosceles triangle can be split into two right-angled triangles and this can sometimes help when finding missing lengths.

EXERCISE 21E

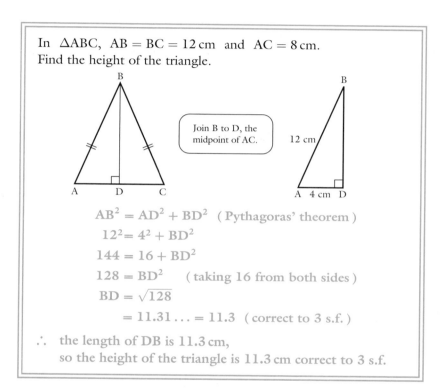

In $\triangle ABC$, $AB = BC = 12$ cm and $AC = 8$ cm.
Find the height of the triangle.

Join B to D, the midpoint of AC.

$$AB^2 = AD^2 + BD^2 \quad (\text{Pythagoras' theorem})$$
$$12^2 = 4^2 + BD^2$$
$$144 = 16 + BD^2$$
$$128 = BD^2 \quad (\text{taking 16 from both sides})$$
$$BD = \sqrt{128}$$
$$= 11.31\ldots = 11.3 \quad (\text{correct to 3 s.f.})$$

$\therefore$ the length of DB is 11.3 cm,
so the height of the triangle is 11.3 cm correct to 3 s.f.

Give your answers correct to 3 significant figures.

1 The sketch shows the front of a ridge tent in which $AB = AC = 150$ cm. $BC = 120$ cm. Find the height of the ridge above the ground.

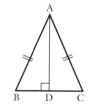

2 A vertical flagpole ED is supported by two stays attached to the pole at B, 1 m from the top and fixed to two points in the ground, A and C, on opposite sides of the pole. If $AB = BC = 7.9$ m, $AC = 5$ m and D is the midpoint of AC find the height of the flagpole.

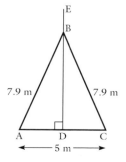

3

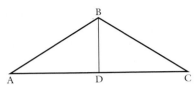

The sketch shows a roof truss for a garage. If AC = 3.5 m and
BD = 1.3 m, find the lengths of the sloping timbers AB and BC.

4 In △ABC, AB = BC = 5.2 cm and AC = 6 cm. Find the height
of the triangle measured from B.

5 In △PQR, PQ = QR = 9 cm and the height of the triangle,
measured from Q, is 7 cm. Find the length of PR.

6 The main feature of a carnival float is made from an equilateral
triangle of side 12 ft. The triangle is fixed on a trailer so that one side
rests on the base of the trailer which is 4 ft above road level.
Will the carnival float pass under a bridge that will take vehicles up to
14 ft high? Give reasons for your answer.

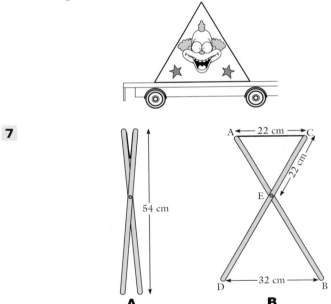

7

Diagram **A** shows a folding stool in its closed postion. When closed
it reaches a height of 54 cm.
In diagram **B** the stool is shown opened out. The two supports AB
and CD pivot at E such that AE = EC = 22 cm, AC = 22 cm and
the distance between the feet of the stool is 32 cm. Find

a the height of the seat AC above the pivot

b the equal lengths EB and ED

c the height of the seat above the ground.

A straight line joining two points on the
circumference of a circle is called a *chord*.
AB is a chord of a circle with centre O.
OA and OB are radii and so are equal.
Hence triangle OAB is isosceles and we
can divide it through the middle into
two right-angled triangles.

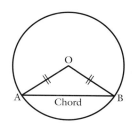

EXERCISE 21F

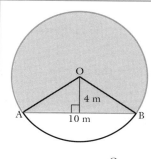

The diagram shows a cross-section
through a tunnel.
AB, which represents the floor of
the tunnel, is 10 m long and is a
chord of a circle, centre O.
If the floor is 4 m from O, find the
radius of the circle.

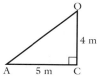

The distance from the centre is the
perpendicular distance so OC = 4 m.
From symmetry AC = 5 m.

$$OA^2 = AC^2 + OC^2 \text{ (Pythagoras' theorem)}$$
$$= 5^2 + 4^2$$
$$= 25 + 16 = 41$$
$$OA = \sqrt{41} = 6.403\ldots$$
$$OA = 6.40 \text{ (correct to 3 s.f.)}$$

The radius of the circle is 6.40 m correct to 3 s.f.

Give your answers correct to 3 significant figures.

1 A circle with centre O has a radius of 5 cm.
AB = 8.4 cm. Find the distance of the chord
from the centre of the circle.

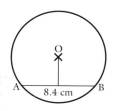

2

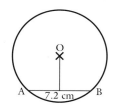

O is the centre of the circle and AB is a chord
of length 7.2 cm. The distance of the chord
from O is 3 cm.
Find the radius of the circle.

3 The sketch shows a hole in a door ready to take a lock. AB is a chord of length 15 mm in a circle, centre O, of radius 14 mm.
Find the distance of the chord from O.

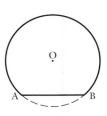

4 The diagram shows the cross-section through a tunnel. AB, which represents the floor of the tunnel, is of length 4 m and the radius of the circle is 3 m. Find

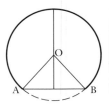

a the distance of O above the floor of the tunnel

b the height of the tunnel.

5 The diagram shows the stage in an arena. The stage is part of a circle, centre P and radius 7.6 m. The front of the stage, which is a chord of the circle, is 9.4 m wide. Find

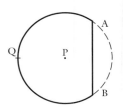

a the distance from P to the front of the stage

b the maximum depth of the stage, i.e. the distance from the front of the stage to Q.

6

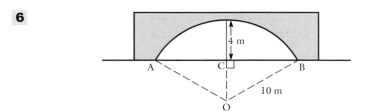

The diagram shows the cross-section of a tunnel which has a maximum height of 4 m above the horizontal base AB. The roof of the tunnel is part of a circle, centre O and radius 10 m.
If C is the midpoint of AB find

a the length of OC

b the length of AC

c the width of the tunnel at its base.

CONVERSE OF PYTHAGORAS' THEOREM

The last question in **Exercise 21D** showed that if AC is the longest side in $\triangle ABC$

and $\widehat{B} < 90°$,

then $AC^2 < AB^2 + BC^2$

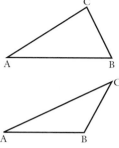

but if $\widehat{B} > 90°$, $AC^2 > AB^2 + BC^2$

We can see that $AC^2 = AB^2 + BC^2$ only when $\widehat{B} = 90°$.

These results give us the converse of Pythagoras' theorem.

> In a triangle, if the square of one side is equal to the sum of the squares of the other two sides, the triangle contains a right angle, and this right angle is opposite the longest side.

EXERCISE 21G

The lengths, in centimetres, of the sides of two triangles are

a 10, 24 and 26 **b** 4.2, 5.6 and 7.2.

Determine whether or not each triangle contains a right angle.

a (If the triangle contains a right angle it is opposite the longest side.

(Longest side)$^2 = 26^2 = 676$
Sum of the squares of the other two sides
$\quad = 10^2 + 24^2$
$\quad = 100 + 576 = 676$

Since the square of the longest side is equal to the sum of the squares of the other two sides the triangle contains a right angle and this is opposite the side of length 26 cm.

b (Longest side)$^2 = 7.2^2 = 51.84$
Sum of the squares of the other two sides
$\quad = 4.2^2 + 5.6^2$
$\quad = 17.64 + 31.36$
$\quad = 49$

The square of the longest side is not equal to the sum of the squares of the other two sides so the triangle does not contain a right angle.

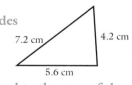

> We can also deduce that the angle opposite the longest side is greater than 90° because the square of that side is greater than the sum of the squares of the other two sides.

In each question from **1** to **6** the lengths, in centimetres, of the three sides of a triangle are given. Find whether or not the triangle contains a right angle.

Give reasons for your answers.

1 6, 7, 9   **3** 4.7, 5.9, 7.5   **5** 9, 39, 40

2 21, 72, 75   **4** 3.9, 5.2, 6.5   **6** 15, 36, 39

7 A gardener pegs out the ground to sow a rectangular lawn which is to measure 45 m by 35 m. When he checks a diagonal he finds it to be 57.0 m correct to the nearest tenth of a metre. What would you expect the length of the other diagonal to be compared with the one already measured – that is, is it longer or shorter?

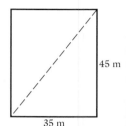

45 m

35 m

8 Jim buys a flatpack wardrobe which, when assembled, measures 2000 mm high by 900 mm wide by 580 mm deep. Before fitting the doors he checks the lengths of the diagonals of the front to see that the wardrobe is standing vertically and that the doors will fit squarely. The first diagonal measures 2193 mm. Should the doors fit squarely when he hangs them? Justify your answer.

9 Lucy is making a frame for a wedding photograph. The frame is to measure 30 cm by 20 cm. When she checks it for squareness she finds that the length of one diagonal is 36.3 cm.
Is the frame square (i.e. are the corners all 90°)?

10

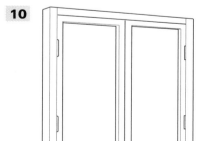

A carpenter is making a rectangular window frame that is to measure 5 ft by 4 ft.
He checks the length of one diagonal and finds it to be 6 ft 4.8 in, correct to the nearest tenth of an inch.
Is the window square?
Justify your answer.

MIXED EXERCISE

EXERCISE 21H

In questions **1** to **6** find the length of the missing side. If any answers are not exact give them correct to 3 significant figures.

If you notice a 3,4,5 triangle or a 5,12,13 triangle, you can use it to get the answer quickly.

1 Find AC.

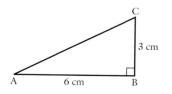

3 Find PR.

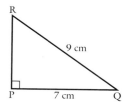

2 Find LM.

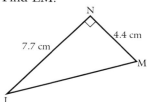

4 Find YZ.

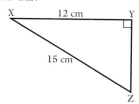

5 In $\triangle$ABC, $\hat{B} = 90°$, AB = 1.25 m, CA = 8.25 m. Find BC.

6 In $\triangle$PQR, $\hat{Q} = 90°$, PQ = 65 cm, QR = 60 cm. Find PR.

7 In $\triangle$ABC, AB = 1.8cm, BC = 8 cm and AC = 8.4 cm.
Does this triangle contain a right angle ? If your answer is 'yes', state which angle it is and justify your answer.

8 In $\triangle$DEF, DE = 42mm, EF = 56 mm and DF = 70 mm.
Is DEF a right-angled triangle ? If your answer is 'yes' state which angle is 90° and justify your answer.

9 Find the length of

a a diagonal of a square of side 10 cm

b an edge of a square whose diagonals are of length 10 cm.

10

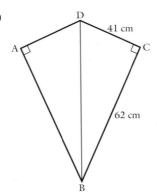

In the kite ABCD, $\widehat{A} = \widehat{C} = 90°$.
DC = 41 cm and BC = 62 cm.
Find the length of the diagonal BD.

11 The diagram shows the side view of a coal bunker.
Find the length of the slant edge.

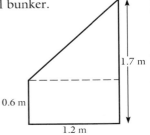

12

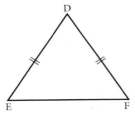

In ΔDEF, DE = DF = 3.4 in
and EF = 3.8 in.
Find the height of the triangle.

13 O is the centre of the circle and AB is
a chord of length 18.2 cm.
The perpendicular distance of the
chord from O is 6.3 cm.
Find the radius of the circle.

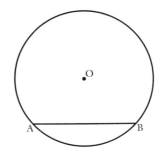

14 A groundsman is marking out a new
tennis court on grass. The length of
the court is 78 ft and the width is 36 ft.
He lays out a rectangle with these
measurements, but wants to check
that his angles are true right angles.

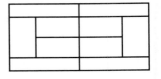

He checks the lengths of the two diagonals which should be the same
length. To the nearest tenth of a foot, how long should they be ?

PRACTICAL WORK

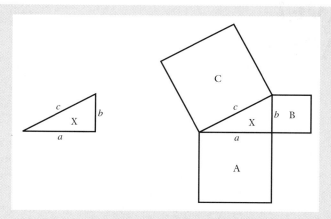

Draw any right-angled triangle and draw the square on each of the three sides. Mark the four areas A, B, C and X as shown in the diagram. Cut out one of each shape and make another three triangles identical to X. Arrange the shapes in two different ways as shown below. Sketch the two arrangements and mark in as many lengths as possible with *a*, *b* or *c*.

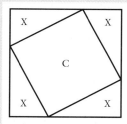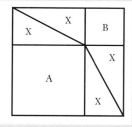

a What can you say about the areas of these two diagrams? Justify your answer.

b If the four triangles marked X are removed from each diagram, what can you say about the areas that remain? What relation does this give for **i** areas A, B and C **ii** lengths *a*, *b* and *c*?

TRAVEL GRAPHS

When Isobel gets to the bus stop she has to travel 1 mile along the bus route to get to school, and she must be in school by **8.50**. She can walk at **4** mph and if she catches the bus, it will travel at an average speed of **20** mph. The times that a bus arrives at the bus stop are unpredictable so Isobel is often undecided whether to wait for a bus or to walk.

- What is the latest time that Isobel can leave the bus stop on foot to be certain of arriving in school on time?

- If she is sure of catching a bus as soon as she gets to the bus stop, what is the latest time she can arrive there and still get to school in time?

- One morning she decides to walk but sees a bus pass her after she has been walking for 5 minutes. How much longer would she have had to wait before the bus arrived? If she had waited, how much earlier would she have got to school than she did by walking?

These questions can be answered by drawing suitable travel graphs. A travel graph is drawn by plotting distance covered against time taken.

FINDING DISTANCE FROM A GRAPH

When we went on holiday in the car we travelled to our holiday resort at a steady speed of 30 kilometres per hour (km/h), that is, in each hour we covered a distance of 30 km.

This graph plots distance covered against time taken and shows that

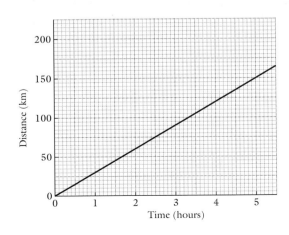

in 1 hour we travelled   30 km
in 2 hours we travelled   60 km
in 3 hours we travelled   90 km
in 4 hours we travelled 120 km
in 5 hours we travelled 150 km.

EXERCISE 22A   The graphs that follow show 5 different journeys. For each journey find

    a the distance travelled

    b the time taken

    c the distance travelled: in 1 hour (questions **1**, **2** and **3**)

                                 or in 1 second (questions **4** and **5**).

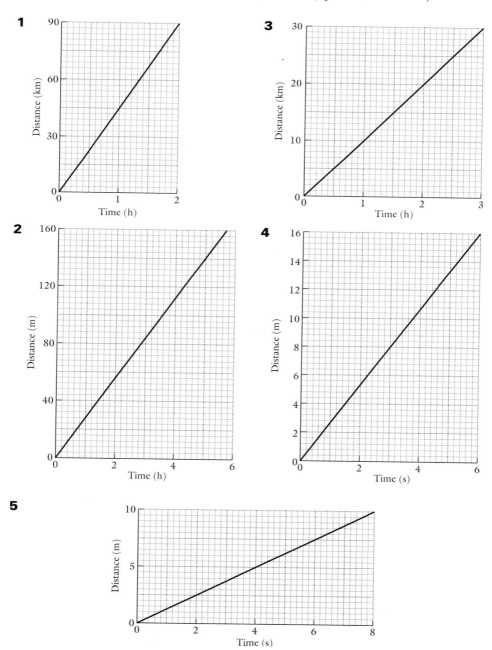

**DRAWING
TRAVEL GRAPHS**

If Peter walks at 6 km/h, we can draw a graph to show this, using 2 cm to represent 12 km on the distance axis and 2 cm to represent 1 hour on the time axis.

Plot the point which shows that in 1 hour he has travelled 6 km. Join the origin to this point and produce the straight line to give the graph shown.

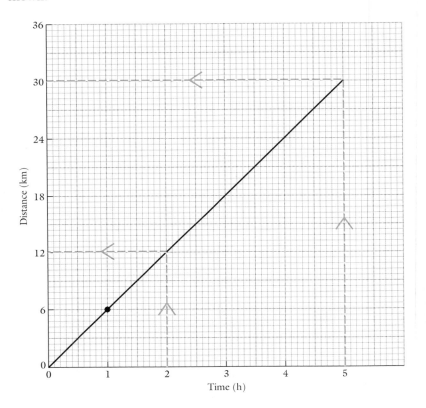

From this graph we can see that in 2 hours Peter travels 12 km and in 5 hours he travels 30 km.

Alternatively we could say that

if he walks 6 km in 1 hour

he will walk 6×2 km, i.e. 12 km in 2 hours

and he will walk 6×5 km, i.e. 30 km in 5 hours

The distance walked is found by multiplying the speed by the time,

i.e.   distance = speed × time

EXERCISE 22B

Draw a travel graph to show Sally's journey of 150 km in 3 hours. Plot distance along the vertical axis and time along the horizontal axis. Let 4 cm represent 1 hour and 2 cm represent 50 km.

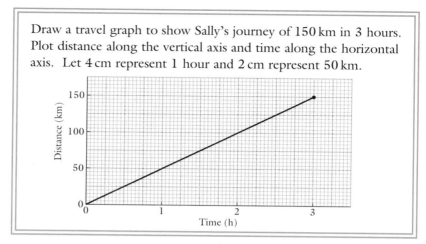

Draw travel graphs to show the following journeys. Plot distance along the vertical axis and time along the horizontal axis. Use the scales given in brackets.

1 60 km in 2 hours (4 cm ≡ 1 hour, 1 cm ≡ 10 km)

2 180 km in 3 hours (4 cm ≡ 1 hour, 2 cm ≡ 50 km)

3 100 km in $2\frac{1}{2}$ hours (2 cm ≡ 1 hour, 2 cm ≡ 25 km)

4 75 miles in $1\frac{1}{4}$ hours (8 cm ≡ 1 hour, 2 cm ≡ 25 miles)

5 240 m in 12 sec (1 cm ≡ 1 sec, 2 cm ≡ 50 m)

6 Alan walks at 5 km/h. Draw a graph to show him walking for 3 hours. Take 4 cm to represent 5 km and 4 cm to represent 1 hour. Use your graph to find how far he walks in
   a $1\frac{1}{2}$ hours   **b** $2\frac{1}{4}$ hours.

7 Julie can jog at 10 km/h. Draw a graph to show her jogging for 2 hours. Take 1 cm to represent 2 km and 8 cm to represent 1 hour. Use your graph to find how far she jogs in
   a $\frac{3}{4}$ hour   **b** $1\frac{1}{4}$ hours.

8 Jo drives at 35 mph. Draw a graph to show her driving for 4 hours. Take 1 cm to represent 10 miles and 4 cm to represent 1 hour. Use your graph to find how far she drives in
   a 3 hours   **b** $1\frac{1}{4}$ hours.

9 John walks at 4 mph. Draw a graph to show him walking for 3 hours. Take 1 cm to represent 1 mph and 4 cm to represent 1 hour. Use your graph to find how far he walks in
   a $\frac{1}{2}$ hour   **b** $3\frac{1}{2}$ hours.

The remaining questions should be solved by calculation.

10 An express train travels at 200 km/h. How far will it travel in

    a 4 hours   **b** $5\frac{1}{2}$ hours?

11 Ken cycles at 24 km/h. How far will he travel in

    a 2 hours   **b** $3\frac{1}{2}$ hours   **c** $2\frac{1}{4}$ hours?

12 An aeroplane flies at 300 mph. How far will it travel in

    a 4 hours   **b** $5\frac{1}{2}$ hours?

13 A bus travels at 60 km/h. How far will it travel in

    a $1\frac{1}{2}$ hours   **b** $2\frac{1}{4}$ hours?

14 Susan can cycle at 12 mph. How far will she ride in

    a $\frac{3}{4}$ hour   **b** $1\frac{1}{4}$ hours?

15 An athlete can run at 10.5 metres per second.
How far will he travel in   **a** 5 sec   **b** 8.5 sec?

16 A boy cycles at 12 mph. How far will he travel in

    a 2 hours 40 min   **b** 3 hours 10 min?

17 A Boeing 747 travels at 540 mph. How far does it travel in

    a 3 hours 15 min   **b** 7 hours 45 min?

18 A racing car travels around a 2 km circuit at 120 km/h. How many laps will it complete in

    a 30 min   **b** 1 hour 12 min?

CALCULATING THE TIME TAKEN

Georgina walks at 6 km/h so we can find out how long it will take her to walk 24 km; as she walks 6 km in 1 hour, it will take her $\frac{24}{6}$ hours to walk 24 km, i.e. 4 hours.

In the same way if Georgina walks 15 km, it will take her $\frac{15}{6}$ hours, i.e. $2\frac{1}{2}$ hours.

This demonstrates that

$$\text{time} = \frac{\text{distance}}{\text{speed}}$$

1 How long will Zena, walking at 5 km/h, take to walk

    a 10 km   **b** 15 km?

2 How long will a car, travelling at 80 km/h, take to travel

    a 400 km   **b** 260 km?

3 How long will it take David, running at 10 mph, to run

    a 5 miles   **b** $12\frac{1}{2}$ miles?

4 How long will it take an aeroplane flying at 450 mph to fly

    a 1125 miles   **b** 2400 miles?

5 A cowboy rides at 14 km/h. How long will it take him to ride

    a 21 km   **b** 70 km?

6 A rally driver drives at 50 mph. How long does it take him to travel

    a 75 miles   **b** 225 miles?

7 An athlete runs at 8 m/s. How long does it take him to cover

    a 200 m   **b** 1600 m?

8 A dog runs at 20 km/h. How long does the dog take to travel

    a 8 km   **b** 18 km?

9 A liner cruises at 28 nautical miles per hour. How long will it take to travel

    a 6048 nautical miles   **b** 3528 nautical miles

10 A car travels at 56 mph. How long does it take to travel

    a 70 miles   **b** 154 miles?

11 A cyclist cycles at 12 mph. How long will it take him to cycle

    a 30 miles   **b** 64 miles?

12 How long will it take a car travelling at 64 km/h to travel

    a 48 km   **b** 208 km?

AVERAGE SPEED

Russell Compton left home at 8 a.m. to travel the 50 km to his place of work. He arrived at 9 a.m. Although he had travelled at many different speeds during his journey he covered the 50 km in exactly 1 hour.

We say that his *average speed* for the journey was 50 kilometres per hour, or 50 km/h. If he had travelled at the same speed all the time, he would have travelled at 50 km/h.

Judy Smith travelled the 135 miles from her home to London in 3 hours. If she had travelled at the same speed all the time, she would have travelled at $\frac{135}{3}$ mph = 45 mph. We say that her average speed for the journey was 45 mph.

In each case:
$$\text{average speed} = \frac{\text{distance travelled}}{\text{time taken}}$$

This formula can also be written:
$$\text{distance travelled} = \text{average speed} \times \text{time taken}$$

and
$$\text{time taken} = \frac{\text{distance travelled}}{\text{average speed}}$$

A useful way to remember these relationships is from the triangle:
(Cover up the one you want to find.)

Suppose that a car travels 35 km in 30 min, and we want to find its speed in kilometres per hour. To do this we must express the time taken in hours instead of minutes,

i.e.
$$\text{time taken} = 30 \text{ min} = \tfrac{1}{2} \text{ hour}$$

Then
$$\text{average speed} = \frac{35}{\frac{1}{2}} \text{ km/h} = 35 \times \frac{2}{1} \text{ km/h}$$
$$= 70 \text{ km/h}$$

Great care must be taken with units. If we want a speed in kilometres per hour, we need the distance in kilometres and the time in hours. If we want a speed in metres per second, we need the distance in metres and the time in seconds.

EXERCISE 22D

Find the average speed for each of the following journeys.

1 80 km in 1 hour

2 120 km in 2 hours

3 60 miles in 1 hour

4 480 miles in 4 hours

5 80 m in 4 sec

6 135 m in 3 sec

7 150 km in 3 hours

8 520 km in 8 hours

9 245 miles in 7 hours

10 104 miles in 13 hours

11 252 m in 7 sec

12 255 m in 15 sec

Tony drives 39 km in 45 minutes. Find his average speed.

$$45\,\text{min} = \frac{45}{60}\,\text{hour} = \frac{3}{4}\,\text{hour}$$

First, convert the time taken to hours.

$$\text{average speed} = \frac{\text{distance travelled}}{\text{time taken}}$$

$$= \frac{39\,\text{km}}{\frac{3}{4}\,\text{hour}}$$

$$= 39 \times \frac{4}{3}\,\text{km/h} = 52\,\text{km/h}$$

Find the average speed in km/h for a journey of

13 40 km in 30 min

15 48 km in 45 min

14 60 km in 40 min

16 66 km in 33 min

Find the average speed in km/h for a journey of 5000 m in $\frac{1}{2}$ hour.

$$5000\,\text{m} = \frac{5000}{1000}\,\text{km} = 5\,\text{km}$$

We need distance in kilometres.

$$\text{average speed} = \frac{\text{distance travelled}}{\text{time taken}}$$

$$= \frac{5\,\text{km}}{\frac{1}{2}\,\text{hour}}$$

$$= 5 \times \frac{2}{1}\,\text{km/h} = 10\,\text{km/h}$$

Find the average speed in km/h for a journey of

17 4000 m in 20 min

19 40 m in 8 sec

18 6000 m in 45 min

20 175 m in 35 sec

Find the average speed in mph for a journey of

21 27 miles in 30 min   **23** 25 miles in 25 min

22 18 miles in 20 min   **24** 28 miles in 16 min

The following table shows the distances in kilometres between various places in the United Kingdom.

	London	Bradford	Cardiff	Leicester	Manchester	Oxford	Reading	
Bradford	320							
Cardiff	250	332						
Leicester	160	160	224					
Manchester	310	55	277	138				
Oxford	90	280	172	120	230			
Reading	64	320	192	164	264	45		
York	315	53	390	174	103	290	210	

Use this table to find the average speeds for journeys between

25 London, leaving at 1025, and Manchester, arriving at 1625

26 Oxford, leaving at 0330, and Cardiff, arriving at 0730

27 Leicester, leaving at 1914, and Oxford, arriving at 2044

28 Reading, leaving at 0620, and London, arriving at 0750

29 Bradford, leaving at 1537, and Oxford, arriving at 1907

30 Cardiff, leaving at 1204, and York, arriving at 1624

31 Bradford, leaving at 1014, and Reading, arriving at 1638.

Problems frequently occur where different parts of a journey are travelled at different speeds in different times and we wish to find the average speed for the whole journey.

Consider for example, Roger May, who travels the first 50 miles of a journey at an average speed of 25 mph and the next 90 miles at an average speed of 30 mph.

Using   time in hours $= \dfrac{\text{distance in miles}}{\text{speed in mph}}$

gives   time to travel 50 miles at 25 mph $= \dfrac{\text{distance}}{\text{speed}}$

$$= \dfrac{50 \text{ miles}}{25 \text{ mph}} = 2 \text{ hours}$$

time to travel 90 miles at 30 mph $= \dfrac{\text{distance}}{\text{speed}}$

$$= \dfrac{90 \text{ miles}}{30 \text{ mph}} = 3 \text{ hours}$$

∴   the total distance of 140 miles is travelled in 5 hours

i.e.   average speed for whole journey $= \dfrac{\text{total distance}}{\text{total time}}$

$$= \dfrac{140 \text{ miles}}{5 \text{ hours}} = 28 \text{ mph}$$

Note: never add or subtract average speeds.

Often it is more convenient to enter this information in a table like the one given below. You may need to do some calculations like those shown above, before you can complete the table.

	Speed in mph	Distance in miles	Time in hours
First part of journey	25	50	2
Second part of journey	30	90	3
Whole journey		140	5

We can add the distances to give the total length of the journey, and add the times to give the total time taken for the journey.

Average speed for whole journey $= \dfrac{\text{total distance}}{\text{total time}}$

$$= \dfrac{140 \text{ miles}}{5 \text{ hours}} = 28 \text{ mph}$$

1 I walk for 24 km at 8 km/h, and then jog for 12 km at 12 km/h. Find my average speed for the whole journey.

2 A cyclist rides for 23 miles at an average speed of $11\frac{1}{2}$ mph before his cycle breaks down, forcing him to push his cycle the remaining distance of 2 miles at an average speed of 4 mph. Find his average speed for the whole journey.

3 An athlete runs 6 miles at 8 mph, then walks 1 mile at 4 mph. Find his average speed for the whole journey.

4 A woman walks 3 miles at an average speed of $4\frac{1}{2}$ mph and then runs 4 miles at 12 mph. Find her average speed for the whole journey.

5 A motorist travels the first 30 km of a journey at an average speed of 120 km/h, the next 60 km at 60 km/h, and the final 60 km at 80 km/h. Find the average speed for the whole journey.

6 Phil Sharp walks the 2 km from his home to the bus stop in 15 min, and catches a bus immediately which takes him the 9 km to the railway station at an average speed of 36 km/h. He arrives at the station just in time to catch the London train which takes him the 240 km to London at an average speed of 160 km/h. Calculate his average speed for the whole journey from home to London.

7 A liner steaming at 24 knots takes 18 days to travel between two ports. By how much must it increase its speed to reduce the length of the voyage by 2 days? (A knot is a speed of 1 nautical mile per hour.)

If we are given a graph representing a journey, which shows the distance travelled plotted against the time taken, we can get a lot of information from it. The worked example in the next exercise shows how we can extract such information.

The graph below shows the journey of a coach which passes three service stations A, B and C on a motorway. B is 60 km north of A and C is 20 km north of B. Use the graph to answer the following questions.

a At what time does the coach leave A?

b At what time does the coach arrive at C?

c At what time does the coach pass B?

d What is the average speed of the coach for the whole journey?

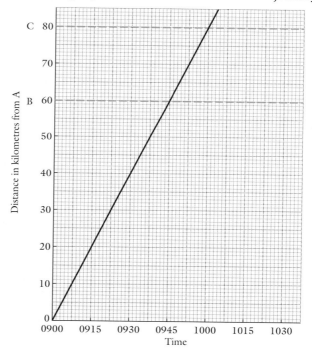

a The coach leaves A at 0900.

b Find the point on the graph level with C, then use a ruler as a guide to find the point on the time axis that is immediately below this point.

It arrives at C at 1000.

c It passes through B at 0945.

d Distance from A to C = 80 km.

Time taken to travel from A to C is 1000 − 0900, i.e. 1 hour.

$$\text{Average speed} = \frac{\text{distance travelled}}{\text{time taken}}$$

$$= \frac{80\,\text{km}}{1\,\text{hour}} = 80\,\text{km/h}$$

1 The graph shows the journey of a car through three towns, Axeter, Bexley and Canton. Axeter is 100 km south of Bexley and Canton is 60 km north of it. Use the graph to answer the following questions.

a At what time does the car

  i leave Axeter

  ii pass through Bexley

  iii arrive at Canton?

b How long does the car take to travel from Axeter to Canton?

c How long does the car take to travel

  i the first 80 km of the journey?

  ii the last 80 km of the journey?

d What is the average speed of the car for the whole journey?

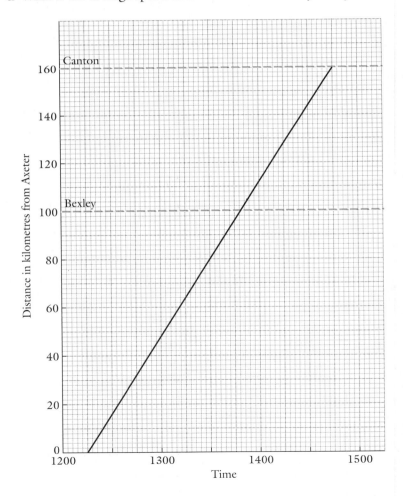

2 The graph shows the journey of an express train which starts from A and passes through stations at B and C on the way to its destination at D.

 a How far is it

  i from A to B

  ii from B to C

  iii from C to D?

 b How long does the journey take

  i from A to D

  ii from B to C?

 c Find the average speed for the whole journey.

 d Where is the train at 1100?

 e What time is it when the train is 20 miles short of C?

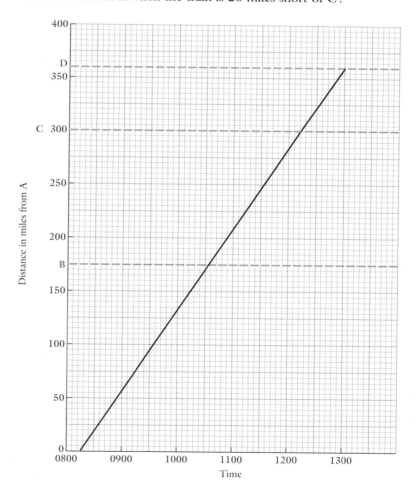

3 A coach leaves Newcombe at noon on its journey to Lee via Manley. The graph shows its journey.

   a How far is it

      i from Newcombe to Manley

      ii from Manley to Lee?

   b How long does the coach take to travel from Newcombe to Lee?

   c What is the coach's average speed for the whole journey?

   d How far does the coach travel between 1.30 p.m. and 2.30 p.m.?

   e After travelling for $1\frac{1}{2}$ hours, how far is the coach from

      i Newcombe

      ii Manley?

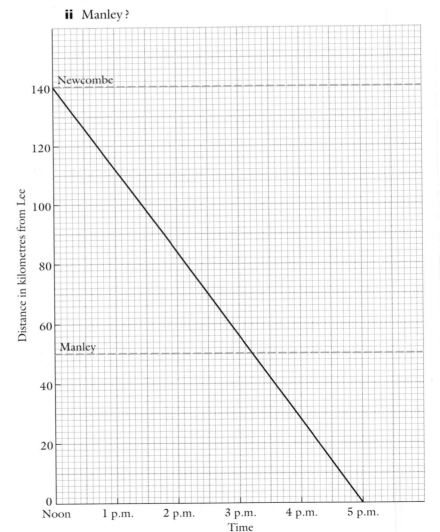

4 Mr Brown used the family car to transport his children from their home to the nearest mainline railway station and then returned home. The graph shows his journey.

  a How far is it from home to the station?

  b How long did it take the family to get to the station?

  c What was the average speed of the car on the journey to the station?

  d How long did the car take for the return journey?

  e What was the average speed for the return journey?

  f What was the car's average speed for the round trip?

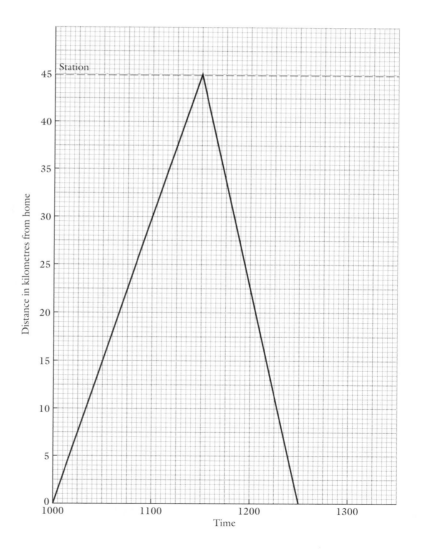

5 The graph shows the journey of a car through three service stations A, B and C, on a motorway.

a Where was the car at
  i 0900
  ii 0930?

b What was the average speed of the car between
  i A and B   **ii** B and C?

c For how long does the car stop at B?

d How long did the journey take?

e What was the average speed of the car from A to C?
Give your answer correct to 1 significant figure.

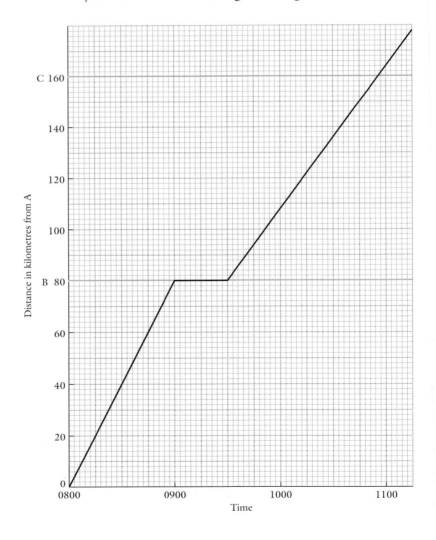

6 The graph shows Bill's journey on a sponsored walk.

   a How far did he walk?

   b How many times did he stop?

   c What was the total time he spent resting?

   d How long did he actually spend walking?

   e How long did the walk take him?

   f What was his average speed for the whole journey?

   g Over which of the four stages did he walk fastest?

   h Over which two stages did he walk at the same speed?

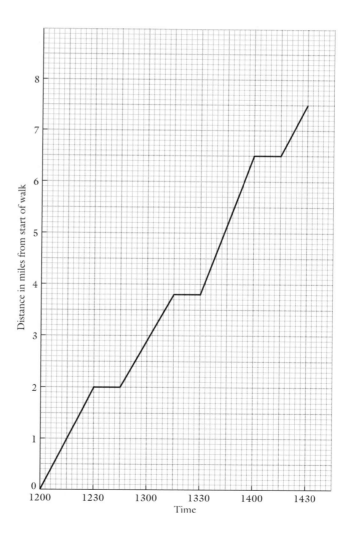

The graph shows Mrs Webb's shopping trip to the nearest town on a bicycle.

a How far is town from home? **b** How long did she take to get to town?

c How long did she spend in town?

d What was her average speed on the outward journey?

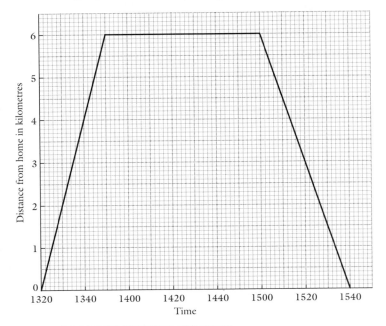

a (The greatest distance the graph rises from 0 (home) is 6 km.)

It is 6 km from home to town.

b (She arrived in town when she stopped going further away from home. This is where the graph stops going up.)

Mrs Webb left home at 1320 and arrived in town at 1350.
The journey therefore took 30 minutes.

c (She left for home when the graph starts to go down.
i.e. she stayed in town for the time that the graph is parallel to the time axis.)

She arrived in town at 1350 and left at 1500. She therefore spent 1 hour and 10 min there.

d On the outward journey:

$$\text{average speed} = \frac{\text{distance travelled}}{\text{time taken}} = \frac{6\,\text{km}}{\frac{1}{2}\,\text{hour}}$$

$$= \frac{6}{1} \times \frac{2}{1}\,\text{km/h} = 12\,\text{km/h}$$

7 The graph shows the journey of a train from Newpool to London and back again. Use the graph to answer the questions that follow.

a How far is Newpool from London?

b How long did the outward journey take?

c What was the average speed for the outward journey?

d How long did the train remain in London?

e At what time did the train leave London, and how long did the return journey take?

f What was the average speed on the return journey?

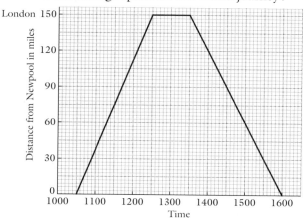

8 The graph represents the journey of a motorist from Leeds to Manchester and back again. Use this graph to find

a the distance between the two cities

b the time the motorist spent in Manchester

c his average speed on the outward journey

d the average speed on the homeward journey (including the stop).

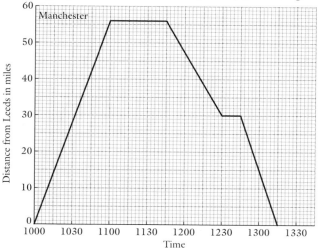

9 The graph below shows Judith's journeys between home and school.

   a At what time did she leave home in   **i** the morning   **ii** the afternoon?

   b How long was she in school during the day?

   c How long was she away from school for her midday break?

   d What was the average speed for each of these journeys?

   e Find the total time for which she was away from home.

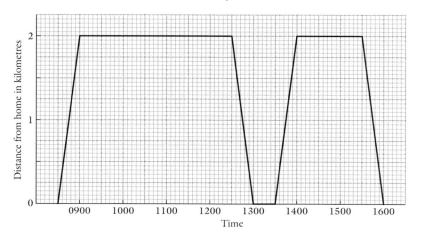

10 The graph below shows the journeys of two cars between two service stations, A and B, which are 180 km apart. Use the graph to find

   a the average speed of the first motorist and his time of arrival at B

   b the average speed of the second motorist and the time at which she leaves B

   c when and where the two motorists pass

   d their distance apart at 1427.

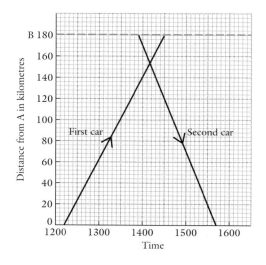

11 The school fête is always held on the first Saturday in July. The graphs show the journeys to school of three pupils on the day of the fête last year.

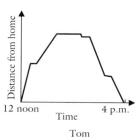

Andrew   Kate   Tom

Use these graphs to answer the following questions.

a Who got to the fête first?

b Who stayed there longest?

c Who left for home first?

d Who took the longest time to get home?

e Who had the slowest journey to the fête?

f What did Kate do when she got to school?

12 Jane leaves home at 1 p.m. to walk at a steady 4 mph towards Cornforth, which is 6 miles away, to meet her boyfriend Tim. Tim leaves Cornforth at 2.00 p.m. and jogs at a steady 6 mph to meet her. Draw a graph for each of these journeys taking $4 \text{ cm} \equiv 1$ hour on the time axis and $1 \text{ cm} \equiv 1$ mile on the distance axis. From your graph find

a when and where they meet,

b their distance apart at 2.10 p.m.

13 Solve the problems set at the beginning of the chapter.

MIXED EXERCISE

EXERCISE 22G

1 Jenny runs at $12\frac{1}{2}$ km/h. Draw a graph to show her running for $2\frac{1}{2}$ hours. Use your graph to find

a how far she has travelled in $1\frac{3}{4}$ hours

b how long she takes to run the first 20 km.

2 A ship travels at 18 nautical miles per hour. How long will it take to travel **a** 252 nautical miles **b** 1026 nautical miles?

3 Find the average speed in km/h of a journey of 48 km in 36 min.

4 The graph shows John's walk from home to his grandparents' home.

   a How far away do they live?

   b How long did the journey take him?

   c What was his average walking speed?

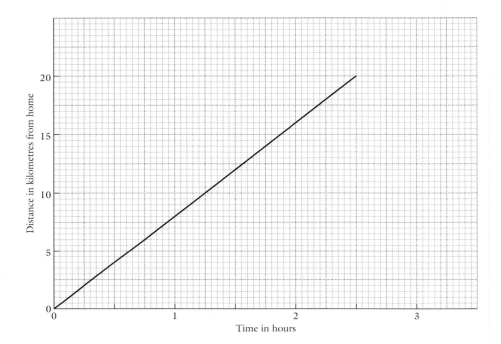

5 I left London at 1147 to travel the 315 miles to York. If I arrived at 1717, what was my average speed?

6 I walk $\frac{2}{3}$ mile in 10 min and then run $\frac{1}{3}$ mile in 2 min. What is my average speed for the whole journey?

7 The graph opposite shows Paul's journey in a sponsored jog from A to B. On the way his sister, who is travelling by car in the opposite direction from B to A, passes him.

   a How far does Paul jog?

   b How long does he take?

   c How much of this time does he spend resting?

   d What is his average speed for the whole journey?

   e What is his sister's average speed?

   f How far from A did Paul and his sister pass each other?

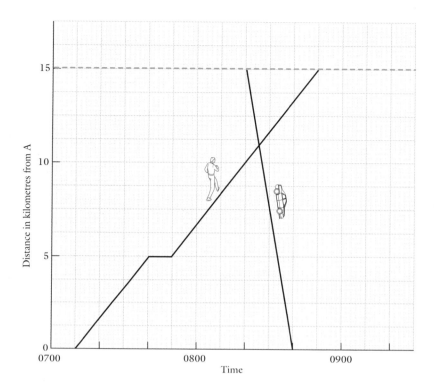

PRACTICAL
WORK

a Describe in words your journey to or from school. For example, walk, wait at the bus stop, bus ride, walk.

b Time the different parts of the journey.

c Use a map to work out the distance travelled for each part of the journey.

d Draw a graph of your journey. You will have to assume that any bus, car or train you use travels at a steady speed from where you start to where you get off.

e Work out the average speed for the various parts of the journey and for the whole journey.

f Think of ways of making the graph more realistic. For instance, how does the varying speed of the bus affect the graph?

g Draw a travel graph to show alternative ways of making the journey.

h What extra information can you get from your graph?

INVESTIGATION

Work in a group for this investigation.

Paul walked from A to B, a distance of 10 miles, at 4 mph.

Jenny cycled from A to B at 7 mph.

Paul left A at 9 a.m. and Jenny left A at 9.30 a.m.

Jenny passed Paul at C, x miles from A.

a In terms of x, how long did Paul take to reach C?

b In terms of x, how long did Jenny take to reach the same place?

c Use the information given to form an equation and then solve it. Hence find when and where Jenny passed Paul.

d Draw a travel graph to represent the given information, and then use it to find when and where Jenny passed Paul.

e Write a brief report on the advantages and disadvantages of the two methods for finding when and where Jenny passed Paul.

SUMMARY 5

VOLUMES

A solid with a constant cross-section is called a *prism*.

The *volume of a prism* is given by

area of cross-section × length

The *volume of a cylinder* is given by $V = \pi r^2 h$

The *density* of a material is the mass of one unit
of volume of the material, for example,
the density of silver is $10.5\,\text{g/cm}^3$, that is, $1\,\text{cm}^3$ of silver weighs $10.5\,\text{g}$.

ENLARGEMENT

When an object is enlarged by a scale factor 2, each line on the image is
twice the length of the corresponding line on the object.

The diagram shows an
enlargement of a triangle,
with centre of enlargement
X and scale factor 2.
The dashed lines are guidelines.
$XA' = 2XA$

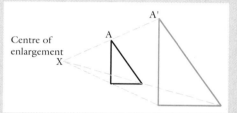

When the scale factor is less than one, the image is smaller than the
object. (Note that the image is still called an enlargement.)

This diagram shows an enlargement
with scale factor $\frac{1}{4}$ and centre of
enlargement O.
$OA' = \frac{1}{4}OA$

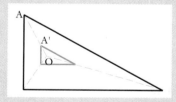

**ANGLES OF
ELEVATION AND
DEPRESSION**

If you start by looking straight ahead,
the angle that you turn your eyes
through to look *up* at an object
is called the angle of elevation,
the angle you turn your eyes
through to look *down* at an object
is called the angle of depression.

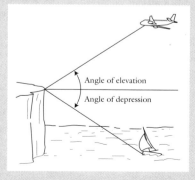

**THREE-FIGURE
BEARINGS**

A three-figure bearing of a point A from a point B
gives the direction of A from B as a clockwise angle
measured from the north.

For example, in this diagram, the bearing of
A from B is 140°.

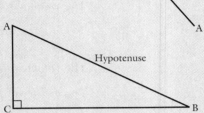

**PYTHAGORAS'
THEOREM**

Pythagoras' theorem states that
in any right-angled triangle ABC
with $\widehat{C} = 90°$, $AB^2 = AC^2 + BC^2$

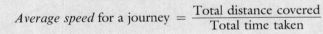

Conversely if, in a triangle the square of the longest side is equal to the
sum of the squares of the other two sides, then the angle opposite the
longest side is a right-angle.

**DISTANCE,
SPEED AND TIME**

The relationship between distance, speed and time is given by

$$\text{Distance} = \text{Speed} \times \text{Time}$$

which can also be expressed as $\text{Speed} = \dfrac{\text{Distance}}{\text{Time}}$

or as $$\text{Time} = \dfrac{\text{Distance}}{\text{Speed}}$$

A useful way to remember these relationships is
from the triangle: (cover up the one you want to find)

Average speed for a journey $= \dfrac{\text{Total distance covered}}{\text{Total time taken}}$

**REVISION
EXERCISE 5.1
(Chapters 18
and 19)**

1 Find the capacity, in litres, of a cylinder with

    a radius 2.5 cm and height 5 cm

    b diameter 7.4 cm and height 6.8 cm.
    (Give each answer correct to 3 significant figures.)

2 The diagram shows the cross-section of
a metal girder which is 2 m long. Find

    a the area of cross-section in cm²

    b the volume of the metal used to
cast the girder in
      i cm³   **ii** m³.

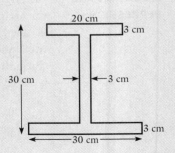

3 The sketch shows a wooden table lamp before the electrical fittings are attached. It has been made from two pieces of wood. The base is a circular disc of radius 8 cm and 3 cm thick, while the column is a cylinder of radius 2.5 cm and height 38 cm.
Find the total amount of wood in one lamp.

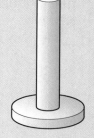

4

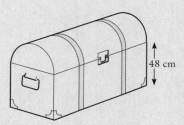

The diagram shows a wooden chest. The internal measurements of the base are 120 cm by 48 cm and the chest is 48 cm deep. The lid is half of a cylinder.
Find, in cubic centimetres, the capacity of the chest

a excluding the lid

b including the lid.

5

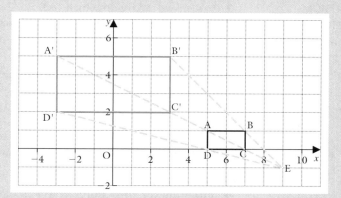

A'B'C'D' is the image of rectangle ABCD, and E is the centre of enlargement.

a Write down the coordinates of the centre of enlargement.

b Find the scale factor.

6 Copy the diagram onto squared paper using 1 cm as 1 unit. Draw
AA′, BB′, CC′ and DD′ and extend the lines until they meet.
Write down the coordinates of the centre of enlargement and find the
scale factor.

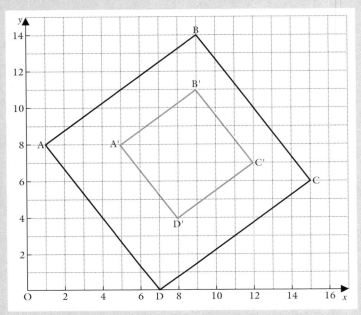

7 Draw axes for x and y from 0 to 15 using 1 cm as 1 unit. Draw
△ABC with A(2, 1), B(4, 2), and C(1, 5). With the origin as the
centre of enlargement and scale factor 3, draw the image of △ABC.

**REVISION
EXERCISE 5.2
(Chapters 20
to 22)**

1 Construct △ABC in which AB = 72 mm, AC = 88 mm and
BC = 54 mm. Measure and write down the size of

  a the largest angle   **b** the smallest angle.

2 From a point on the ground 220 m from the base of a multi-storey
block of flats the angle of elevation of the top is 33°.
Make a scale drawing and use it to find the height of the block.

3 From the control tower, the bearing of an aircraft is 085°. Draw a
freehand sketch showing this information. Find the bearing of the
control tower from the aircraft.

4 From the top of a church tower the Post Office is 300 m on a
bearing of 137°, and the City Hall is 180 m on a bearing of 245°.

  a Using a scale of 1 cm ≡ 30 m, make a scale drawing and use it
  to find the distance of the City Hall from the Post Office.

  b What is the bearing of   **i** the City Hall from the Post Office
                                 ii the Post Office from the City Hall?

5 a

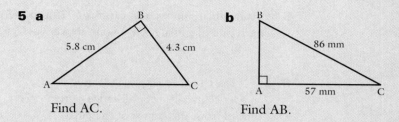

Find AC.

b

Find AB.

6 a In △ABC, $\widehat{B} = 90°$, AC $= 17.4$ cm and AB $= 12.9$ cm.
Find BC.

b In △DEF, DE $= 5.4$ cm, EF $= 9$ cm and DF $= 7.2$ cm.
Does the triangle contain a right angle? If so, which angle is it?

7 The graph shows the journey of an athlete in a race.

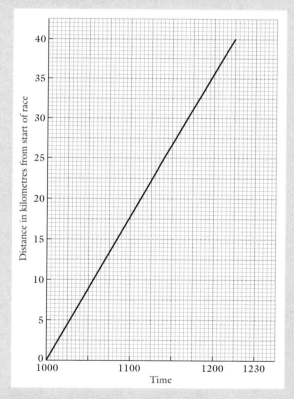

a What was the length of the race?

b How long did the athlete take?

c What was his average speed for the whole journey?

d How far did he travel in the first $1\frac{1}{4}$ hours?

e Did the athlete stop at any time during the race?

f Did the athlete travel at more than one speed?

8 A rectangular cupboard measures 172 cm by 63 cm. What is the length of a diagonal? Give your answer correct to 1 decimal place.

9 a AB is a chord of length 9.8 cm in a circle, centre O. The distance of the chord from the centre of the circle is 3.7 cm. Find, correct to 3 significant figures, the radius of the circle.

b A second chord, CD, is 4.3 cm from O. Find the length of CD.

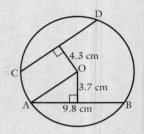

10 The graph shows the journey of a car from Amberley to Brickworth and on to Coldham.

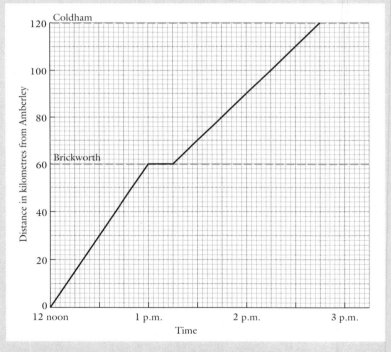

a Where was the car at   **i** 12.30   **ii** 2.15?

b What was the average speed of the car between
   i Amberley and Brickworth   **ii** Brickworth and Coldham?

c For how long does the car stop at Brickworth?

d How long did the journey take including the stop?

e What was the average speed of the car for the whole journey? Give your answer correct to the nearest whole number.

1 a Find the volume of a rectangular block measuring 12 cm by 8 cm by 5 cm.

  b Find the volume of a cylindrical can whose diameter is 3.5 cm and whose height is equal to the diameter of its base. Give your answer correct to 3 significant figures.

2 The diagram shows the cross-section of a metal pipe support that is 28 cm long. Find

  a the area of cross-section

  b the volume of the block.

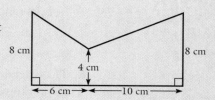

3 Copy the diagram onto squared paper using 1 cm as 1 unit. Draw AA′, BB′ and CC′ and extend the lines until they meet. Give the coordinates of the centre of enlargement and the scale factor.

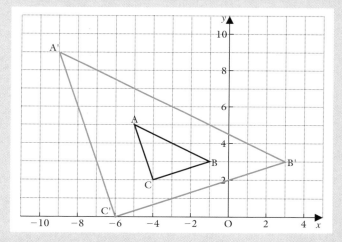

4 Draw axes for *x* and *y* from 0 to 12 using 1 cm as 1 unit. Draw △ABC with A(0, 2), B(3, 1), and C(4, 5). Taking the origin as the centre of enlargement and a scale factor of 2, draw the image of △ABC. Mark the image A′B′C′. Write down the coordinates of the vertices of the image triangle.

5 Construct △DEF in which DE = 67 mm, DF = 94 mm and $\hat{D} = 66°$. Measure and write down the length of EF.

6 From the top of Canary Wharf, which is 244 m high, the angle of depression of a bus on the road below is 38°. Make a scale drawing, using 1 cm ≡ 20 m. Use your drawing to find the distance of the bus from the base of the tower.

7 a A car travels at 72 km/h. How far will it travel in

    i $1\frac{1}{2}$ hours   **ii** 20 minutes   **iii** 1 hour 40 minutes?

  b How far will Rob, walking at 6 km/h, take to walk

    i 12 km   **ii** 20 km?

  c Pete cycles at 12 mph. How long will it take him to cycle

    i 18 miles   **ii** 30 miles?

8 The diagram shows the cross-section through a lean-to conservatory which is attached to the back-wall of a house. Calculate the length of the slant edge of the roof.

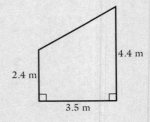

9 The diagonals of a kite ABCD meet at E.
AB = BC = 9.5 cm,   AC = 8.2 cm
and   BD = 19.7 cm.

Find the length of   **a** BE   **b** CD.

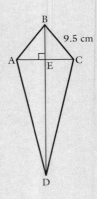

10 The graph opposite represents the bicycle journeys of three school friends, Audrey, Betty and Chris, from the village in which they live to Buckwell, the nearest main town, which is 30 km away.
Use the graph to find

  a their order of arrival at Buckwell

  b Audrey's average speed for the journey

  c Betty's average speed for the journey

  d Chris's average speed for the journey

  e where and when Chris passes Audrey

  f how far each is from town at 2 pm

  g how far Betty is ahead of Chris at 2.15 p.m.

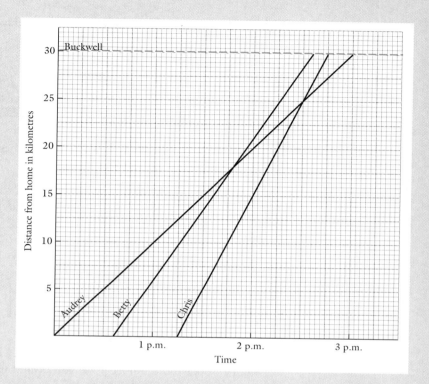

REVISION
EXERCISE 5.4
(Chapters 1 to 22)

1 a Express as a single expression in index form

   i $5^2 \times 5^7$ **ii** $4^9 \div 4^6$

b Write in standard form   **i** $47\,000$ **ii** $0.000\,008\,2$

c Write as ordinary numbers   **i** 6.93×10^4 **ii** 8.44×10^{-3}

2 Find   **a** $4\frac{1}{4} - 2\frac{2}{3}$ **c** $\frac{2}{3}$ of £39   **e** $3\frac{1}{8} \div 3\frac{3}{4}$

         b $4\frac{1}{5} - 5\frac{1}{2} + 2\frac{3}{10}$ **d** $6\frac{2}{3} \times 1\frac{7}{8}$ **f** $\frac{3}{4}$ of $72\,\text{kg}$

3 a The perimeter of a triangle is 45 cm and the lengths of the sides are in the ratio $4 : 5 : 6$. Find the lengths of the three sides.

b What does 1 cm represent on a map with a map ratio of $1 : 50\,000$?

4 a The area of a square is $70\,\text{cm}^2$. Find the length of a side. Give your answer correct to 3 significant figures.

b The area of a triangle is $0.41\,\text{m}^2$. The length of the base is $1.64\,\text{m}$. Find its height.

c The sketch shows the end wall of a building. Find its area.

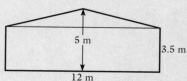

5 a Find the area of a quadrant of a circle of radius 12 cm.

b Hence find the area of the shaded part of the diagram, the outline of which is formed from two quadrants.

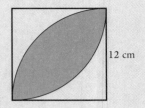

12 cm

6 Show on a sketch the range in which each of these measurements lies.
   a 5 m rounded to the nearest metre.

   b 95 cm rounded up to the next complete centimetre.

   c 35 minutes rounded down to the nearest minute.

7 A rectangular box is to have a capacity of $0.25\,\text{m}^3$. It is to have square ends and be three times as long as it is wide.

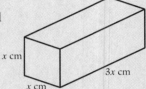

x cm
$3x$ cm
x cm

a Form an equation in x.

b Use a graphics calculator to plot a graph to solve this equation.
   Take values of x from 0 to 1 and values of y from -1 to 1 using a scale of 0.1 for x and for y. If you do not have a graphics calculator available, use trial and improvement to solve this equation.

c Hence find the dimensions of the box, giving all measurements correct to 3 significant figures.

8

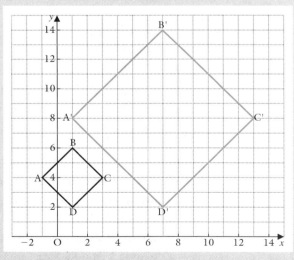

Copy the diagram using 1 cm as 1 unit. Find the coordinates of the centre of enlargement and the scale factor.

9 a I think of a number, add 9 and double the result. This gives 42. What is the number?

b Find the range of possible values of x for which the pair of inequalities $3x + 1 \geqslant 3$ and $4x - 4 < 3$ are true. Illustrate your answer using a number line.

c Solve the equation $8.2x = 14$, giving your answer correct to 3 significant figures.

10 Eddie makes a frame to mount his child's tapestry. The sides of the frame measure 48 cm by 36 cm. When he checks the diagonal he finds it is 60 cm. Is the frame square? Justify your answer.

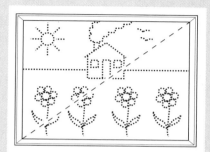

REVISION EXERCISE 5.5 (Chapters 1 to 22)

1 Draw up a possibility space showing all the possible combinations of scores if two dice are rolled together. The scores on each dice are added. Use this possibility space to find the probability of getting

a 5 or less

b 10 or more

c at least 8

d a score that is exactly divisible by 3.

2 a Express £2.50 as a fraction of £12.50

b Express
  i $\frac{17}{20}$ as a decimal

  ii $\frac{17}{60}$ as a percentage correct to 1 decimal place.

c Bendthorpe First XI played 42 matches last season. They won 16 matches, drew 8 and lost the remainder.
  i What fraction of the matches did they lose?
  ii What percentage (to the nearest whole number) did they win?
  iii Express the fraction that they drew as a decimal correct to 2 significant figures.

3 a Find the size of each exterior angle in a regular polygon with 24 sides. What is the size of each interior angle?

b What name do you give to this shape?

c Is this polygon regular? Justify your answer.

d What is the sum of the interior angles of a polygon with five sides?

e Find the value of x.

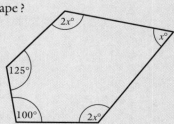

4 a Multiply out the brackets and simplify
  i $3(4x-2)$ **iii** $10-4(x-4)$
  ii $3x+4(2x+3)$ **iv** $3x-x(2-x)$

b

n	1	2	3	4	5
u_n	−1	0	1	2	3

Find a formula for u_n in terms of n.

5 a The table gives the coordinates of three points on a line. Find the equation of the line.

x	−1	3	5
y	−1	1	2

b $A(-1, a)$ and $B(b, 7)$ are two points on the line $y = 4x - 9$. Find a and b.

6 A gardener encloses a rectangular piece of ground as an allotment by using $100\,\text{m}$ of fencing and a straight wall as one side.

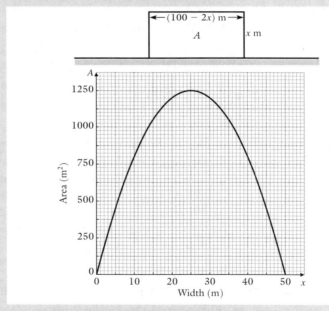

The graph shows the relationship between the area enclosed, $A\,\text{cm}^2$, and its width, $x\,\text{m}$. Use the graph to find

a the width of the rectangle when its area is $1000\,\text{m}^2$

b the maximum area that can be enclosed and the corresponding value of x

c the area enclosed when the width is $20\,\text{m}$.

7 **a** Solve the simultaneous equations $5x + 2y = 22$

$$3x + 2y = 14$$

b The sum of two numbers is 136 and their difference is 10. Find the numbers.

8 **a** A cupboard is 1.5 m wide, 30 cm deep and 120 cm high. Find its volume in   **i** cm^3 **ii** m^3.

b A child's paddling pool is an open cylinder of diameter 2.6 m which is 30 cm deep. It is filled with water to a depth of 18 cm.
   i How many litres of water are there in the pool?
   ii How many more litres are needed to fill the pool?

9 **a** Construct $\triangle ABC$ in which $AB = 115$ mm, $\widehat{A} = 37°$ and $\widehat{B} = 58°$. Measure and write down the length of
   i AC   **ii** BC.

b From a point on the top of a building 55 m high, the angle of depression of a small boat on the river below is 53°. Make a scale drawing using $1\,cm \equiv 5\,m$. Use your drawing to find the distance of the boat from the base of the building.

10 Les sets out on his bike on a journey of 12 km. He has cycled 10 km at 15 km/h when his bicycle suffers a puncture. As a result, he pushes his bike the rest of the distance at 5 km/h.

a How long does he cycle before the breakdown?

b How far does he have to push the bike and for how long?

c Find the total time for his journey.

d Find his average speed for the whole trip.

11 The diagram shows three aircraft, A, B and C.

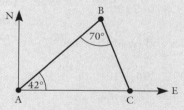

a What is the bearing of B from A?

b What is the bearing of B from C?

c If C is 4 km from A, draw a scale diagram to find how far B is from A.

INDEX

Acute angle   4
Alternate angles   6
Angle
  of depression   383, 441
  of elevation   380, 441
  of rotation   222
Angles on a straight line   4
Area   4
  of a circle   178, 220
  of a parallelogram   136, 220
  of a rectangle   4
  of a square   4
  of a triangle   140, 220
  of compound shapes   151
Average speed   422, 442

Bar charts   298
Bearing   387, 442
Bivariate data   157
Brackets   188, 232, 325

Calculator, using   33
Capacity   4
Centre of enlargement   365, 441
Centre of rotation   210, 213, 222
Chord of a circle   409
Circle   168, 220, 409
  area of   178, 220
Circumference   168, 170, 220
Congruence   5, 365
Constructing triangles and quadrilaterals
  376
Continuous data   291, 294, 340
Conversion graphs   272
Coordinates   6
Correcting to a number of decimal places
  3
Correlation   162, 221
Corresponding angles   6
Curve, equation of   285
Curved graphs   279
Cylinder, volume of   358

Decimals   73, 241
Decimals, division by   2
Decimals, multiplication of   2
Density   361, 441
Diameter   168
Direct proportion   100, 107
Directed numbers   6, 221
  division of   191
  multiplication of   190
Discrete data   291, 340
Distance   418, 420, 422, 442
Dividing by a fraction   56
Division in a given ratio   96

Enlargement   365, 372, 441
  centre of   365, 441
Equation   7, 232, 238, 241, 243, 310,
  324, 336
  of a curve   285
  of a line   258
Equations, simultaneous   312, 336
Equivalent fractions   2
Equilateral triangle   5
Estimates   31
Exterior angles of a polygon   116, 118,
  120

Factor   1
Formula   7, 184, 195
Fractions   48, 63, 65, 67, 70, 71, 73, 106,
  236, 238, 243
  addition   2
  changing to a decimal   3
  dividing by   56, 106, 236, 336
  multiplying   48, 50, 106, 236, 336
  subtraction   2
Frequency polygon   304
Frequency tables   294

Gradient of a straight line   261, 264, 267
Graphics calculator   269, 334

Graphs   256, 332
  conversion   272
  curved   279, 332
  straight lines   256, 338
Guidelines for enlargement   365

Hypothesis   156, 221
Hypotenuse   396

Image   203
Imperial units of length and mass   3
Index   18
Index, zero   22
Indices   1, 18, 19, 20, 105, 193
Inequalities   246, 248, 250, 251, 337
Intercept, y 267
Interior angles   6
Interior angles of a polygon   115, 121, 122
Intersection, point of   268
Isosceles triangle   5, 407

Like terms   7, 185, 189
Line of best fit   162, 221
Line symmetry   6
Line, equation of   258, 338
Line, plotting a   260
Lines parallel to the axes   270

Map ratio   98, 107
Mean   7
Median   7
Metric units of length and mass   3
Mid-class values   304
Mirror line   203
Mixed numbers, multiplying   53
Mixed operations   1, 59
Modal group   303, 340
Multiple   1
Multiplying fractions   48, 50
  mixed numbers   53

nth term of a sequence   198
Negative indices   20
Negative numbers   6, 221

Object   203
Obtuse angle   4
Order of rotational symmetry   210

Parabola   288, 339
Parallel lines   271
Parallelogram   5
Parallelogram, area of   136
Percentage   3, 70, 71, 73, 74, 75, 77, 80, 106
  increase and decrease   80, 107
Perigal's dissection   406
Perpendicular height   141
Perpendicular lines   271
Plotting a line   260
Point of intersection   268
Polygon   113, 220
  exterior angles of   116, 118, 120, 220
  interior angles of   115, 121, 122, 220
  regular   114, 120, 220
Polynomial equations   337
Possibility space   42
Power   18
Prime numbers   1
Prism   353
  volume of   354
Probability   7, 105
  that an event does not happen   38
Proportion, direct   100, 107
Pythagoras' theorem   398, 442
  converse of   411, 442

Quadrilateral   5
Questionnaire   166

Radius   168
Range   7, 303, 340
Ratio   89, 93, 107
Rectangle   5
  area of   5
Rectangular numbers   1
Reflection   202, 203, 222
Reflex angle   4
Regular polygon   114
Relative frequency   7

Representative fraction   98, 107
Rhombus   5
Right angle   4
Right-angled triangle   396
Rotation   202, 222
   centre of   222
Rotational symmetry   6, 210
   order of   210
Rounding continuous values   292

Scale   378
Scale drawing   376
Scale factor   369, 372, 441
Scatter graph   159, 221
Sequence, nth term of   198
Significant figures   26, 27, 105
Simultaneous equations   312, 319, 336
Speed   418, 420, 422, 442
Spreadsheet   200, 289, 331
Square   5
   area of   5
Square numbers   1
Square root   145, 221, 327, 337
Standard form   24, 105
Standard form, changing to   25
Straight line graphs   256, 338
   gradient of   261, 264, 267
Supplementary angles   4

Table, constructing from a formula
   283
Tessellate   128
Three-figure bearing   387, 442
Time   418, 420, 422, 442
Translation   202, 207, 222
Trapezium   5
Travel graph   416
Trial and improvement   329, 337
Triangle, area of   140
   equilateral   4
   isosceles   4, 407
   right-angled   396
Triangular numbers   1

Uniform cross-section   353

Vertically opposite angles   4
Volume   4
   of a cuboid   4, 351
   of a cylinder   358, 441
   of a prism   354, 441

y-intercept   267

Zero index   22